ORIGINS OF INTELLIGENCE

ORIGINS OF INTELLIGENCE

SUE TAYLOR PARKER

and

MICHAEL L. MCKINNEY

THE EVOLUTION OF COGNITIVE DEVELOPMENT IN MONKEYS, APES, AND HUMANS

THE JOHNS HOPKINS UNIVERSITY PRESS
Baltimore and London

© 1999 The Johns Hopkins University Press
All rights reserved. Published 1999
Printed in the United States of America on acid-free paper
9 8 7 6 5 4 3 2 1

The Johns Hopkins University Press
2715 North Charles Street
Baltimore, Maryland 21218-4363
www.press.jhu.edu

Library of Congress Cataloging-in-Publication Data will be found
at the end of this book.
A catalog record for this book is available from the British Library.

ISBN 0-8018-6671-5

CONTENTS

PREFACE

Origins of Intelligence takes its name from Piaget's *Origins of Intelligence in Children,* his seminal work on the developmental origins of intelligence in children. It also echoes another meaning of *origins* expressed in Darwin's *Origin of Species,* the revolutionary book on evolutionary origins. The title reflects the audacity of the undertaking: to trace the evolution of cognitive development in our species.

As suggested by its title and subtitle, *Origins of Intelligence: The Evolution of Cognitive Development in Monkeys, Apes, and Humans* uses Piagetian and neo-Piagetian frameworks to compare cognition and cognitive development in living monkeys and apes. Although imperfect, this strategy yields a consistent body of comparative data that allows us to identify those cognitive characteristics unique to humans and those shared by various primate taxa. This, in turn, provides data for reconstructing the evolutionary history of cognitive development in our species through cladistic analysis. This analysis shows that the stages of human cognitive development recapitulate the stages of cognitive evolution in a series of ancestors, who evolved new stages of cognitive development through terminal extension of previous stages, a form of heterochrony known as *peramorphosis* or adultification. Obviously, this analysis leads to the conclusion that human intelligence did not evolve through neoteny or juvenilization, as others have proposed.

The audacity of this neorecapitulationist conclusion is matched only by the opprobrium psychologists and anthropologists have laid at the door of recapitulation models and the criticism laid at the door of Piagetian models and, in some quarters, evolutionary models of heterochrony (in many cases by those unfamiliar with the literature in these fields). We persist in this approach because it yields a detailed, consistent, comprehensive explanation of cognitive evolution unmatched by any other approaches. In the history of science, arguments are won through competing models rather than claims of disproof. We will be convinced by anyone who can provide an equally powerful and comprehensive explanation for the same phenomena.

ACKNOWLEDGMENTS

We thank our employers, Sonoma State University and the University of Tennessee at Knoxville, for their support. We thank the following colleagues for their helpful comments on various chapters at various stages of composition: Barnard Baars, Elizabeth Bates, Jonathan Coddington, Terrance Deacon, Kurt Fischer, John Gittleman, Claire Kopp, Jonas Langer, Constance Milbrath, Robert Mitchell, Ken McNamara, Patrizia Poti', Sulamith Potter, Nathan Rank, Brian Shea, and Andrew Wilson. We especially thank Jonas Langer, Claire Kopp, and an anonymous reviewer for reading the manuscript in its entirety and making many helpful suggestions. Sue Parker thanks students in her primate cognition classes at Sonoma State University. We thank Judith Gregg, Sam Schmidt of the Johns Hopkins University Press, and Sherry Hawthorne of Wythe Technical Associates for editorial assistance. We thank our spouses for their unflagging support and encouragement and our editor, Ginger Berman, for making it all possible.

INTRODUCTION

Origins of Intelligence demonstrates that cognitive development in human infants and children first parallels and then exceeds that of our closest living relatives, great apes and monkeys. It shows when, and in which common ancestors, new higher-level cognitive abilities must have evolved. It also reviews the evolution and development of the brain. Finally, it shows what selective pressures may have favored the emergence of these new abilities. The detailed, comprehensive coverage of primate cognitive development and the reconstruction of the evolution of cognitive development from comparative data distinguish *Origins of Intelligence* from other attempts to reconstruct the evolution of human mentality.

The evolution of human intelligence has long fascinated anthropologists and psychologists. For many years evolutionary reconstruction was hampered by lack of both comparative data and appropriate methodology. In the past decade, however, as these data have proliferated, reconstruction of the evolution of human intelligence has become a growth industry. Recent volumes focusing on the evolution of primate intelligence include Parker and Gibson's *"Language" and Intelligence in Monkeys and Apes* (1990); Donald's *The Origins of the Modern Mind* (1991); Barkow, Cosmides, and Toobey's *The Adapted Mind* (1992); Gibson and Ingold's (1993) *Tools, Language, and Cognition in Human Evolution;* King's *Information Continuum* (1994); Parker, Mitchell, and Boccia's *Self-Awareness in Animals and Humans* (1994); Byrne's *The Thinking*

Ape (1995); Vauclair's *Animal Cognition* (1996); Mithen's *The Prehistory of the Mind* (1996); Noble and Davidson's *Human Evolution, Language, and Mind* (1996); Russon, Bard, and Parker's *Reaching into Thought* (1996); and Tomasello and Call's *Primate Cognition* (1997).

Origins of Intelligence is similar to other books on primate cognition in its use of frameworks from developmental psychology to compare the intelligence of monkeys, apes, and humans (Byrne, 1995; Mithen, 1996; Tomasello & Call, 1997). It is unique, however, in using a single comprehensive developmental framework to trace the stages of cognitive development in monkeys and apes as compared with those of human children. Like other volumes, it uses comparative data to speculate on the evolution of cognition (Byrne, 1995; Donald, 1992; Mithen, 1996; Tomasello & Call, 1997), incorporating paleontological and archeological data to do so (Mithen, 1996; Noble & Davidson, 1996). It is unique in using a systematic methodology to reconstruct the evolution of the stages of mental development and changes in developmental rates during ape and hominid evolution.

Like Mithen, and unlike most others, we conclude that the stages of development of human cognition roughly recapitulate the stages of their evolution. However, whereas Mithen uses developmental recapitulation as a heuristic device for speculating on the evolution of human mentality, we derive our recapitulation model from our comparative developmental data.

Origins of Intelligence differs from the other volumes in its systematic use of cladistic and heterochronic analysis of comparative data to reconstruct the evolution of human cognitive development. Finally, it differs from these volumes in its discussion of brain evolution and development.

THE PLAN OF THIS BOOK

Origins of Intelligence is a book about the evolution of cognitive development in monkeys, apes, and humans. We call our approach to this topic "comparative developmental evolutionary studies" because it combines three fields of study: (1) developmental psychology, (2) biological anthropology and comparative psychology, and (3) evolutionary biology. Comparative developmental evolutionary studies also draw on paleoanthropology and archeology (Parker, 1990a). This integrative approach allows us to compare primate abilities and to reconstruct their evolution.

This emerging field is represented by a loose society of researchers from around the world. Some investigators are developmental and/or comparative psychologists, some are biological anthropologists, others are ethologists, and others evolutionary biologists and paleontologists. Some focus exclusively on comparative studies of cognition; others focus on evolutionary reconstruction. In the past quarter century, they have published related works in a variety of journals (*Behavioral and Brain Sciences, Evolutionary Anthropology, Folia primatologica, Human Evolution, International Journal of Primatology, Journal of Comparative Psychology, Journal of Human Evolution, Primates,* and others) and at least 20 edited and single-authored books (Antinucci, 1989; Boysen & Capaldi, 1993; Byrne, 1995; Chevalier-Skolnikoff & Poirier, 1977; Gardner, Gardner, & Van Cantfort, 1989; Gibson & Ingold, 1993; Mitchell & Thompson, 1986; Parker & Gibson, 1990; Parker, Miles, & Mitchell, in press; Parker et al., 1994; Patterson & Linden, 1981; Povinelli & Eddy, 1996; Premack, 1976; Russon et al., 1996; Savage-Rumbaugh et al., 1993; Tomasello & Call, 1997; Whiten & Byrne, 1988; de Waal, 1983; Whiten, 1991), many of them published since 1990.

Although incomplete, data on primate cognitive development at the close of the twentieth century constitute an impressive corpus compared to what was available a century ago. In any case, comparative data are sufficiently advanced to justify a midterm report. The purpose of this report is to summarize and integrate a large body of comparative data in a comprehensive framework, but also to identify gaps and to suggest fruitful directions for further research.

The book is the product of authors schooled in two evolutionarily oriented disciplines: Michael McKinney is a paleontologist who has specialized in the theory of the evolution of development across animal taxa. Sue Parker is a biological anthropologist who has specialized in comparative studies of human and nonhuman primate cognitive development.

This book is divided into two parts. Part I compares cognitive development in human infants and children and monkeys and apes; Part II describes the evolution of cognitive development. Part I was written by the first author; Part II was coauthored (chapters 8 and 13 are by McKinney; chapters 9, 10, and 11 are by Sue Parker; and chapter 12 is coauthored).

Chapter 1 discusses epistemological and methodological issues involved in comparative developmental studies of primate cognition. After exploring issues of anthropomorphism and anecdotalism, it presents a taxonomy of different kinds of studies addressing issues of setting, rearing, sampling, and instruments of measurement.

Chapters 2 through 5 compare primate cognitive development topically by domain and subdomain. Chapter 2 reviews the development of physical cognition in human children, apes, and monkeys in the subdomains of object concept, causality, and space. Chapter 3 reviews the development of logical mathematical knowledge in the subdomains of logic, classification, number, seriation, and conservation. Chapter 4 reviews social cognition in the subdomains of imitation and pretend play, theory of mind, and self-awareness. Chapter 5 reviews the development of language in human infants and children and symbol use in apes.

Chapter 6 reviews species differences and similarities in cognition. This review reveals that the highest-level ability great apes achieve is generally equivalent to that of 2- to 4-year-old children in the preoperations period, whereas the highest level that monkeys achieve is roughly equivalent to that of 1-year-old children in the sensorimotor period. It also reveals that the development of monkeys and apes follows the same sequence as the development of human infants and children up to the sensorimotor or symbolic and intuitive subperiods of the preoperations period, respectively. Finally, it reveals that great apes complete the development of the sensorimotor and preoperations periods substantially later than human children do, at about age 3 and 8 years as compared to 2 and 4 years, respectively, whereas monkeys complete their development at a much earlier age than humans do. Significant differences in the rate of development of various abilities in monkeys and apes are also revealed.

Chapter 7 places species-typical patterns of cognitive development in the context of life histories. In addition to age, it uses dental development as a baseline to compare rates of cognitive development in monkeys, apes, fossil hominids, and humans. This approach demonstrates that the longer the period of dental development, the greater the number of subperiods of cognitive development traversed by the species. It also shows that longer periods of dental development also correlate with larger brain size, longer immaturity, and life span. Finally, it addresses the question of innateness. Although use of the comparative method of evolutionary reconstruction implies a genetic basis for the character states that are compared, it does not imply the hard-wired modular model of neoinnatists. Rather, it is more compatible with the model of epigenetic construction of succeeding levels of cognitive development (Elman et al., 1996; Karmiloff-Smith, 1995).

Part II focuses on comparative approaches to the evolution of cognitive development in hominoids. Chapter 8 introduces the concept that evolution often occurs through changes in the timing and rate of development. It shows that, just as bodily organs and behaviors evolve, so do life cycles and developmental patterns. It then describes the various kinds of changes in timing and rate of development that can occur during the evolution of ontogeny. It explains how two major outcomes of such changes, underdevelopment (juvenilization) and overdevelopment (adultification), can result from common and predictable patterns of change in developmental timing and rates.

Chapter 9 introduces and then uses comparative methodology to reconstruct the evolutionary history and common ancestry of the patterns of cognitive development identified in the preceding chapters. This analysis allows us to see the pattern of the evolution of cognitive development in great apes and humans. It shows that the periods and stages of ape and human intellectual development evolved through addition of new cognitive stages. Comparative analysis of the various developmental patterns also reveals that the timing of the development of the various periods shifted to an earlier age during human evolution. Finally, it shows that the rate of development of these periods accelerated during human evolution. These processes have resulted in some recapitulation of cognitive evolution during human cognitive development. This is the crux of our argument.

Chapter 10 introduces our favorite scenarios for the adaptive significance of hominoid cognition. Specifically, it hypothesizes that the new cluster of symbolic-level abilities manifested by living great apes arose in their common ancestor as an adaptation for apprenticeship in extractive foraging with tools on a variety of embedded foods. Chapter 11 proposes methods for generating and testing models of adaptive significance. It then uses these methods to compare various alternative models that have been presented to explain the adaptive significance of primate cognition.

Chapter 12 reviews comparative data on brain size and function, and brain development in monkeys, apes, humans, and early hominids. It presents new data on the implications of developmental constraints on brain development for understanding brain evolution.

The last chapter returns to larger questions regarding the significance of the extension of development in evolution. It reviews data on the frequency and distribution of extension and juvenilization in the fossil record. It discusses implications these modes of heterochronic evolution have for progressive change in life on earth and places trends in the evolution of human brains and intelligence in a larger context.

PART I
COGNITIVE DEVELOPMENT IN HUMAN AND NONHUMAN PRIMATES

Because some similarities between apes and humans are homologous,
that is, rooted in our common evolutionary past, it can be questioned
whether descriptions of humanlike behavior in apes based on appropriate
descriptions of hominoid-hominid continuities are truly anthropo-
morphic because these behaviors are shared by the two and thus are not
unique to humans.

MILES, 1997, P. 387

CHAPTER 1
COMPARATIVE DEVELOPMENTAL STUDIES
OF PRIMATE COGNITION

Ever since Darwin (1930) proclaimed that human intelligence arose through natural selection and since Huxley (1959) proclaimed the genealogical affinity between humans and great apes, anthropologists, biologists, and psychologists have sought a serviceable framework for comparing human and nonhuman primate intelligence for the purpose of reconstructing the evolution of human mentality. The quest has persisted, despite many vicissitudes, from that day to this.

The first step in reconstructing the evolution of cognitive development in primates is to compare the mental abilities of monkeys, apes, and humans using a consistent framework. The idea of using the stages of cognitive development of human children as a basis for comparing the mentalities of monkeys and apes dates at least to the turn of the last century. At that time, lack of knowledge about cognitive development, both of humans and of monkeys and apes, rendered this approach impractical. Indeed, this research program has become feasible only in the last quarter of the twentieth century, as a comprehensive model of human cognitive development emerged and primate studies burgeoned.

This research program in comparative developmental studies of primate cognition became feasible with the emergence of comprehensive models of human cognitive development, particularly those of Piaget and the neo-Piagetians. Overall, these models show that human infants and children develop intellectually from birth through adolescence, undergoing three major transitions during this time. The first transition, from the sensorimotor period to the symbolic subperiod of the pre-operations period, occurs at about 2 years of age. The second transition, from the preoperations period to the concrete operations period, occurs at about 6 years of age. The third transition, from concrete operations to formal operations, occurs at about 12 years of age (Case, 1985, 1987). Finer gradations, described in terms of *stages of development*, occur within each of these periods. These major transitions represent comprehensive reorganizations of thinking that occur more or less synchronously across the domains of physical, logical-mathematical, symbolic, and social cognition. At the core of Piagetian theory is the idea that developmental stages are sequentially constructed and reconstructed at higher levels through assimilation of experience to schemes and through accommodation of those schemes to experience (Baldwin, 1897, 1906; Piaget, 1966, 1970; Piaget & Inhelder, 1969).

These models provide a consistent framework for organizing and presenting a wide array of comparative studies of primates. First, they offer a comprehensive, integrated picture of cognitive development; second, they trace intellectual development to prelinguistic early infancy (the sensorimotor period) and therefore are appropriate for studies of non-linguistic creatures; and, third, either directly or indirectly, they have shaped modern research on cognitive development in both human and nonhuman primates.

A BRIEF HISTORY OF COMPARATIVE PSYCHOLOGY

The idea of reconstructing the evolution of intelligence traces to Darwin and Darwin's designated successor in comparative studies of animal intelligence, George Romanes. Romanes's goal was to establish the continuity between animal and human intelligence (Richards, 1987). Writing extensively on this subject, he defined the categories of mental activity, with a particular focus on instinctive activities that were transmitted through heredity (Romanes, 1882, 1888, 1898). Romanes defined *intelligence* as the ability to infer unperceived qualities from perceived qualities (Richards, 1987).

Romanes's published accounts of animal behavior are in an anecdotal, natural history tradition. He described the intelligent behaviors of a captive cebus monkey (Romanes, 1882) and a captive chimpanzee (Romanes, 1888). In one of the earliest studies of numerical knowledge in primates, he described the ability of a female chimpanzee to respond to verbal commands to pick up from one to six straws, a description that inspired Yerkes's later studies (Yerkes & Yerkes, 1929). Romanes, who died in 1894 at the age of forty-six, named Lloyd Morgan as his literary executor to see to the publication of two volumes of his metaphysical essays (Richards, 1987).

Morgan, a student of Thomas Huxley, represented the emerging behaviorist tradition. In 1892, Morgan had published a criticism of Romanes's notion that a dog could have abstract ideas. He called for an objective empirical basis for comparative studies. At this time, at a meeting of the International Congress of Experimental Psychology, Morgan first proclaimed his canon "that in no case is an animal activity to be interpreted as the outcome of the exercise of a higher psychical faculty, if it can be fairly interpreted as the outcome of the exercise of one which stands lower in the psychological scale," later published in his *Introduction to Comparative Psychology* (Morgan, 1895; Richards, 1987, p. 381).

Whereas Romanes saw instinct as "conscious reflex action," a cognitive act involving conscious interpretation of sensations, Morgan saw instinct and intelligence in behavioral and neurological terms. He argued that animals learn through trial and error and imitation while humans reason by rational inference. Both men agreed that instinct was a product of natural selection and perhaps the inheritance of habitual intelligent actions. Whereas Romanes believed in the continuity between animal and human intelligence, Morgan believed that the animal and the human mind differed fundamentally (Richards, 1987).

Morgan's ideas inspired Edward Thorndike's experimental research into animal learning, which, in turn, began the reaction against anthropomorphism and anecdotalism in comparative psychology. This reaction culminated in the domination of the learning theory paradigm in the United States. This behaviorist revolution may also be traced to Jennings and Loeb in the United States and Bekterev and Pavlov in the Soviet Union (Roback, 1964).

At the turn of the nineteenth century, before the rise of behaviorism, comparative psychology and developmental psychology in the United States were closely connected. At that time, both fields were profoundly influenced by Darwinism. Whereas Romanes and Morgan were Darwin's successors in the comparative study of animal behavior, William James, Granville Stanley Hall, and James Mark Baldwin were his successors in the study of human psychology.

Granville Stanley Hall, a founding figure in American developmental psychology, was an enthusiastic Darwinian. He was chief of one of the first psychological laboratories, founder of the *American Journal of Psychology*, author of more than four hundred books and articles, president of Clark University, and official host to Sigmund Freud's one trip to the United States (Kessen, 1965). Today Hall is remembered primarily as the proponent of the idea that children recapitulate the stages of their ancestor's evolution in the development of their play and their fears (Hall, 1897, 1904; Kessen, 1965). Hall's ideas and methodology were tellingly attacked by Edward Thorndike, William James, Hugo Munsterberg, and James Mark Baldwin. The reaction against Hall's recapitulation theory and his creative interpretation of anecdotes turned both developmental and comparative psychology away from the evolutionary concerns inspired by Darwin's theory.

James Mark Baldwin, another founding figure in American developmental psychology, was caught in the cross-fire between the philosophical approach of the nineteenth century and the empirical approach of the twentieth. Like Hall, he was an avid evolutionist who envisioned a comparative developmental psychology that could reconstruct the evolution of intelligence. Also like Hall, he believed in parallels or concordances in evolution and development. Baldwin elaborated his model in three major works on human development and its evolution: *Mental Development in the Child and the Race* in 1894 (revised in 1906), *Social and Ethical Interpretations in Mental Development* in 1897, and *Development and Evolution* in 1902.

Although Baldwin established the *Psychological Review* in 1894 in collaboration with James McKeen Cattell, his influence waned after he was forced to resign from the Johns Hopkins University in 1909 in the wake of a sex scandal (Richards, 1987). Despite this scandal, and despite swimming upstream against the emerging tide of behaviorism, his ideas profoundly influenced the work of Jean Piaget and George Herbert Mead, among others.

The close connection between comparative and developmental psychology that had been forged out of their mutual interest in evolution dissolved under the impact of behaviorism. The uniquely American brand of psychology known as *behaviorism* arose out of critiques of Morgan and Thorndike and the growing interest in Pavlov's work on conditioning, culminating in J. B. Watson's manifesto (J. B. Watson, 1913). Behaviorism, which dominated American psychology for fifty years (from the second decade of the twentieth century through the fifties), eclipsed the interconnected strands of developmental and comparative psychology envisioned by Baldwin and other early psychologists. Behaviorists were logical positivists who emphasized the discontinuities between the mentality of man and animals, distrusted the notion of consciousness, and sought to discover universal laws of learning. These views naturally turned psychologists away from questions of evolution and species differences (Beach, 1965).

The reemergence of developmental psychology in the United States began with a wave of European immigrants who were hired into major universities in the thirties. Prominent among these were Wolfgang Köhler, Kurt Lewin, Heinz Werner, and Erik Erikson. These scholars introduced the works of Gestaltists, Freudians, and Piagetians. Piaget's work was popularized by J. McVicker Hunt and John Flavell as well as Jerome Bruner. A second, more effective, challenge to behaviorism came in the sixties with the cognitive revolution.

An early shot in the cognitive revolution was an influential paper by Karl Lashley at a Hixon Conference in 1949, but the revolution was officially announced at the Symposium on Information Theory at the Massachusetts Institute of Technology in 1956 (H. Gardner, 1987). It was effectively promulgated in the sixties through the Center for Cognitive Studies that George Miller and Jerome Bruner began at Harvard (Parker, 1990a). This interdisciplinary center included linguists and developmental psychologists. It stimulated the earliest child language studies, including those of Roger Brown, Patricia Greenfield, and Dan Slobin. It also stimulated the information-processing approach to developmental psychology characteristic of some major neo-Piagetians, including Robbie Case, Robert Sternberg, and Robert Siegler. These and other neo-Piagetians, including Jonas Langer, eventually had a strong influence on comparative studies of primate cognition.

STUDIES OF PRIMATE COGNITION

The first American researcher in primate mentality was a Yale professor, Robert M. Yerkes, who started an ape colony on a private estate in Montecito, California, in 1914. (The colony was later moved to Atlanta, Georgia, and named the Yerkes Regional Primate Research Center.) Yerkes was an evolutionist as well as a behaviorist (Reed, 1987). He was strongly influenced both by Romanes and by Wolfgang Köhler's landmark studies of ape cognition at the Anthropoid Research Station of the Prussian Academy of Sciences in the Canary Islands. Köhler's (1927) research on problem solving in chimpanzees was guided by Gestaltist notions of the form of actions in response to problems and of good errors based on reasonable interpretations of situations. Köhler noted,

"One would like to have a standard for the achievements of intelligence described here by comparing with our experiments the performances of human beings . . . and above all, human children of different ages," (Köhler, 1927, p. 238).

Following Yerkes, research into primate cognition in the United States took two very different forms. One was natural history reports of the development of home-reared infant apes and human infants (C. Hayes, 1952; Kellogg & Kellogg, 1933). The other was controlled laboratory studies of learning in primates, especially the hardy laboratory monkeys, rhesus macaques, which began with Harry Harlow's laboratory at the University of Wisconsin in 1932 (and today stands next to the Wisconsin Regional Primate Research Center).

Most of the next generation of U.S. researchers in primate cognition worked in these and other primate research centers established by Congress in 1960 under the aegis of the National Institutes of Health (NIH). These included such major figures from Harlow's laboratory as William Mason and, later, Steven Suomi. (Mason later moved to University of California at Davis, and Suomi, to the NIH.) Emil Menzel Jr. worked at the Delta Regional Primate Center with William Mason and Hans Kummer. Duane Rumbaugh came to Yerkes Regional Primate Research Center from the San Diego Zoo. David Premack also worked at Yerkes in the fifties. (Of these early researchers, only Beatrix and Allen Gardner, the first ape language researchers, worked entirely outside the primate research centers.) This generation pioneered studies of cognitive abilities across a variety of domains: tool use, space, number, logic, and communication. They did not focus on development, nor did they use models from developmental psychology.

The move toward comparative developmental studies using these models came with the third generation of researchers in primate cognition. Many of these were students or postdoctoral researchers who were trained in the behaviorist tradition under the second generation of influential comparative psychologists (Kim Bard, Sarah Boysen, Dorothy Fragaszy, Robert Mitchell, and Sue Savage-Rumbaugh).

In the seventies and eighties the ranks of this generation of U.S. researchers expanded to include biological anthropologists (Dorothy Cheney, Suzanne Chevalier-Skolnikoff, H. Lyn Miles, Sue Parker, and Daniel Povinelli) and psycholinguists (Elizabeth Bates, Patricia Greenfield, and Michael Tomasello). Further broadening came through the influence of cognitive researchers from Canada (Mirielle Mathieu, Claude Dumas, and Anne Russon), Great Britain (James Anderson, Richard

Byrne, and Andrew Whiten, the latter two trained in developmental psychology), Holland (Frans de Waal), Germany (Jurgen Lethmate), Italy (Francesco Antinucci, Francesco Natale, Patrizia Poti', Giovanna Spinozzi, and Elizabetta Visalberghi), Spain (Juan Carlos Gomez), France (Jacques Vauclair), Switzerland (Christophe and Hegwig Boesch), and Japan (Tetsuro Matsuzawa and Shoji Itakura).

This broadening in the disciplinary bases of comparative studies turned researchers' attention back to questions that had interested Darwin's successors. Anthropologists and zoologists brought an increasing focus on evolutionary reconstruction; psycholinguists brought an increasing emphasis on parallels between language acquisition in apes and human infants; and developmental psychologists brought an increasing interest in parallels in cognitive development.

The first wave of explicitly developmental studies came from ape language researchers who analyzed the symbolic abilities of chimpanzees, orangutans, and gorillas, and, more recently, bonobos or pygmy chimpanzees. These studies have used a variety of symbol systems, including American Sign Language (R. A. Gardner, Gardner, & Van Cantfort, 1989; Miles, 1983; Patterson, 1980; Premack, 1976; Rumbaugh, 1977; Terrace, 1979). Although plagued by controversy, these language studies have contributed substantially to our knowledge of the mentality of apes. (Chapter 5 reviews these studies.)

Stimulated in part by Allison Jolly's (1972) discussion of Piagetian stages of development in monkeys in her popular textbook, *The Evolution of Primate Behavior*, the next wave of comparative developmental studies, beginning in the late seventies, focused on (Piagetian) sensorimotor intelligence in monkeys and apes (Antinucci, 1989; Chevalier-Skolnikoff & Poirier, 1977; Dore & Dumas, 1987; Parker, 1990a). Following this, beginning in the eighties, comparative developmental studies began to focus on social cognition, including imitation, communication, deception, and "theory of mind" (Mitchell & Thompson, 1986; Whiten, 1991; Whiten & Byrne, 1988). Closely connected studies of self-awareness, which began in the seventies by Gallup (1970), have recently focused more explicitly on development (Parker, Mitchell, & Boccia, 1994; Povinelli, Rulf, Landau, & Bierschwale, 1993).

Perhaps the earliest mention of Piagetian concepts in the primate literature is a brief discussion of object permanence in an article by Allison Jolly (1964) on lemur problem solving. Beginning in the seventies, several papers were published on the object concept in monkeys and apes (Mathieu, Bouchard, Granger, & Hersovitch, 1976; Vaughter,

Smotherman, & Ordy, 1972; Wise, Wise, & Zimmerman, 1974). These were followed by articles on the development of this and other selected sensorimotor-period series in macaques (Parker, 1977) and great apes (Chevalier-Skolnikoff, 1976, 1977; Redshaw, 1978). Additional Piagetian studies of macaques and great apes were published in the seventies and eighties (Antinucci, 1989; Antinucci, Spinozzi, Visalberghi, & Volterra, 1982; Braggio, Hall, Buchanan, & Nadler, 1979; Chevalier-Skolnikoff, 1983; Hallock & Worobey, 1984; Mathieu & Bergeron, 1981; Wood, Moriarty, Gardner, & Gardner, 1980). (See Dore & Dumas, 1987, and Parker, 1990a, for reviews.) The pace of this research has accelerated in the nineties.

THE QUESTION OF ANTHROPOMORPHISM

Historically, the use of frameworks from developmental psychology in comparative studies of primate cognition has been dismissed as anthropomorphic. Learning theorists, cognitive ethologists, cognitive psychologists, and others have objected that comparative studies based on human developmental psychology are unscientific because they are distorted by anthropomorphic interpretations, sometimes known as Clever Hans effects.

This criticism has been especially strong in the domain of ape language studies (Sebeok & Rosenthal, 1981). More recently, the same criticism has been leveled at studies of social cognition that employ anthropomorphic terms (Kummer & Dasser, 1990). A more serious charge is that anthropomorphism blinds investigators to the nature and significance of species-typical behaviors by imposing human tasks on nonhuman species (Mithen, 1993).

Anthropomorphism operates on several levels in comparative studies. First, it springs from human cognitive and social organization. It is a consequence of our capacity to understand other minds and our adaptation for teaching language to our infants (Miles, 1997).

Second, anthropomorphism has played an important role in generating hypotheses about primate abilities. Anthropomorphic identifications (overt or covert) between humans and animals have inspired most comparative studies (Mitchell, Thompson, & Miles, 1997). Given the nature of the human psyche, such identification is probably inevitable.

Third, *pragmatic anthropomorphism*—treating an ape like a child—has served as a methodological tool for eliciting symbolic behaviors in ape language studies and may in fact be a requirement for stimulating such behavior (Miles, 1997). This can be seen, for example, in the demonstration that the same individual, the chimpanzee Nim, showed linguistic abilities similar to those of 2-year-old children in communicative environments similar to those between human mother and infant but failed to do so when subjected to strict conditioning procedures (O'Sullivan & Yeager, 1989) (see chapter 5).

Fourth, in scientific studies *anthropomorphism is the null hypothesis* rather than the conclusion. The null hypothesis is disconfirmed by the elucidation of differences in abilities and their developmental patterns between humans and other species (Parker, 1997). Piagetian and neo-Piagetian frameworks are valuable for comparative studies because they allow us to identify a broad range of cognitive differences, as well as similarities, between humans and great apes and between great apes and monkeys.

Uncritical use of human frameworks could indeed blind investigators to species-typical behaviors and abilities. This could occur, for example, if they limited their investigation to the use of tests in experimental or cross-fostered settings and ignored naturalistic behaviors. Fortunately, however, developmental models are uniquely suited to testing the null hypothesis because they allow interpretive analysis of spontaneous behaviors in natural settings. Their utility for such studies is revealed in the neo-Piagetian analyses of tool use, imitation, and pretend play in wild chimpanzees.

Finally, it is important to realize that some behaviors and abilities in great apes that are considered "humanlike" are actually "apelike." In other words, they are shared characters that derive from the common ancestry of apes and humans; to focus on these elements is to be apeocentric rather than anthropocentric. It is the virtue of developmental frameworks that they allow us to identify such elements and to distinguish them from those that are truly unique to our species.

THE ANECDOTE PROBLEM

Anecdotes are narrative descriptions of behaviors that have been interpreted psychologically (Mitchell, 1997). They have been recorded through ad libitum rather than systematic sampling, sometimes retrospectively (Byrne, 1997). Many have questioned the validity of qualitative descriptions based on uncontrolled observations of behaviors. Nevertheless, anecdotes continue to demand our attention, partly because of their narrative structure and partly because they are often the only records of rare behaviors. The problem with anecdotes lies in the unexamined psychological presumptions that underpin them rather than the descriptions themselves. Unless disciplined to do otherwise, the human mind naturally tends to construct a plausible narrative interpretation of an event, ignoring another interpretation that may be equally plausible (Mitchell, 1997). The standard check on this tendency is to separate description from interpretation (Martin & Bateson, 1993).

PARAMETERS OF COMPARATIVE DEVELOPMENTAL STUDIES

In recent years, primatologists have increasingly turned to models of cognitive developmental stages in human children as frameworks for studying species differences and similarities in mentality. These models provide the only comprehensive and systematic basis for comparing developmental patterns and for distinguishing differing levels of cognitive complexity in primate species. Comparative use of these developmental models raises a variety of methodological issues.

DOMAINS AND MODULES

The first issue is the definition of domains or categories of knowledge. Most developmental psychologists agree that children display several major *domains* of knowledge. These include physical, logical-mathematical, social, and symbolic knowledge. Each of these, in turn, includes several subdomains, for example, space and causality within physical cognition. Operationally, domains of knowledge have been defined by the categories used by researchers in child development. Presumably, these categories reflect some intuitive understanding of the organization of the content of the mind. Karmiloff-Smith (1995) defines *cognitive domains* as sets of mental representations that sustain particular areas of knowledge.

According to the classification of mental structure made by the cognitive scientist Fodor (1983), domains are neo-Cartesian structures characterized by their propositional contents. They are distinguished from "horizontal" mental facilities such as attention, perception, and memory, which function as mechanisms of knowledge acquisition that cut across domains of knowledge. Domains are also distinguished from "vertical" facilities such as aptitudes, propensities, and dispositions, which underlie individual and or species differences in ability (Fodor, 1983).

Modules, in contrast to domains, are autonomous, innately specified, hardwired entities specific to perceptual facilities, as, for example, mechanisms of color vision, shape analysis, and three-dimensional relations (Fodor, 1983, p. 37). Fodor contrasts them with *central processes*, which are defined by the opposing characteristics of cutting across domains, forming through learning, being assembled from nearly equipotential components, and sharing horizontal resources with other cognitive systems. Roughly speaking, modular systems function as input systems, and central systems look at what these systems deliver and use that information to constrain the parameters of the hypotheses they generate. The subsystems of perception such as color perception or form perception are modular systems.[1]

Some use the term *modules* more broadly. Neoinnatists such as Chomsky (1957), Pinker (1994), and Cosmides and Tooby (1991) speak of language and computational modules, for example, thereby extending the concept into the territories of content knowledge. (Chapters 2 and 4 present further discussion of neoinnatist and modularity models.) Some developmental psychologists have challenged Fodor's definition of modules based on data on the construction of modules during development (Elman et al., 1996; Karmiloff-Smith, 1995).

INSTRUMENTS OF MEASUREMENT

The second issue in comparative developmental studies is choice of instruments of "measurement." There are a variety of Piagetian and neo-Piagetian approaches to testing sensorimotor and preoperational cognition. Considerable disagreement exists about the equivalence of performances based on differing testing procedures (Fischer & Bidell, 1991; Fischer, Bullock, Rotenberg, & Raya, 1993; Lourenco & Machado, 1996).

In the domains of physical cognition, perhaps the commonest instrument in comparative developmental studies of primate cognition has been the use of Uzgiris and Hunt's (1975) ordinal scales of development. These scales systematize testing of Piaget's sensorimotor-period series by listing related tasks in the order in which they appear developmentally. (See tables in chapter 2 for a summary of tasks for assessing the five series: object concept, space, means-ends, operational causality, and schemes for relating to objects.)

Use of these scales, however, presents the challenge of identifying classic Piagetian stages within these ongoing sequences. Few investiga-

---

1. Fodor (1983) poses the following questions about cognitive systems: (1) Are its operations domain specific or do they cut across domains? (2) Is its structure innately specified or formed through some form of learning? (3) Is its structure assembled from subunits or mapped directly onto neural substrates? (4) Is its structure hardwired or constructed from nearly equipotential neural mechanisms? (5) Is it autonomous or does it share horizontal resources (memory, etc.) with other cognitive systems?

tors have published the criteria for these extrapolations. Various investigators have sometimes used different tasks as diagnostics for a given Piagetian stage. They may also have used different criteria for scoring tests. Needless to say, standardization of diagnostic tasks is important for refining comparative developmental studies.

Yet another approach to physical cognition has been the interpretation of observational data on naturalistic behaviors in light of Piagetian and neo-Piagetian series and scales. This approach has been used as a basis for testing (e.g., Antinucci, 1989) and for interpretation of naturalistic behaviors (e.g., Parker, 1977; Chevalier-Skolnikoff, 1977; Miles, 1990). The latter approach has been used in colony, rehabilitant, and wild settings, as well as in home settings.

In the domains of logical-mathematical cognition, Langer's (1980, 1986) tasks of spontaneous manipulation of standardized classes of objects have been important instruments (e.g., Antinucci, 1989). These include geometric objects of various shapes and colors whose manipulations reveal the subjects' understanding of logic and classification as well as physical causality. Note that this instrument involves spontaneous rather than conventionally tested responses to the objects.

Other studies in the domains of logical-mathematical cognition have focused on the transition between preoperations and concrete operations periods. These include studies of number, seriation, and transitivity and of conservation (Boysen & Berntson, 1990; Boysen, Berntson, Shreyer, & Quiqley, 1993; Boysen & Capaldi, 1993; Gillan, 1981; Lourenco & Machado, 1996; Premack, 1983). Some of these studies are based on methodologies of neo-Piagetians that facilitate nonoperational solutions to the problems (discussed in chapter 3).

Developmental studies of social cognition began with Piaget's classical studies of imitation and moral development (Piaget, 1928, 1962). Since then, neo-Piagetian developmental studies have continued to focus on these domains. More recently, studies have focused on self-awareness and theory of mind. Wellman's (1990) and Perner's (1991) neo-Piagetian stages, for example, hold considerable promise for comparative studies in this latter domain. Finally, of course, stages of language development in human children have provided a framework for comparative studies of symbolic abilities in great apes (R. A. Gardner et al., 1989; Greenfield & Savage-Rumbaugh, 1990; Savage-Rumbaugh et al., 1993).

It is the existence of instruments for assessing abilities in so many cognitive domains that gives the comparative developmental approach its unique power. We call the use of multiple instruments, for example, all the sensorimotor series scales, the *pattern approach* as contrasted with the more common approach of studying abilities in a single series or domain, which we call the *category approach.*

We recommend the pattern approach because it can reveal different levels of achievement in the various cognitive series or domains within members of the same species. This approach, for example, has revealed that macaques achieve a higher level of understanding of object permanence than of causality. Identification of such differences is central to reconstructing the evolution of cognition (see chapter 9). Unfortunately, many comparative developmental studies have been limited to one developmental series within physical or logical domains, most frequently to the object concept series or causality in tool use, rather than the full sensorimotor series that Piaget described.

BEHAVIORAL MODALITIES

A related methodological issue in comparative developmental studies is the value of distinguishing among modalities or agents of action (e.g., manual, facial, whole body). Whenever possible, it is important to distinguish the modalities in which abilities are manifested within cognitive domains.

It is important to distinguish, for example, imitation in manual-gestural, facial, and vocal modalities because this reveals important species differences. One of the strengths of the Uzgiris and Hunt scales is that they do this. Likewise, it is important for the same reason to distinguish spatial knowledge manifested by the whole body from spatial knowledge manifested through manual manipulation of objects such as Piaget describes. Only by distinguishing different modalities can we detect evolutionary changes in the range and scope of abilities.

TEMPORAL DESIGNS

Another crucial methodological issue is the need for developmental studies as opposed to studies of adult abilities. As we have emphasized, evolutionary reconstruction of hominoid cognitive development is our primary aim. In order to accomplish this, we need systematic comparative data, not only on species differences in adult abilities, but also on the *rates of development of each of the various domains of cognition* in related primate species. Such comparative data depend upon appropriate measures of development. Piagetian and neo-Piagetian stages provide the only comprehensive framework for such comparative developmental studies.

Developmental studies encompass several designs. The first, which might be called *pseudodevelopmental studies*, are comparative studies of the highest levels of cognitive abilities in the adults of related primate species using developmental measures from human infants and children. Pseudodevelopmental studies provide useful information about the evolution of the adult stage of development. The second, which might be called *strictly developmental studies*, provide information about the sequence of development of cognitive abilities and the ages at which the stages are achieved. These may be either cross sectional or longitudinal in design. The third, which might be called *quasi-developmental studies*, provide information about the sequence of development beginning at an arbitrary age. These are studies that began when the subjects were already several months old.

Only strictly developmental studies can provide the information necessary for reconstructing the evolution of development of various abilities, that is, their time of onset and offset and their rate of development. Therefore, when possible, we focus on comparative data on the timing and rate of development of cognitive abilities coming from strictly developmental studies. Both longitudinal studies, which give specific developmental trajectories, and cross-sectional studies, which give larger samples, are useful in such reconstructions. Unfortunately, strictly developmental studies are rare compared to pseudodevelopmental and quasi-developmental studies. The study of cognitive development in primates is just coming into its own. We hope this book will stimulate greater interest in strictly developmental comparative studies.

The ideal methodology in comparative developmental studies would be strictly developmental studies (both longitudinal and cross sectional) of multiple cognitive domains distinguishing modalities of action. This research design is important because there is reason to believe that temporal asynchronies in development may occur among domains and also among sensorimotor modalities within domains. Only by identifying differences in relative timing among domains and modalities that occur between species can we track their role in cognitive evolution. Comparative evidence suggests, for example, that object permanence evolved before understanding of causality in anthropoids. Likewise, imitation in the gestural/manual and facial modalities seems to have preceded imitation in the vocal modality in hominid evolution (Parker, 1996b).

OTHER METHODOLOGICAL PARAMETERS

Identifying species differences in domains and modalities of cognitive development is crucial for reconstructing the evolution of cognitive development. Other methodological issues in comparative developmental studies concern the descriptive modes, settings, rearing conditions, and sample sizes in comparative developmental studies.

REARING CONDITIONS

Comparative studies have been done on individuals raised under a variety of rearing conditions. These conditions range from wild rearing to laboratory and colony rearing in groups, in mother infant pairs, in peer groups, or in isolation to cross-fostering by humans. Some investigators have studied subjects reared in seminaturalistic colonies. Others have studied subjects reared in semi-isolation in laboratories. Still others have studied subjects reared in intimate contact with human caretakers. These differences could be significant, if rearing conditions affect the rate and extent of cognitive development or performance. On the other hand, consistency of results across settings and rearing conditions would be equally significant. Comparative studies of the effects of rearing conditions are an important priority.

SETTINGS

Comparative developmental studies have been performed in a variety of settings. Most studies have been carried out in captivity. Recently, however, some studies have been done in a rehabilitant setting intermediate between captive and wild settings. In this setting, previously captive great apes are provisioned in camps on the edge of a forest. Ideally, developmental research should be done in all of these settings because each setting provides unique insights. Moreover, when complementary data emerge across settings, the convergent evidence is more compelling than evidence from any one setting (de Waal, 1991).

ACCOUNTING FOR PERFORMANCE CONTEXTS

The influence of training on cognitive performance is a potential confound in interpretation of data on human and nonhuman primate cognition. Fischer and Bidell's (1991) distinction among performance levels in studies of human cognition is useful for comparative studies: (1) the *optimal (middle) level*, which is manifested under conditions of contextual support; (2) the *functional (lowest) level*, which is manifested with no contextual support; and (3) the *scaffolded (highest) level*, which is manifested in a context in which an adult or older child performs part of the task. Fischer et al. (1993) elaborate and extend this distinction in the concept of *developmental range:* "We propose that functional and optimal levels define the developmental range of a domain for a child— from skills that the child can produce easily on his or her own to skills that the child can produce only with strong contextual support" (Fischer et al., 1993, p. 104). We appeal to this distinction throughout our discussion, emphasizing the functional and optimal levels of performance in primate cognition as species typical. Scaffolded levels, following from training or enculturation studies, though important, need to be interpreted with caution because the scaffolding is done by humans rather than conspecifics. Indeed, the capacity for elaborate scaffolding is probably a species-specific ability of humans.

DESCRIPTIVE MODES

De Waal (1991) distinguishes between qualitative and quantitative modes of observation. Among quantitative modes he distinguishes several subtypes: experiments, natural experiments, controlled observations, and uncontrolled observations. He particularly recommends the pseudo-experimental procedure of controlled observations that are designed to test hypotheses generated from observations of naturalistic behaviors.

Convergent evidence from qualitative and quantitative studies is more compelling than evidence from either kind of study alone. Qualitative accounts are vital for capturing rare instances and for generating hypotheses, particularly hypotheses regarding the adaptive significance of behaviors. Quantitative accounts are vital for testing these and other hypotheses (de Waal, 1991).

SAMPLING METHODS

Since 1974 when Jeanne Altmann (1974) published her seminal paper on methodology, most primatologists have described their sampling of spontaneous behaviors in terms of the techniques she defined and elaborated. These include ad libitum, focal animal, scan, and sequence or behavior sampling rules. Other techniques include choice of measures (latency, frequency, intensity, or duration) and choice of rules for recording behavior (continuous or time sampling) (Martin & Bateson, 1993).

SAMPLE SIZES

In all cases, comparative cognitive studies have involved small numbers of subjects from a few species. Even smaller numbers of subjects have been studied developmentally. Most of the ape language studies have focused on a single subject. (The studies of the Gardners and associates are notable exceptions.) This fact and the use of human rearing strategies have contributed to criticism of these studies. It is instructive to note that many of the seminal studies in child language acquisition also focused on one or two children.

Because most studies have been done on captive primates, they have focused primarily on the few species on monkeys and apes widely available in research colonies, laboratories, and zoological gardens. Most studies have focused on chimpanzees, cebus, and macaques, but a few studies have been done on bonobos, gorillas, orangutans, vervets, langurs, and squirrel monkeys. In addition, several language studies have assessed the mentality of cross-fostered chimpanzees, bonobos, gorillas, and orangutans. (Notably absent have been studies of gibbons and other lesser apes, baboons, and galagos and other prosimians.) Of the species studied, chimpanzees represent the largest sample size, followed by macaques and then cebus. In the case of macaques and cebus, it is important to note that the data for species (Japanese, stump-tailed, and long-tailed macaques and wedge-capped and other cebus) have been consolidated.

SUMMARY AND CONCLUSIONS ABOUT METHODOLOGICAL ISSUES

This chapter addresses some methodological issues raised by the research program in comparative developmental evolutionary studies. These issues include anthropomorphism, the scope and comprehensiveness of studies, modes, instruments and techniques of data gathering, settings, rearing conditions, sampling, and sample sizes (Table 1-1).

The chapters on cognitive development summarize comparative developmental data topically in parallel fashion across species. Each chapter presents theory and data on human infants and children, followed by parallel data on great apes and monkeys in a given domain of knowledge and its subdomains. The chapter on physical knowledge includes object manipulation, causality, and space. The chapter on logical knowledge includes classification, seriation, number, and conservation. The social knowledge chapter includes imitation and pretend play, theory of mind,

TABLE 1-1 METHODOLOGICAL PARAMETERS AND VARIABLES IN COMPARATIVE DEVELOPMENTAL STUDIES OF PRIMATE COGNITION

Parameters	Variables	Related variables
Settings	Field Rehabilitant camp Colony Laboratory Home	Kind, degree, and agency of socialization
Rearing conditions	Wild Group Mother-infant Peer Cross-fostered	Species, numbers, and kinds of social interactions
Descriptive modes	Qualitative Quantitative	 Uncontrolled Controlled natural experiments Controlled experiments
Age focus	Terminal/adult level Developmental levels	
Temporal designs	Pseudo-developmental Quasi developmental Strictly developmental	Longitudinal vs. cross sectional
Scope/breadth of study	Single domain (categorical approach) Multiple domains (pattern approach)	
Measurement instruments	Uzgiris-Hunt scales Langer object sets Piagetian stages Language acquisition stages	Sensorimotor modalities (auditory-vocal; facial- visual; manual-visual; whole body)
Observational strategies	Naturalistic behaviors Spontaneous behaviors Tested behaviors	

and self-awareness. The chapter on linguistic knowledge is organized in terms of lexical, semantic, and grammatical knowledge.

After presenting the data in each domain by subdomain, we reorder the data by species to show a coherent developmental picture within domains for each of the species for which there are some data. The data are summarized in tables designed to organize and simplify presentation. In the case of nonhuman primates, these are designed to graphically display gaps in the developmental data.

Gaps in the developmental data are only one problem with the data. Another problem is incommensurate data arising from the variety of models and procedures that have been used to collect and analyze the data. Within the strictly Piagetian camp, for example, some investigators have done formal tests, some have interpreted spontaneous behaviors on standardized objects, and others have interpreted naturalistic behaviors in Piagetian or neo-Piagetian terms. Some of those who have done formal testing have used Uzgiris and Hunt's (1975) procedures; others have followed Piaget's own descriptions. Still others have used Langer's (1980, 1986) procedures.

Despite these and other methodological inconsistencies and ambiguities in comparative developmental studies, the patterns that emerge across studies are remarkably consistent. Therefore, we believe existing data place us on the threshold of reconstructing the evolution of cognitive development. Data are sufficient to reveal the outlines of some tantalizing patterns. We believe that these patterns merit discussion and continuing research. Also, we hope that frank discussion of methodological issues will stimulate more systematic and refined research in comparative cognitive development.

SITUATING PRIMATE COGNITION IN NATURAL AND EVOLUTIONARY CONTEXTS

As indicated above, the first step in reconstructing the evolution of primate cognition is to find a suitable framework for comparing species' abilities and to organize existing data within that framework. The second step is to systematically compare the abilities of closely related species with one another and with those of members of an outgroup of more distantly related species. (In the case of the great apes, this means comparing abilities of all the species in all three genera with those of monkeys.) Part I of this book is devoted to these tasks. Part II is devoted to reconstructing the evolution and adaptive significance of these abilities.

Comparing species in the same genus or family most closely related to one another is critical to evolutionary research. By mapping the abilities of sister species onto their family tree, it is possible to determine which ancestor first displayed the new abilities. This evolutionary approach to comparative studies differs fundamentally from the classical approach of comparative psychologists, who compared model organisms of such distantly related species as pigeons, rats, and rhesus monkeys. (Chapter 9 presents further discussion of procedures for evolutionary reconstruction.)

Once we have discovered the origins of these characters, we can try to reconstruct the adaptive significance of these characteristics. Demonstrating the adaptive significance of characteristics is difficult and sometimes impossible. Adaptations are the outcome of selection for particular functions in particular environments. In principle, they are recognizable by specificity of their design features (Dawkins, 1986). Diagnosis is complicated, however, by the fact that conditions change, characters take on new functions, and beneficial outcomes of behaviors are not necessarily the reason for their existence (Williams, 1966).

A first step toward identifying the possible adaptive functions of characters is to see how they operate in the lives animals live in the wild and to try to demonstrate their contributions to fitness (Williams, 1992). Attempts to do this must be speculative because most studies of cognition have been done on captive primates, even on laboratory-reared or home-reared individuals. Nevertheless, we do our best to relate cognitive abilities to descriptions of life in the wild or in naturalistic social groups.

A further step toward identifying possible adaptive functions of characters is to look for analogous characters among distantly related species who could not have inherited the character from a common ancestor with the group under study. Parallel functions in two such distantly related groups represent powerful evidence for adaptation (Coddington, 1988). (Chapter 9 provides examples and discussion.)

One would like to have a standard for the achievements of intelligence described here by comparing with our experiments the performances of human beings . . . and above all human children of different ages.

KÖHLER, 1927, P. 239

CHAPTER 2
DEVELOPMENT OF PHYSICAL COGNITION IN CHILDREN, APES, AND MONKEYS

Of all the domains of primate cognition, physical knowledge is probably the best known. This chapter compares the development of physical knowledge in human and nonhuman primates in the subdomains of object concept, causality, and space. It begins with a review of key stages of physical cognition as they have been formulated by Piaget. We follow this with reviews of neo-Piagetian and neoinnatist formulations. These formulations generally fall into three categories: (1) replications, (2) structural reformulations, and (3) redesigns of specific tests. In this chapter we review a few well-known examples of each of these three kinds of studies that focus on the sensorimotor and preoperations periods. These studies are summarized topically under the relevant domains. The review is organized in chronological terms from sensorimotor to preoperational development within each domain of physical cognition (object, space, and causality). We begin with a brief overview of Piaget's model of development of physical knowledge.

PIAGET'S APPROACH TO PHYSICAL COGNITION IN CHILDREN

Piaget identified four major periods of intellectual development in human children from birth through adolescence. He called these four periods the sensorimotor, preoperations, concrete operations, and formal operations periods because he believed that they are characterized by movement from the strictly practical or motoric understand-

TABLE 2-1 SUMMARY OF PIAGET'S PERIODS OF COGNITIVE DEVELOPMENT

Periods and subperiods	Cognitive domains		
	Physical knowledge	Logical-mathematical knowledge	Social and intra-personal knowledge
Sensorimotor period Early subperiod (birth to 1 year) Late subperiod (1 to 2 years)	Discovery of practical properties of objects, space, time, and causality	Practical construction of logical relationships among objects	Discovery of interper- sonal efficacy through circular reactions and novel schemes through imitation
Preoperations period Symbolic subperiod (2 to 4 years) Intuitive subperiod 4 to 6 years)	Discovery of immediate causes of actions and reactions	Construction of nonreversible classes	Construction of new routines and social roles through imitation and pretend play
Concrete operations period Early subperiod (6 to 9 years) Late subperiod (9 to 12 years)	Discovery of simple mediated causes of actions and reaction	Construction of reversible hierarchical classes	Construction of more complex routines and roles based on rules
Formal operations period Early subperiod (12 to 15 years) Late subperiod (16 to 18 years)	Discovery of simple laws of physics through measurement and control of variables	Discovery of logical necessity through hypothetical deductive reasoning	Construction of universal rules and principles

ing of the sensorimotor period, to the symbolic understanding of the preoperations period, to the concrete conceptual understanding of concrete operations, to the abstract conceptual understanding of physical and logical-mathematical phenomena of the formal operations period. (Table 2-1 outlines Piaget's four periods. Summaries of these developmental periods are found in Piaget and Inhelder, 1969.) Piaget believed that the structures of thought changed radically over these periods, with the most significant shifts occurring at the transition from sensorimotor to symbolic knowledge at 2 years and at the transition from preoperational to concrete operational thought at 6 or 7 years.

According to Piaget, the sensorimotor period begins with simple, un-coordinated schemes such as sucking and grasping and culminates in the understanding that actions are reversible—as, for example, covering and uncovering an object or placing an object inside another object and removing it. Sixth-stage sensorimotor thought also leads to the achievement of symbolic thought, which involves the ability to mentally represent simple events and relationships through a variety of means, words, drawings, and pretend play. The emergence of this ability depends upon the achievement of deferred imitation at the end of the sensorimotor period. Although powerful relative to sensorimotor knowledge, preoperational knowledge is fairly rigid and stimulus bound compared to operational knowledge.

The achievement of concrete operations is a landmark that involves a qualitative shift in intellectual understanding. True operations or concepts that emerge at this point are characterized by mentally reversible and comprehensible actions that allow the child to surmount the influences of perception. These new mental capabilities allow the child to understand that quantities are conserved during spatial temporal transformations (e.g., that the amount of juice remains the same when it is poured into a low, flat dish or into a tall, thin jar). They can understand that the same set of objects can be classified into various classes and subclasses (e.g., by color or shape). They can understand that different-size objects can be seriated, that relations among these objects are transitive (e.g., if D is larger than B and B is larger than A, then D is larger than A), and that straight lines and angles can be constructed, etc. When children achieve concrete operations, they are able to explain the reasons for conservation of quantities in terms of the reversibility of the operations of pouring or shaping or rearranging the elements or, alternatively, in terms of the compensation of increases in one dimension by decreases in the associated dimension. This understanding leads them to have a sense of the necessity of their conclusions (Lourenco & Machado, 1996).

Because our focus is on comparative studies of intellectual development, we focus primarily on Piaget's two earliest periods, the sensorimotor period and the preoperations period.

PIAGET'S SENSORIMOTOR PERIOD

Piaget's researches into the earliest stages of development began with detailed studies of his own three children during the first three years of their lives. On the basis of these studies he identified six developmental stages occurring during what he called the sensorimotor period.

TABLE 2-2 PIAGET'S SENSORIMOTOR SERIES

Stages (approx. months)	Object concept	Causality	Space	Time	Sensori-motor intelligence	Imitation
First stage (0 to 2)	Reflexive adaptations					
Second stage (2 to 3)	No search for hidden objects	Diffuse sense of efficacy	Separate perceptual spaces	Perceptual durations	Primary circular reactions	Sporadic imitation of schemes from own repertoire
Third stage (3 to 8)	Beginning search for hidden object, but only as an extension of own movement	Magico-phenomena-listic causality	Coordina-tion of subjective spaces	Subjective sense of sequence of own actions	Secondary circular reactions: efforts to recreate interesting spectacles	Systematic imitation of schemes in own repertoire
Fourth stage (8 to 12)	Active search for hidden objects without following displace-ments	Beginning externaliza-tion of causality	Transition to objective space, reversing objects in space	Beginning notions of before and after	Coordina-tion of secondary schemes, beginning differentia-tion of means and ends	Imitation of own invisible schemes through simple matching of self and other
Fifth stage (12 to 18)	Sequential searching for hidden object following visible dis-placements	Objectifi-cation of causality, use of instruments	Objectifi-cation of space, dis-placing objects relative to other objects	Memory of sequential order of events	Tertiary circular reactions, search for new means	Imitation of novel schemes through trial-and-error matching
Sixth stage (18 to 24)	Mental represen-tation of sequential displace-ments of objects	Mental represen-tation of simple causal relations	Mental represen-tation of invisible dis-placements	Mental represen-tation of short series of events	Mental represen-tation of tertiary circular reactions, insight	Deferred imitation of novel schemes

These six stages progress concurrently across six different series: object, space, time, and causality in the physical domain; imitation in the social domain; and "sensorimotor intelligence," which straddles these subdomains and possibly the domain of logic. They are described in *The Origins of Intelligence in Children* (Piaget, 1952), *The Construction of Reality in the Child* (Piaget, 1954), and *Play, Dreams and Imitation in Childhood* (Piaget, 1962). Table 2-2 summarizes the six stages of the sensorimotor period across the six series.

According to Piaget's model, all six stages in the sensorimotor period develop in concert in all the series because they depend upon the same underlying structures, the *circular reactions*. These circular reactions contribute to the development of the other series because they are instruments of *assimilation* and *accommodation*, which cause the infant to repeatedly explore properties of objects, space, time, and causality. Imitation is a special form of circular reaction in which the infant repeats the model's actions rather than her own actions.

Mental representation of simple object relations in space and time, which arises during the sixth stage, is both the culmination of the sensorimotor period and the transition into the symbolic subperiod of the preoperations period. Piaget argues that the sixth-stage achievement of mental representation occurs through interiorization of trial-and-error accommodation through imitation of actions on objects, for example, mental representation of placing a chain inside a matchbox allows the child to solve the problem through insight (i.e., through interiorized mental actions on objects resulting in understanding of the solution to a problem without overt trial-and-error groping) (Piaget, 1952).

As we will see below, the sensorimotor period has been widely studied, and the stages Piaget described have been repeatedly replicated (e.g., Corman & Escalona, 1969; Konner, 1976; Uzgiris & Hunt, 1975). Even so, researchers have discovered that when tasks are simplified, human infants can display some sensorimotor abilities at an earlier age than Piaget described.

PIAGET'S PREOPERATIONS AND CONCRETE OPERATIONS PERIODS

Piaget's research on the two subsequent periods of development, the preoperations and the concrete operations periods, tracks development in some of the same domains as his earlier work, for example, time, causality, and space, but also follows development in logical-mathematical reasoning (discussed in chapter 3).

Whereas Piaget relied heavily on observations of spontaneous behavior in his studies of the sensorimotor period and in his studies of imitation and symbolic play during the symbolic subperiod (Piaget, 1962), he relied on interviewing in his study of games with rules and moral judgment (Piaget, 1965b). Other of Piaget's early investigations of intellectual development in young children involved purely verbal forms of the clinical method in which he posed questions about the nature of thoughts; of dreams; of living things; and of the sun, the moon, and the weather (Piaget, 1928, 1929).

Following these early studies, Piaget and his collaborators shifted to the so-called clinical-critical method, which emphasized the use of physical materials to pose problems for preschool and school-aged children to solve, such as his famous conservation problem, in which he asked children whether the amount of liquid was the same after being poured into a taller, thinner jar or a shorter, wider dish. Following their solutions, children were asked a series of questions about the reasons for their answers and even challenged with alternative interpretations to assess the stability of their ideas (Lourenco & Machado, 1996). Piaget used this problem-oriented methodology, for example, in his investigations of space (Piaget & Inhelder, 1967), geometry (Piaget, Inhelder, & Szeminska, 1960), classification (Inhelder & Piaget, 1964), number (Piaget, 1965a), and conservation of discontinuous quantities (Piaget & Inhelder, 1974).

Piaget described two subperiods of the preoperations period: the symbolic subperiod, from 2 to 4 years, and the intuitive subperiod, from 4 to 6 years. As suggested by its name, the first subperiod is dominated by the emergence and elaboration of semiotic functions or symbolic capacities, including language, symbolic or pretend play, dreaming, and drawing. The second subperiod is characterized by mastery of the symbolic functions and the emergence of immature—intuitive—forms of concepts or preconcepts.

In Piaget's early work, his formulation of physical thought during the preoperations period is couched primarily in terms of the disabilities children show rather than in terms of the emerging abilities they show. In contrast, operational thought is characterized by the capacity for mental reversible transformations, compensation and decentration, conservation of quantities, and dynamic imagery. Preoperational thought is characterized by irreversible transformations, centration, nonconservation of quantities, and static imagery.

Overall, Piaget argues that preoperational children are perception bound; their attention is always centered on a particular state of an object and on their own particular point of view. Symbolic thought, which grows out of imitation and symbolic play, confuses types with individuals. It operates by analogy rather than deduction and hence is transductive. Like symbolic thought, from which it springs, intuitive thought is an extension of sensorimotor adaptation. "Just as [sensorimotor intelligence] assimilates objects to response-schemata, so intuition is always in the first place a kind of action carried out in thought" (Piaget, 1966, p. 137). Nevertheless, it is more complex than symbolic thought in that it concerns complex configurations.

Concrete operational children, in contrast, are largely freed from perceptual constraints. By this time, "thought is no longer tied to particular states of the object, but is obliged to follow successive changes with all their possible detours and reversal; and it no longer issues from a particular viewpoint of the subject, but coordinates all the different viewpoints in a system of objective reciprocities" (Piaget, 1966, p. 142). As Piaget says, "An operational system derives its content from a series of abstractions of the subjects' actions, and not from particular features or properties of objects" (Piaget & Inhelder, 1967, p. 484). Finally, during concrete operations, logical-mathematical knowledge combines with and regulates physical knowledge so that laws of space, time, and causality are deducible.

NEO-PIAGETIAN APPROACHES TO PHYSICAL COGNITION

Neo-Piagetian studies of physical cognition fall under three general categories: replications, focal test redesigns, and systematic reformulation. We briefly characterize each of these approaches before summarizing some specific studies.

NEO-PIAGETIAN REPLICATION STUDIES

Neo-Piagetian replication studies are aimed at systematic replication of Piaget's observations and experiments on large populations of subjects. In their large-scale replication study of the sensorimotor period, for example, Uzgiris and Hunt (1975) translated Piaget's descriptions of stage-typical behaviors into ordinal sequences. These are not demarcated in terms of Piaget's six stages. (Consequently, various investigators who have translated their sequences into Piagetian stages have used different criteria.) Uzgiris and Hunt note that they abandoned the stages because they did not reflect the full variety of actions they observed in infants. They also abandoned Piaget's stages because they were tied to specific ages. Uzgiris and Hunt found greater variation in the ages at which children passed through these sequences than was indicated by Piaget.

This and other replications of sensorimotor-period tasks based on Piaget's criteria have yielded results broadly consistent with his results. One significant difference that emerges from the Uzgiris and Hunt study is that the rate of development across the scales is not exactly synchronous, as Piaget had postulated. With some caveats about statistical methodology, Uzgiris and Hunt note two dissociations. First, the rate of development of vocal imitation exceeded that of gestural schemes. Second, the rate of development of object permanence, construction of object relations in space, and the schemes for relating to objects exceeded that of means-ends. The disparities vary across development but tend to increase across the sequence. (These findings take on new significance when viewed in the context of developmental asynchronies in nonhuman primate species—see chapters 6 and 9.)

SPECIFIC TEST REDESIGNS

Rather than replicating or systematically reformulating Piaget's models, many studies have redefined specific Piagetian tasks. In most cases these redesigns have been based on simplified experiments or new technologies, particularly the use of virtual reality and videotape analysis. Many of the infant studies use a *two-test design* in which infants are habituated to a scene and then tested with two variants, a possible and an impossible variant. If the infant looks longer at the impossible event, this is taken as evidence that a conceptual understanding has been violated.

SYSTEMATIC REFORMULATIONS

Several neo-Piagetians have proposed systematic reformulations of Piaget's model across the developmental span or some segment of it. Notable among these is Robbie Case's model (Case, 1985, 1992a; Case & Okamoto, 1996). As we have indicated, Piaget's description of the pre-operations period is couched primarily in terms of young children's disabilities relative to the abilities they display in the concrete operations period. As in the case of gaps in the formulation of the sensorimotor period, Piaget and his colleagues had begun to address this limitation in his model at the end of his life (Piaget & Garcia, 1991). Following this lead, Robbie Case has developed an integrated model for mental development in children 4 to 10 years of age. This period embraces the end of the relational period and the bulk of the dimensional period. Case proposes that children display "central conceptual structures" that cut across physical, logical, and social domains (Case, 1996).

Case (1996) shows that children undergo parallel development in the social, numerical, and spatial domains from 2 to 5 years during the relational period. He also notes that they undergo a major transition from relational to dimensional thinking between the ages of 4 and 6 years in all three domains. This transition occurs when children are able to integrate two separate structures of first-order symbolic relations into a new system of second-order symbolic relations. In the domain of space, this new second-order structure allows them to understand maps, Cartesian coordinates, and perspective. In the domain of number, it allows them to understand number systems, for example, the base 10 number system. In the social domain, it allows them to understand the relationship between two story lines. This is Case's reformulation of Piaget's transition from the preoperations to the concrete operations period.

COMPARATIVE STUDIES OF SENSORIMOTOR INTELLIGENCE
ROAD MAP FOR COMPARATIVE STUDIES

In the following sections of this chapter, we present comparative data on the development of physical cognition in humans, monkeys, and apes. Each section focuses on a particular cognitive domain. We focus on circular reactions, objects, space, and causality because they have been important in comparative studies of primate mentality in the physical domain. In addition to summarizing Piaget's stages for each of these domains, we also summarize alternative formulations by neo-Piagetians and neoinnatists.

SENSORIMOTOR INTELLIGENCE AND CIRCULAR REACTIONS IN HUMAN CHILDREN

The little-known sensorimotor intelligence series is central to all the others. It is characterized by three sequentially developing *circular reactions*, that is, repeated schemes. The *primary circular reactions*, which occur during the second stage, beginning at about 2 months of age, are characterized by repeated actions on the self or on environmental objects that can be assimilated to simple schemes, for example, repeated sucking on the nipple or on own thumb or repeated touching or grasping own hands.

The *secondary circular reactions*, which occur during the third stage, beginning at about 3 months of age, are characterized by repeated actions that lead to interesting spectacles in the environment, for example, repetition of the shaking scheme with a rattle or repetition of the kicking scheme, which fortuitously activates a mobile. These reactions depend primarily upon the development of hand-eye coordination and voluntary prehension. They are secondary in that they focus on recreating phenomena outside the self through the self's actions.

During the fourth stage, beginning about 8 months of age, infants begin to coordinate their secondary schemes in intentional acts, for example, letting go of one object in order to grasp another. "It is this coordination between a distinctly different action serving as means (= setting aside the obstacle) and the final action (= grasping the object) which we shall consider as the beginning of the fourth stage" (Piaget, 1952, p. 217). In other words, schemes become mobile or freely combinable. This stage is also characterized by application of familiar schemes in new situations. Using the *derived secondary reactions* at the end of the fourth stage, the child begins to explore new objects through serial application of his familiar schemes, sometimes discovering new secondary circular reactions.

Finally, the *tertiary circular reactions*, which occur during the fifth stage, at about 12 months of age, develop out of the derived secondary circular reactions. These reactions are very similar except that they focus on the outcome of the actions rather than the actions themselves. They are characterized by systematic varying of the position or the intensity of schemes on objects with the aim of creating varying effects of objects on other objects or in force fields, for example, dropping toys from varying heights or dropping various objects into water from the same height. They are tertiary in that they focus on the outcome of systematic variations of secondary schemes. Strictly speaking, these reactions occur only during initial experimentation. The same kind of systematic trial-and-error groping toward a solution is involved in solving such problems as using the stick as an instrument, which arises during the fifth stage in the causal series.

SENSORIMOTOR INTELLIGENCE AND CIRCULAR REACTIONS IN MONKEYS AND APES

In one of the largest-scale replications of Piaget's sensorimotor-period stages, Uzgiris and Hunt (1975) devised ordinal scales of development for Piaget's sensorimotor series listing related tasks in the order in which they appear developmentally. These include the schemes for relating to

TABLE 2-3   SCHEMES FOR RELATING TO OBJECTS

Step	Object relations schemes	Month of age
1	Incidental use of an object in the exercise of a scheme, e.g., mouthing	2
2	Momentary attention to object used in exercise of a scheme, e.g., visual inspection	3
3	Systematic use of objects in the exercise of schemes, e.g., hitting	4
4	Beginning differentiation of schemes through interaction with objects, e.g., shaking	5
5	Beginning focus on the properties of objects, examining	6
6	Selective application of schemes in relation to properties of objects	7
7	Acquisition of new schemes as a result of studying properties of objects, e.g., throwing and dropping	8 to 9
8	Beginning social use of objects	10
9	Nonverbal reference to objects during social interaction, e.g., showing	14
10	Verbal labeling of objects	18

Note. Adapted from Uzgiris and Hunt (1975) (scale 6).

objects scale (outlined in Table 2-3), which includes some items from Piaget's sensorimotor intelligence series.

SENSORIMOTOR SCHEMES FOR RELATING TO OBJECTS SCALE

The schemes for relating to objects scale includes some items from Piaget's sensorimotor intelligence series but without explicit reference to circular reactions. Although it has rarely been used in comparative studies, we are including this scale because it is useful for comparing motor patterns of various primate species.

This scale is important because it explicates several classes of manual schemes on single objects: (1) simple holding, mouthing, and visual inspecting schemes; (2) simple hitting, shaking, and waving schemes; (3) letting go schemes of dropping and throwing; (4) socially instigated actions, and (5) showing and naming schemes. When repeated, actions in category 1 correspond to primary circular reactions; actions in category 2 correspond to secondary circular reactions; and those in category 3 correspond to tertiary circular reactions in Piaget's model.

The sensorimotor object relations scale neglects actions on two objects, such as stacking; placing in, on, and under; and pushing and/or pulling one object with another. This class of object interactions is important in comparative studies of object manipulation. Anecdotal evidence suggests that this class of object manipulation is rare in great apes as compared to human children (Vauclair, 1982; Mignault, 1985) and virtually nonexistent in monkeys (excepting cebus).

Miles (1990) tested Chantek, the signing orangutan, on the object relations scale when he was 4 years and 7 months of age. He passed items 1 to 7 on this scale, but failed item 8 (appreciating gravity in playing with items). In addition to Miles's study, there is a cross-sectional study of object manipulation in wild orangutan infants (2½ to 4½ years) and juveniles (5 to 8½ years) (Bard, 1995). This Piagetian-inspired study distinguished four categories of object manipulation: single actions, goal-directed sequences, tool use, and invention of new means through mental combinations. Overall, the infants displayed a greater percentage of single actions, and juveniles displayed a higher percentage of goal-directed actions. Only the two juveniles displayed evidence of invention of new means. Although the highest percentage of manipulations occurred in foraging, the highest percentage of goal-directed sequences occurred during locomotion.

There is also a tabulation of action schemata and their frequencies in three young cebus and two young long-tailed macaques (Natale, 1989b). This tabulation reveals that virtually all the schemata used by the young cebus and macaques fall into the first category of simple holding, mouthing, and visual inspection (about 45%). The cebus also showed a moderate frequency of banging (12%) and a low frequency of throwing (0.36%), hitting (0.88%), and rolling (0.96%). Work on adult macaques reveals that rubbing, sliding, and rolling objects under the hand are the most common patterns of object manipulation (Torigoe, 1985). (These schemes are important in cleaning food.) More recently, Poti' (1996) has published a description of the development of object manipulation schemes in chimpanzees (see below).

Various published descriptions of infant gorillas and chimpanzees reveal that these great apes traverse roughly the same course of development of object manipulation schemes as human infants do. First, they frequently engage in simple holding, mouthing, and visual inspecting schemes. Second, they engage in simple hitting, shaking, and waving schemes when they are slightly older. Third, they later begin to engage in letting go schemes of dropping and throwing. Finally, they engage in some object showing (Parker, personal observation, in press). Only cross-fostered great apes seem to display socially instigated actions and

TABLE 2-4 DEVELOPMENTAL SENSORIMOTOR STAGES OF SCHEMES FOR
RELATING TO OBJECTS IN MONKEYS, GREAT APES, AND HUMANS

Species	Age in months				
	Stage 2	Stage 3	Stage 4	Stage 5	Stage 6
Humans[a]	2	4 to 5	5 to 7	8 to 10	14 to 18
Chimpanzees[b]	?	?	?	?	?
Bonobos	?	?	?	?	?
Gorillas	?	4.5	16	24	39
Orangutans	?	?	?	?	?
Gibbons	?	?	?	?	?
Baboons	?	?	?	?	?
Macaques	0.75 to 1.75	No SCR	1.25 to 6	None	None
Cebus	?	?	?	?	None

Notes. From Uzgiris and Hunt (1975) (scale 6). SCR = Secondary circular reactions.
    ? = No data available.
[a] Uzgiris and Hunt (1975).
[b] Parker (1977).

naming schemes that are common in human children. Schemes in the
second and third categories are important in agonistic displays.

Fortunately, the sensorimotor object relations scale can be applied
fairly easily to existing data on spontaneous behaviors. We have taken
developmental ages for macaques for this scale from Parker (1977). We
have also done this for gorillas from Parker (in press) in separate scales in
order not to conflate it with other scales done by Redshaw (see Table 2-4).

Schemes relating to objects are critical to describing and understanding species differences in cognition, first, because they are species-
typical patterns that have evolved for specific functions and, second, because for each species they constitute the basic repertoire of actions on
the world. This repertoire constrains the kinds of relationships individuals are able to construct. Other factors such as the goals of actions and
the rewarding contingencies also bear on construction, however. For example, Uzgiris and Hunt (1975) note that a turning point in human development occurs when infants begin looking at the effects their schemata
have on objects. In contrast, great apes seem to focus less on these
effects. Mignault (1985) notes that young chimpanzees manipulate objects for the sake of manipulating them, without trying to discover the
properties of objects. Poti' (1996) makes a similar point.

Likewise, Poti' (1996) found distinctive patterns of object manipulation in chimpanzees as compared to human children. Specifically, chimpanzees showed "global manipulations," that is, manipulation of two or more objects as one unit with one part of the body. Human children, in contrast, showed "analytic manipulations," that is, manipulation of one object with one act as in placing one object on top of another sequentially or bringing two objects together holding one in each hand. Chimpanzees also showed a tendency to hold constructions rather than placing them on a substrate. Finally, chimpanzees put objects on either side of their bodies on the ground when they used a substrate. Human infants, in contrast, increasingly used substrates and placed objects in front of themselves.

CIRCULAR REACTIONS AND THE SENSORIMOTOR INTELLIGENCE SERIES IN MONKEYS AND APES

Piaget's sensorimotor intelligence series focuses on the development of circular reactions. These schemes were neglected by Uzgiris and Hunt and other neo-Piagetians and, consequently, by primatologists who have used these secondary sources. This is unfortunate because the comparative study of circular reactions is useful for diagnosing and understanding differences among human and nonhuman primates (Parker, 1993).

As previously indicated, secondary circular reactions are characterized by repetition aimed at re-creating contingent effects of actions on the environment. Tertiary circular reactions are characterized by repetition with variation aimed at creating contingent variations in the behavior of the objects acted on (directly or indirectly). (Behaviors are classified as tertiary circular reactions during their initial exploratory phase only.) Accurate classification of repeated actions as circular reactions can be problematic because determining the goal of repeated actions is difficult. Investigators may have difficulty distinguishing circular reactions from other repeated actions.

All monkeys and apes studied so far show such primary circular reactions as thumb sucking, visual following of the hand, and hand-hand touching (Chevalier-Skolnikoff, 1983; Parker, 1977). Of the monkeys so far studied, only cebus clearly show secondary circular reactions, that is, repetition of schemes in order to re-create a contingent effect. (Antinucci, 1989). Great apes all show secondary circular reactions of this kind (Chevalier-Skolnikoff, 1977, 1983; Mathieu and Bergeron, 1981). Great apes display a more elaborated array of derived secondary reactions involving serial application of object manipulation schemes than

TABLE 2-5 DEVELOPMENTAL STAGES OF PIAGET'S SENSORIMOTOR INTELLIGENCE SERIES (CIRCULAR REACTIONS) IN MONKEYS, GREAT APES, AND HUMANS

Species	Age in months				
	Stage 2	Stage 3	Stage 4	Stage 5	Stage 6
Humans[a]	2	3	8	12	18
Chimpanzees[b]	2 to 5.1	5 to 6	6 to 11½	11½ to 36 or 42	?
Bonobos	?	?	?	?	?
Gorillas[c]	2.8	3 to 4.5	4.5 to 10	39 (semi-TCRs)	39
Orangutans[b]	1½ to 2	2½ to 6	6 to 11½	11½ to 36	24 to 48
Gibbons	?	?	?	?	?
Baboons	?	?	?	?	?
Macaques[d]	0.5 to 3	1 to 4 (no SCR)	2½ to 5	no TCR	None
Langurs[e]	0.5 to 0.75	0.5 to 2.25	2.25 to 4	4	None
Cebus[f]	1	9 to 12 SCR	12	No TCR	None

Notes. SCR = Secondary circular reactions. TCR = Tertiary circular reactions.
  ? = No data available.
[a]Uzgiris and Hunt (1975). [d]Parker (1977).
[b]Chevalier-Skolnikoff (1983). [e]Chevalier-Skolnikoff (1983).
[c]Parker (in press). [f]Antinucci (1989).

macaques (Parker, personal observation). Some investigators have reported tertiary circular reactions in great apes (Chevalier-Skolnikoff, 1977, 1983). Given the difficulty in diagnosing true tertiary circular reactions, this remains an important subject for research.

Table 2-5, which summarizes the development in this series in apes and monkeys, reveals the following patterns: First, humans and great apes (excluding bonobos, who have yet to be studied) go through the same sensorimotor stages in the same sequence within each series. Second, in the sensorimotor intelligence series, macaques fail to show both secondary and tertiary circular reactions and sixth-stage insight, but they do show some fourth-stage coordinations. Third, cebus show secondary, but not tertiary, circular reactions or sixth-stage insight. Fourth, macaques and langurs complete their highest stages by about 6 months of age, whereas cebus complete theirs at 12 months or later.

Fifth, chimpanzees and gorillas show primary and secondary circular reactions, rich derived secondary circular reactions, and poorly developed or incomplete tertiary circular reactions, the latter developing only when they are about 3 years old. According to Piaget, imitation depends upon circular reactions and therefore develops concurrently or slightly after they do. Unfortunately, these results are not strictly comparable to those in the Uzgiris-Hunt scale since they were derived through observation.

If we skip ahead to chapter 4 to look at the imitation series, we see that, like circular reactions, imitation is generally slow to develop in great apes. The two series do not, however, show the direct association we might expect given Piaget's idea that imitation depends upon circular reactions

COMPARATIVE STUDIES OF THE OBJECT CONCEPT

THE OBJECT CONCEPT IN HUMAN CHILDREN

THE OBJECT CONCEPT SERIES

The object concept series, which is the best known and best studied of Piaget's sensorimotor-period series, culminates in the sixth stage with the achievement of object permanence. It originates during the first and second stages, from birth to 2 months. During the first stage, the infant recognizes objects only in the sense of assimilating them to such simple schemes of looking, sucking, and grasping. During the second stage, he begins to accommodate to them and anticipate their behavior. During the third stage, from 3 to 7 months, the infant extends his accommodation to objects by visually following them or by reaching for them after they drop from his hand. If an object is partially covered as he reaches for it, he will continue until he grasps it, but if it is completely covered he will stop his movement.

During the fourth stage, from 8 to 12 months, the infant begins actively searching for an object after it is covered and removes the cloth or screen and grasps it. Yet if the same object is placed under an adjacent cloth in his sight, the infant shows a *typical reaction*, searching for the object under the cloth where he originally found it. Later, if he looks under the second cloth and fails to find it, he shows a *residual reaction*, failing to search for it under the first cloth. If it is placed under two cloths, he will search under the first, but not under the second. Not until the fifth stage, beginning at about 12 months, does the infant search under the last cloth under which he has seen the object hidden. If, however, the object is hidden in the hand or in a container as it is moved under two or more cloths, the child shows the same problems he previously showed with a visibly displaced object. In this case, the child

TABLE 2-6 DEVELOPMENT OF VISUAL PURSUIT AND THE PERMANENCE
OF OBJECTS

Step	Object concept schemes	Month of age
1	Visually follows slow-moving object through a complete arc	1
2	Glance lingers at the point at which a slow-moving object disappeared	2
3	Searches for a partially hidden object	4 to 5
4	Glance returns to point at which a slow-moving object disappeared	5 to 8
5	Finds an object hidden under a single screen	7
6	Directly finds an object sequentially hidden under two screens	7
7	Directly finds an object hidden under one of three screens	7
8	Finds an object hidden under superimposed screens	9 to 10
9	Searches in container and then under the screen for an object hidden in an invisible displacement under a single screen	11 to 13
10	Searches in container and then under the correct screen for an object hidden in an invisible displacement under two screens	13
11	Searches in container and then under the two screens for an object hidden in an invisible displacement under two screens after observing invisible displacements under different screens	14

Note. Adapted from Uzgiris and Hunt (1975) (scale 1).

searches for the object in the hand or the container in which he saw the object placed rather than looking under the cloth (the residual reaction). If he finds it under a cloth, it is by chance.

Finally, during the sixth stage, beginning at about 18 months, the child searches for the invisibly displaced object under the last cloth under which the hand or the container was placed and, if he fails to find it there, systematically searches for it in reverse order. Piaget argues that this depends upon the child's mentally representing the invisible displacements of the object.

The large-scale replication study by Uzgiris and Hunt (1975) also included a study of the object concept series (Visual Pursuit and Permanence of Objects scale). Their results are summarized in Table 2-6.

NEOINNATIST STUDIES OF THE OBJECT CONCEPT

The object concept series has received more critical attention from neoinnatists than any other series in the sensorimotor period. We briefly review one such study by Rene Baillargeon and her colleagues.

Baillargeon and her colleagues (Baillargeon, Spelke, & Wasserman,

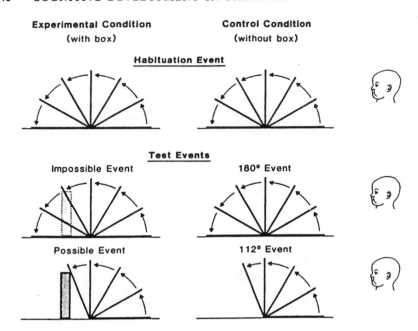

FIGURE 2-1
Experimental and control conditions on object permanence test
(Baillargeon, 1987). © 1987 by the American Psychological Association.
Reprinted with permission.

1985; Baillargeon, 1987) studied object understanding in a series of experiments on 3½-, 4½-, and 5½-month-old infants. These experiments
compared infants' habituation to possible and "impossible events." One
test event (the possible event) showed a rotating screen that was stopped
by an object; the other test event (the impossible event) showed a rotating screen that went through and was not stopped by an object (Figure
2-1). (The control events involved a rotating screen without potential
blockage.)

Before they were tested, infants were segregated into slow and fast
habituators. Fast-habituating Infants reliably looked for a longer period
at the impossible event than at the possible event, whereas slow-habitu-
ating infants failed to show any difference. Baillargeon argues that these
experiments demonstrate that infants have object permanence long before they engage in manual search for a covered object at 8 to 12 months

during the fourth stage of Piaget's sensorimotor object permanence series.[1]

Fischer and Bidell (1991) note that Baillargeon neglects to mention Piaget's description of indicators of object permanence at 3 to 4 months in the third stage of the series. These indicators include interrupted prehension (attempt to grasp an object that has been dropped) and visual search for the disappeared object. If these third-stage behaviors are considered, Baillargeon's results accord more closely with Piaget's.

THE OBJECT CONCEPT IN MONKEYS AND APES

The sensorimotor object concept series has been studied in more species than any other sensorimotor series. In the seventies several investigators reported that monkeys achieve the sixth stage. Parker (1977) reported that a stump-tailed macaque showed early sixth-stage behavior of searching under one cloth for an invisibly hidden object. Wise, Wise, and Zimmerman (1974) reported that rhesus monkeys attain the sixth stage. Vaughter et al. (1972) reported that squirrel monkeys achieve the sixth stage. Likewise, Mathieu et al. (1976) reported that chimpanzees and cebus attained the sixth stage whereas woolly monkeys did not. Most other investigators have reported the sixth stage in great apes (Hallock & Worobey, 1984; Redshaw, 1978; Wood et al., 1980)

More recently in larger-scale studies, Antinucci et al. (1982) and Natale (Natale, 1989c, 1989d) found that Japanese macaques and long-tailed macaques both failed to achieve the sixth stage of object permanence as measured by systematic search for invisibly displaced objects. De Blois and Novak confirmed this finding on a large sample of rhesus monkeys (De Blois & Novak, 1994).

Given the discrepant results on sixth-stage object permanence in monkeys, some explanation should be sought. Differences regarding the achievement or nonachievement of the sixth-stage understanding of invisible displacements probably stem from differences in methodology.

---

1. In a recent critique of the research by Baillargeon et al., Bogartz et al. (in press) present an alternative explanation for the results of these experiments. They show that the results can be explained in terms of the looking time the infant requires to perceptually process novel stimuli. They demonstrate that Baillargeon et al. failed to control for relevant sources of novelty in their two-test design. Bogartz et al. argue that the relationships between the events on the test trials and the habituation period were the significant variables in Baillargeon's experiment. By controlling for these variables using an "event set x event set multitest design," they were able to show that the results Baillargeon et al. reported were artifacts of their experimental design.

TABLE 2-7 DEVELOPMENTAL SENSORIMOTOR STAGES IN OBJECT PERMANENCE SERIES IN MONKEYS, GREAT APES, AND HUMANS

Species	Age in months				
	Stage 2	*Stage 3*	*Stage 4*	*Stage 5*	*Stage 6*
Humans[a]	2	4 to 8.5	7 to 8	7 to 10	11 to 21
Chimpanzees[b]	3	6	9	12	18
Bonobos	?	?	?	?	?
Gorillas[c]	1.5	3.5	7.5	7 to 9	9 to 10
Orangutans	?	?	?	?	?
Gibbons	?	?	?	?	?
Baboons	?	?	?	?	?
Macaques[d]	0.75	1 to 1.5	1.75	3.5	None or partial (?)
Cebus[e]	1	2 to 3	3 to 7.5	8 to 12	None
Squirrel monkeys	?	?	?	?	None
Woolly monkeys	?	?	?	? Partial	None

Note. Object permanence series is scale 1 in Uzgiris and Hunt (1975). ? = No data available.
[a]Uzgiris and Hunt (1975).
[b]Hallock and Worobey (1984), Wood et al. (1980).
[c]Redshaw (1978).
[d]Antinucci, Spinozzi, Visalberghi and Volterra (1982).
[e]Antinucci (1989).

Given the controlled testing of the later studies, it seems likely that their results are correct. It may be that repeated testing in the earlier studies resulted in success from associational learning rather than insight.

Table 2-7 summarizes data on the developmental ages for stages 2 to 6 in the sensorimotor object concept series in humans, chimpanzees, gorillas, macaques, and cebus monkeys from the studies discussed above. Although the object concept series has been studied in more primate species than any other series, developmental data are lacking for bonobos and orangutans, gibbons, and all monkeys except macaques and cebus monkeys. Figure 2-2 shows a baby stump-tailed macaque looking for a covered object.

Table 2-7 reveals several patterns in the development of the object concept series. First, all the subject species go through the same sequence in their development in this series. Second, humans and great apes complete all six stages, whereas the monkeys fail to achieve the sixth stage. Third, whereas humans and great apes are similar in their developmental rates in this series, gorillas develop more rapidly than chimpanzees and humans. (Until larger developmental sample sizes are available, however, this conclusion must be tentative.) Fourth, macaques complete their five stages by their 4th month of age, and cebus complete their fifth stage by 12 months of age, about the same age as chimpanzees.

COMPARATIVE STUDIES OF CAUSALITY

KNOWLEDGE OF CAUSALITY IN HUMAN CHILDREN

PIAGET'S SENSORIMOTOR CAUSALITY SERIES

During the first two stages of development of sensorimotor causality, the infant experiences a diffuse sense of efficacy and phenomenalism asso-

FIGURE 2-2
Infant stump-tailed macaque, "JP," searching for a hidden object (photograph by Shirley Strumm).

TABLE 2-8   MEANS FOR ACHIEVING DESIRED EVENT (MEANS-END)

Steps	Means-end schemes	Month of age
1	Hand-eye coordination; watching own hand movements	2
2	Repetition of schemes that produce an interesting effect, e.g., thrashing arms and legs to activate a mobile	3
3	Use of schemes for more than one end, e.g., attempt visually directed grasping at toy with two hands	3 to 4
4	Successful visually directed grasping at toy	4
5	Differentiation of schemes, e.g., drops one object in order to grasp another	8
6	Anticipatory adaptation of means to ends, e.g., pulling a support to obtain a toy	8
7	Use of some form of locomotion to retrieve a toy	9
8	Anticipatory adaptation of means to ends, e.g., pulling a support to obtain a toy only if toy is on support	10
9	Beginning construction of alternative means to a given end, e.g., pulling on a string tied to a toy on a table	12
10	Further construction of alternative means to a given end, e.g., pulling on a string tied to a toy that is not in view	13
11	Use of objects as extension of body as means, e.g., use of stick to reach object	15 to 18
12	Showing of foresight in a problem situation, e.g., putting a necklace in a narrow-necked container	19 to 29
13	Recognition of hindrances toward ends, e.g., avoiding placing solid ring on peg	22

Note. Adapted from Uzgiris and Hunt (1975) (scale 2).

ciated with simple schemes of assimilation and accommodation such as sucking, grasping, and bringing hand to eye. During the third stage, as the infant masters visually directed prehension, he begins to find pleasure in repeating the visual and auditory phenomena he inadvertently produces through his various schemes of shaking, hitting, banging, etc. Piaget calls this form of causality *magico-phenomenalistic* because the child acts as if his actions are sufficient to cause effects in the environment without reference to the intermediaries or mechanisms necessary for the effects; for example, he continues to try to elicit movement from a mobile by shaking his arm even when it is no longer connected to the string attached to the mobile. When his own actions coincide with an interesting outcome, the infant does not distinguish among the causes of actions of others on his body, his actions on objects, and interactions among objects. His magico-phenomenalistic secondary circular reactions occur with both physical and social objects, for example, in his attempts to elicit imitation from his mother by repeating an action that previously elicited imitation.

During the fourth stage, as she begins to apply familiar schemes to new situations and to differentiate means from ends, the infant begins to understand that objects can be centers of causality independent of her own activity but only as she activates them through her actions of touching or pushing. During this stage, she begins to push away or remove obstacles in order to grasp a toy. She places the adult's hand on or next to toys to make them activate the toy, suggesting that she sees the hand as an independent source of causality; "efficacy, originally concentrated in the movement itself, becomes decentralized, objectified and spatialized by being transferred to intermediaries" (Piaget, 1952, p. 299).

In the fifth stage, she begins to develop and experiment with new means for achieving goals involving an understanding of the causal links between actions of objects, for example, using a support to pull an object toward herself or using a stick to rake in an out-of-reach object. The use of the stick develops from an efficacious outcome of touching or hitting the object with the stick into an understanding of the dynamics of the action of the stick on the movement of the object. The child gropes toward solutions to such problems through directed trial and error, which is a form of tertiary circular reaction. Through these experi-

TABLE 2-9 OPERATIONAL CAUSALITY

Step	Causality schemes	Month of age
1	Hand eye coordination; watching own hand movements	2
2	Repetition of schemes that produce an interesting effect, e.g., thrashing arms and legs to activate a mobile	3
3	Cessation of an interesting spectacle evokes a "procedure," e.g., shaking arm when mobile stops	4
4	Beginning appreciation of other centers of causality, e.g., touches mother's hand after seeing her shake an object	5
5	Further appreciation of other centers of causality, e.g., hands mechanical toy to mother after demonstration	12 to 15
6	Further objectification of causality seen in attempts to activate objects, e.g., tries to activate mechanical toy after demonstration	18
7	Greater objectification of causality seen in attempts to activate objects, e.g., tries to activate mechanical toy without demonstration	21

Note. Adapted from Uzgiris and Hunt (1975) (scale 4).

ments she comes to understand the objective nature of causality and to see herself and other objects as subject to forces, as, for example, in rolling a ball rather than pushing it and in letting herself glide on a swing.

In the sixth stage, the child becomes capable of mentally representing the causes of visible outcomes and anticipating the outcome of possible actions, so, for example, she is able to act by insight.

Uzgiris and Hunt (1975) also replicated Piaget's study of the development of sensorimotor knowledge of causality. Their study identified two separate series, called *means-ends* (Table 2-8) and *operational causality* (Table 2-9).

NEOINNATIST STUDIES OF SENSORIMOTOR CAUSALITY
IN CHILDREN

Spelke and her associates (Spelke, Breinlinger, Macomber, & Jacobson, 1992) devised so-called impossible events to test infants' understanding of causality. In the test conditions, infants were shown an object in a particular position. The object was covered by a screen and then placed in a second position, which was possible or impossible. They have found that infants of 3 or 4 months expect objects to behave in accord with the principles of continuity and solidity. These infants, for example, spent more time looking at an object that had been placed in a position it could only have reached by going through a barrier (impossible event), than they did at an object that achieved its position without doing so (possible event). In contrast, Spelke et al. (1992) found that infants this age did not expect objects to behave in accord with inertia and gravity.

Spelke et al. (1992) interpret these results to mean that infants of 3 or 4 months are capable of active representation of certain aspects of causality. They argue that these abilities are part of a central core of physical knowledge that cannot have arisen from peripheral experience with object perception or action. They also argue that these abilities appear in essentially mature form at this age. These interpretations differ from Piaget's interpretation of the origins of representation, the immature form of early thought, and the action origins of causality.

Again, as in previous studies showing precocious abilities, these studies are based on the two-test habituation design. Like the other studies, they focus on perception rather than action on objects. Spelke et al. (1992) argue that object perception and causal reasoning are closely linked abilities that may reflect the same capacity. They believe that this is also Piaget's view. We believe that their Gibsonian interpretation contradicts Piaget's view that perception and operational thought are separate realms (Piaget & Inhelder, 1971).

PIAGET'S PREOPERATIONAL CAUSALITY

In their book *Understanding Causality*, Piaget and Garcia (1974) trace the development of understanding of transmission of forces, compositions of forces, inertia, work, and transformations of matter (e.g., from fluid to gas) in 5- to 12-year-old children. Piaget argues that it is only with the achievement of concrete operations that children understand causality in the sense of understanding the relationship among forces. During preoperations their understanding of causal and logical-mathematical relations are undifferentiated; hence their thinking is prelogical and precausal. During concrete operations, their understanding of causality is facilitated by the emergence of logic, which allows them to transcend the immediately observable effects to deduce transformations.

In 1991, Piaget and Garcia (1991) reported on a study of causality that partially fills the gap between Piaget's description of the development of causality in the sensorimotor period and later in the intuitive subperiod. In this study, children aged 1 to 4 years old were placed before an array of implements and a set of three out-of-reach objects on a board, which they were instructed to get. The implements included a straight stick, a rake, and a stick with a hook at the end. The goal objects were a toy cat, a toy monkey in a open box made of clear plastic (the monkey had hookable tail looped up on its head), and a toy dog in a clear plastic box opening to the top and the back.

Piaget and Garcia distinguish three stages of development in this study:

Stage I, at the end of the sensorimotor period from about 1 to 1¾ years —
The children reached for the goal objects and picked up and manipulated the instruments, rolling, sweeping, and/or banging them on the board without intending to use them as tools to get the objects. At an intermediate stage they pushed and/or pulled the goal objects with a tool, intending to achieve them, and in some cases succeeding in raking in the toy cat.

Stage II, during the symbolic subperiod from about 2 to 4 years —
Children easily used a tool to achieve the cat and succeeded in getting the monkey out the box by the hook (but without clear evidence of planning) but failed to get the dog from its box. The goals were clear, but they were achieved primarily through trial and error.

Stage III, during the intuitive subperiod from about 3 to 4½ years —
Children understood the specific functions of each implement, and they
were able to coordinate three schemes necessary to achieve the goal of
obtaining the dog toy ("push out, move laterally, and pull to self"). They
were able to plan their actions.

KNOWLEDGE OF CAUSALITY IN MONKEYS AND APES

SENSORIMOTOR CAUSALITY

The sensorimotor causality series has attracted the interest of primatologists. Perhaps interest in this series followed from Köhler's (1927)
studies of primate cognition focusing on physical knowledge, particularly tool use. Although Köhler's study was informed by a Gestalt perspective, it is easily translated into Piagetian terms, as Piaget himself
noted. The fifth stage focuses specifically on the development of two
tasks, use of a support to draw in an object and use of a stick as a tool
to draw in an object. Indeed, Piaget's focus on use of a stick may have
been influenced by Köhler's study, which he cites. The sixth stage corresponds to Köhler's description of insightful use of a tool. (Piaget cites
Köhler, who defined *insight functionally* in terms of a sudden, smooth
curve of action to solve a problem without trial and error.)

It is important to specify exactly what Piaget meant by fifth-stage
use of a stick to retrieve an object. Success or failure in achieving the
object is not the crucial factor distinguishing fifth-stage tool use from
fourth-stage object use. Rather, understanding of the dynamic relationship between the position and movement of the stick relative to that of
the goal object is the key. In the fifth stage the child discovers these relationships throughout what Piaget calls "directed trial-and-error groping." In other words, the fifth-stage child focuses on the relationship between the actions of the stick and the actions of the goal object. A child
or a nonhuman primate who gets an inaccessible object simply by hitting at it with a stick is not showing fifth-stage abilities. The occasional
successful use of a stick by macaques and baboons does not necessarily
constitute fifth-stage understanding of causality (Natale, 1989a).

It is important to note, however, that only three species of macaques
— stump-tailed macaques (Parker, 1977), Japanese macaques (Antinucci,
Spinozzi, Visalberghi, & Volterra, 1982), and long-tailed macaques (Antinucci, 1989) — have been studied explicitly from this perspective. There
is evidence that another species in this genus, lion-tailed macaques, use
and even prepare tools (Westergaard, 1988). Several captive monkeys of

TABLE 2-10 DEVELOPMENTAL SENSORIMOTOR STAGES IN CAUSALITY SERIES ("MEANS-END") IN MONKEYS, GREAT APES, AND HUMANS

| Species | Age in months | | | | |
	Stage 2	Stage 3	Stage 4	Stage 5	Stage 6
Humans[a]	2	3 to 4	8 to 9	10 to 15	19 to 22
Chimpanzees[b]	3	3	6 to 9	22 by tool use; 12 by use of a string	By 30
Bonobos	?	?	?	?	?
Gorillas[c]	1½	2½	4.5 to 7	36 by tool use	?
Orangutans	?	?	?	?	?
Gibbons	?	?	?	?	?
Baboons[d]	?	?	2	6	?
Macaques[e]	0.5 to 1	1.3	2.3	?	None
Cebus[e]	1½ to 2	2 to 2½	3 to 4	9 to 18 / 9[d] = Support / 18 = Stick	None

Note. Causality series is scale 2 in Uzgiris and Hunt (1975). ? = No data available.
[a]Uzgiris and Hunt (1975).   [c]Redshaw (1978).
[b]Hallock and Worobey (1984), Mathieu   [d]Westergaard (1992, 1993).
  and Bergeron (1981).   [e]Antinucci (1989); Westergaard (1988).

this species spontaneously used browse to extract syrup from an apparatus designed to allow probing. Some of these individuals subtracted side branches of the browse to render it better suited for syrup dipping. This almost certainly qualifies as fifth-stage causality. The youngest lion-tailed macaque to use tools in this study was 1.7 years old.

In a parallel study of anubis baboons, Westergaard (1992, 1993) found that they too used browse and paper towels to extract syrup from a container. These actions arose from actions of rubbing two objects together, which began at about 2 months of age and culminated in tool use at about 6 months of age. These would correspond to fourth- and fifth-stage causality, respectively.

Uzgiris and Hunt's (1975) means-ends series focuses directly on tool use. Table 2-10 on means-ends gives available developmental ages for sensorimotor causal knowledge in humans, monkeys, and apes based on Piagetian and neo-Piagetian studies. It also includes data from other compatible studies of tool use cited above. Within the fifth stage of that scale we focus on use of a support and use of a stick as the criterial behavior.

TABLE 2-11 DEVELOPMENTAL SENSORIMOTOR STAGES OF "OPERATIONAL CAUSALITY" IN MONKEYS, GREAT APES, AND HUMANS

Species	Age in months				
	Stage 2	Stage 3	Stage 4	Stage 5	Stage 6
Humans[a]	2	3	5	12 to 18	18 to 21
Chimpanzees[b]	?	6	9 to 12	22	?
Bonobos	?	?	?	?	?
Gorillas[c]	1.5	1.5 to 3	4.5 to 8.4	9.4	?
Orangutans	?	?	?	?	?
Gibbons	?	?	?	?	?
Baboons	?	?	?	?	?
Macaques	?	?	?	?	?
Cebus	?	?	?	?	?

Note. Operational causality is scale 4 in Uzgiris and Hunt (1975). ? = No data available.
[a]Uzgiris and Hunt (1975).
[b]Hallock and Worobey (1984).
[c]Mathieu and Bergeron (1981), Redshaw (1978).

Table 2-10 on sensorimotor means-ends reveals the following patterns: First, all species studied seem to traverse the same sequence of stages. Second, data are lacking on the development of the sixth stage in great apes. It is clear from Köhler's work and from the symbol use of great apes, however, that they are capable of mental representation of tool use. Third, using intelligent tool use as a criterion for the fifth stage we see that great apes and cebus monkeys definitely achieve this level. We also see that several macaque species fail to achieve it, whereas one succeeds. Baboons may also succeed. Fourth, tool use develops later in great apes and cebus monkeys than it does in human infants and earlier in lion-tailed macaques and baboons. Fifth, fifth-stage tool use develops later than fifth-stage object permanence and fifth-stage spatial knowledge in great apes.

Their operational causality series, in contrast to the means-ends series, focuses on the understanding of *agency* in causality, that is, what another individual (or mechanical device) can do to cause an effect independently of the child's own actions on that object. Table 2-11 on sensorimotor operational causality is too sketchy to interpret. Anecdotal reports suggest that great apes are less interested in agency in causality than human infants are. It is interesting to note that, like the fifth stage in the means-ends series, the fifth stage in operational causality series may occur later in chimpanzees than in humans, but earlier in gorillas than in humans.

PREOPERATIONAL CAUSAL KNOWLEDGE IN PRIMATES

Preoperational knowledge of causality in monkeys and apes has been assessed primarily in terms of tool use and tool production. In his classic study, Köhler (1927) described how one of his chimpanzees, Sultan, was able to solve a problem by constructing a longer tool by inserting a smaller stick into the end of a larger tube. This task probably exceeds the achievements of the fifth stage of the sensorimotor period since it seems to be based on insight. Schiller (1952, 1957) found that only chimpanzees older than 5 years were able to solve this problem.

Advanced Tool Use. In recent years it has become increasingly clear that wild chimpanzees engage in tool use of greater complexity than previously thought. Specifically, they engage in use of *composite tools* "consisting of two or more tools having different functions that are used sequentially and in association to achieve a single goal" (Sugiyama, 1997, p. 23). There are at least eight examples falling into five categories: (1) use of *tool sets;* (2) use of *anvils* and *wedges* to stabilize anvils; (3) use of picks to extract meat after hammering open nuts; (4) use of digging sticks to open insect nests, followed by dipping wands or sticks or fishing sticks; and (5) use of a sponge after smashing fiber with pestle or before use of a stick to retrieve it. Use of compound tools is rare, but indicative of the peak intellectual abilities of chimpanzees. All cases involve individuals who have achieved at least the juvenile stage. Interestingly, all known cases have been in western and central populations (Sugiyama, 1997).

At least one adult chimpanzee, Kate, on Baboon Island in Gambia used a tool set containing several kinds of tools (e.g., small and large chisel, bodkin [awl], and dip stick) to extract honey (Brewer & McGrew, 1990; McGrew, 1992). (A *tool set,* which is a set of tools sequentially involved in achieving a goal, contrasts with a *tool kit,* which refers to the total set of objects used as tools for a variety of purposes [McGrew, 1992].)

Wild chimpanzees in West Africa use anvils to hold the nuts they crack with sticks and stones. Moreover, some individuals use wedges to stabilize their anvils (*metatools*) (Matsuzawa, 1996). This is apparently the most complex form of tool use seen in great apes. See Figure 2-3 for a photograph of a chimpanzee cracking nuts, and Figure 2-4 for Matsuzawa's diagrammatic representation of hierarchical levels of complexity involved in these various forms of tool use.

FIGURE 2-3
A western chimpanzee cracking a nut on an anvil
(photograph by Tetsurso Matsuzawa).

Preliminary data on the development of these forms of tool use in wild chimpanzees suggest that use of a hammer stone first occurs at about 3½ years; use of a hammer and an anvil first occurs in chimpanzees at about 6½ years as compared to about 2½ years in local human children, whereas use of a wedge to support the anvil first occurs at about 9½ years as compared to 6½ years in humans (Matsuzawa, 1994).

Recent experiments comparing tool use in humans, chimpanzees, and cebus monkeys have focused on a form of tool use that exceeds the demands of sixth-stage sensorimotor understanding of causality. The trap tube task involves using a stick to get a goal object inside a clear plastic tube. The goal object is placed near a drop trap into which it will fall if pushed from the near side. Subjects showed various responses to the task: (1) performing random trials, (2) using rule-based strategies of inserting the stick in the end farthest from the goal object, (3) checking to see the location of the stick relative to the trap, and (4) anticipating the result as indicated by looking and then choosing the correct end for insertion. Figure 2-5 depicts the trap tube.

Five chimpanzees, Sarah, aged 34, Darrell, 13, Kermit, 12½, Sheba, 11½, and Bobby, 5½, were tested on two related trap tube tasks. In the first task, in which the trap was in the center of the tube, two of the five chimpanzees, Sheba and Darrell, solved the task after repeated trials using a representational strategy. In the second task, in which the trap was displaced to one side, only Sheba was able to solve the problem using a representational strategy. Human children 3 years old and older are able to solve this problem virtually immediately after inspecting the setup (Limongelli, Boysen, & Visalberghi, 1995). It is notable that cebus monkeys, who achieve fifth-stage causality (Natale, 1989a), are unable to solve the trap tube task by cognitive strategies (Visalberghi & Limongelli, 1996).

The one cebus subject, Roberta, who was able to solve the trap tube task, did so by adopting the strategy of inserting the stick into the opening farthest from the reward (Visalberghi & Limongelli, 1994). Her lack of understanding of causality was revealed when she continued to apply this strategy under other conditions for which it was inappropriate, that is, when the trap was inverted or absent (Visalberghi & Limongelli, 1994). It is notable that cebus are also unable to make compound tools, although they are able to use one stick to rake in a longer stick (Parker & Poti', 1990).

Tool Making. Visalberghi et al. compared the ability of six cebus monkeys, five chimpanzees, four bonobos, and one orangutan to modify

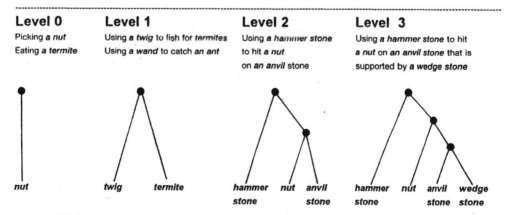

Examples of observed behavior

Level 0	Level 1	Level 2	Level 3
Picking *a nut*	Using *a twig* to fish for *termites*	Using *a hammer stone*	Using *a hammer stone* to hit
Eating *a termite*	Using *a wand* to catch *an ant*	to hit *a nut*	*a nut* on *an anvil stone* that is
		on *an anvil* stone	supported by *a wedge stone*

| *nut* | *twig*   *termite* | *hammer   nut   anvil*
stone   stone | *hammer   nut   anvil   wedge*
stone   stone   stone |

FIGURE 2-4
Analysis of hierarchical levels of tool use in chimpanzees (Matsuzawa, 1996). Reprinted with the permission of Cambridge University Press.

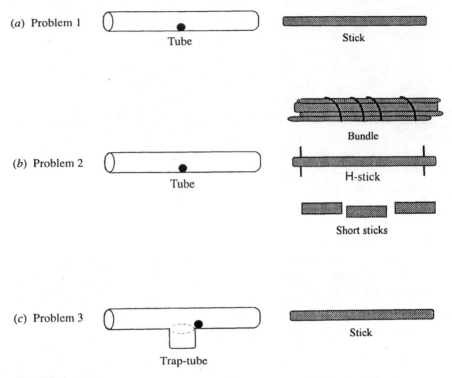

FIGURE 2-5
Experimental apparatuses presented as tasks: (*a*) Problem 1, tube task;
(*b*) Problem 2, complex conditions; (*c*) Problem 3, trap tube task. (Visalberghi
and Limongelli, 1996).

objects for use as tools in the tube task (Visalberghi, Fragaszy, & Savage-Rumbaugh, 1995). The cebus, aged 2½ to 17 years, were housed at Washington State University. The great apes, aged 3 to 20 years, were housed together at the Georgia State Language Research Center. The youngest cebus was 2½ years old, and the youngest great ape was 3 years old. Subjects were presented in sequence with three potential tools for solving the tube task: (1) a straight stick, (2) a bundle of sticks, and (3) an H-shaped stick. Although all the subjects succeeded in the second and third tasks, differences in their performances suggest that they did so by different means. Whereas all the great apes improved significantly across blocks of trials, the cebus continued to make errors throughout. This suggests that at least some of the great apes understood the cause-and-effect relations.

At least two studies have succeeded in eliciting the flaking or fracturing of stones and use of those flakes to cut open a rope tied around a box containing a desired object. The first study was done on a 5½-year-old captive orangutan, Adang, who learned to flake a tool from a core using a hammer stone by observing a model. Before exposure to the model, Adang had been trained to open a box by cutting through a rope with a stone flake (Wright, 1972). The second study was done on Kanzi, the symbol-trained bonobo, (Savage-Rumbaugh & Lewin, 1994). Kanzi found the technique of knocking stones together difficult. He spontaneously invented the technique of throwing stones on a hard surface to fracture them.

Neo-Piagetian Analysis of Advanced Tool Use. In Case's (1985) terms, most of these forms of tool use seem to involve coordination of two or more relationships to create one interrelational structure (Russon & Galdikas, 1992). Unifocal coordinations are typical of children 1½ to 2 years old who cannot combine relationships; bifocal coordinations, of children 2 to 3 years old who can combine two relationships; and elaborated coordinations, of children 3½ to 5 years old who can combine three relationships. In Piaget's terms, unifocal and bifocal coordinations that occur during the relational stage are typical of the symbolic or early subperiod of the preoperations period, and elaborated coordinations bridge the relational and dimensional stages, that is, the symbolic subperiod and the intuitive subperiod of preoperations. In the following section, we tentatively analyze the preceding tool-using performances in terms of this framework.

Inserting one stick inside another to create a longer tool to reach an object is more complex than use of one stick. It seems to involve a bifocal coordination, first between the two sticks, and then between the compound stick and the goal object. Use of a hammer and anvil seems to involve a bifocal coordination, first between the nut and anvil, and then between the nut and hammer in a single interrelational task structure. Use of a wedge to support an anvil adds yet another coordination, suggesting an elaborated coordination. The fact that this behavior seems to be manifested by only late juveniles and adults, and by only a few of them, suggests that this lies near the limit of the cognitive abilities of chimpanzees.

Use of a tool set also seems to involve an elaborated coordination, first between the chisels and the dirt encasing the bees' nest, then between the bodkin and the casing, and then between the dip stick and the honey. Production of a flake to use to cut open a rope so as to get into a box may involve an elaborated coordination, first between the rock and

TABLE 2-12 SUMMARY OF RELATIONAL (SENSORIMOTOR AND PREOPERATIONAL) CAUSAL KNOWLEDGE IN CHIMPANZEES AND HUMANS

Case's stages	Piagetian equivalent	Tasks	Achievement age (months)	
			Humans[a]	Chimpanzees[b]
Unifocal coordination	Sensorimotor sixth stage/ preoperations symbolic	Simple tool use, e.g., raking in an out-of-reach object	18 to 24	40
Bifocal coordination	Preoperations symbolic	Intermediate tool use, e.g., joining two sticks together to rake in an object	24 to 40	78
Elaborated coordination	Preoperations intuitive	Advanced tool use, e.g., using a wedge under anvil before hammering	40 to 60	114

[a] Matsuzawa (1994).
[b] Matsuzawa (1994).

the flake, then between the rock and the hard surface, then between the flake and the rope, and then between the rope and the box.

Insightful solution to the trap tube task apparently involves coordination of a relationship between the goal object and the trap, between the tool and tube, and between the tool and the trap. If this analysis is correct, these forms of tool use involve relational coordinations typical of human children aged 36 to 60 months of age. See Table 2-12 for summary of preoperational causal knowledge in chimpanzees.

In summary, many chimpanzees and orangutans seem to display forms of tool use suggesting bifocal coordinations typical of children of 2 to 3 years. At least a few chimpanzees seem to display forms of tool use suggesting relational coordinations typical of children 3 to 5 years old in the transition between the symbolic and the intuitive subperiods of preoperations. The few developmental reports suggest that full understanding of causality in tool use develops as late as 8 or 10 years.[2]

2. Another study of causality apparently demonstrates symbolic levels of spatial knowledge. Using videotaped presentations, Premack and Woodruff (1978) tested the ability of Sarah, a 14-year-old symbol-trained chimpanzee, to choose an appropriate solution to a series of problems. These problems included trying to escape from a locked cage and trying to wash the floor of the cage. After seeing each problem depicted by a trainer in a videotape, Sarah was presented with representations of

COMPARATIVE STUDIES OF SPACE

SPATIAL KNOWLEDGE IN CHILDREN

PIAGET'S SENSORIMOTOR SPATIAL SERIES

According to Piaget, during the first two stages of the sensorimotor period, infants experience sensory images as more or less separate and uncoordinated "spaces": visual, auditory, buccal, and tactile, except for simple coordinations such as thumb sucking and visual following. No single coordinated space exists.

During the third stage the infant begins to coordinate these spaces as he develops visual control over prehension and lowers his hand to reach for objects he drops. In other words, the infant locates objects in relation to his own body and prehension, but not in relation to one another.

During the fourth stage the infant begins to construct notions of extended space as he extends interrupted prehension to tactually search for an object he has dropped. He constructs notions of near and far and back and front by using his hands to move objects close to and away from his eyes and to turn them over, and by looking at objects from different perspectives by moving his head. He also constructs notions of in front and behind as he pushes over obstacles and reaches around and over obstacles.

During the fifth stage the child constructs a notion of the relationship among objects as opposed to the relationship between his own actions and objects: "The most typical and important behavior pattern in this regard consists in the child's experimental study of visible displacements: carrying objects from one place to another, moving them away and bringing them near, letting them drop or throwing them down to pick them up and begin again, making objects roll and slide along a slope, in short conducting every possible experiment with distant space

---

possible solutions in the form of still photographs of four alternative devices, for example, a key, a hose, a cord, and a paper lighter. In each case, she chose the appropriate items. Then she was presented with a more complex series of alternative solutions, for example, three keys—one intact, one twisted, and one broken—or three hoses—one attached to the facet, one not attached, and one attached but cut. She made only one error in this series, indicating that she understood that the key must be intact, that the hose must be attached and intact, etc. Obviously, using the videotaped and photographic representations of behaviors and physical settings and possible tools involves symbolic-level abilities. In sum, most of these abilities are indicative of at least symbolic-level understanding of causality typical of the early preoperations period in children 2 to 4 years old. This seems to be present in chimpanzees by 3 years of age.

TABLE 2-13 CONSTRUCTION OF OBJECT RELATIONS IN SPACE

Step	Spatial scheme	Month of age
1	Slow alternate glancing between two objects	2
2	Rapid alternate glancing between two objects	3
3	Visually localizes sound source	3 to 5
4	Grasps object directly within reach	4 to 5
5	Reconstructs trajectory of fallen object, looks where it came to rest	6
6	Visually searches for objects that have fallen	7
7	Rotates objects in space, recognizes other side	9
8	Uses one object as a container for another	9
9	Builds a tower of blocks	15
10	Anticipates effects of natural forces, e.g., compensates for gravity in using a string to suspend an object	13 to 15
11	Deduces spatial locations of objects after invisible displacement, e.g., searches in container and then under the screen for an object hidden after observing invisible displacements	14
12	Deduces spatial location after invisible displacements of object in successive locations	17
13	Remembers usual locations of objects	18
14	Deduces spatial location after invisible displacements in reverse order of displacements	

Note. Adapted from Uzgiris and Hunt (1975) (scale 5).

as well as near space" (Piaget, 1952, p. 209). Also characteristic of the later part of the fifth stage is creation of relations among objects, for example, by placing objects in hollow containers and removing them or dumping them out.

Finally, during the sixth stage the child begins to mentally represent changes in the location of objects and hence becomes capable of reconstructing the path of objects that roll out of view and of using detours to find such objects. She also begins to view herself as an object in space and to understand that she must move out of the way to open a door, for example.

Uzgiris and Hunt also replicated Piaget's study of the development of sensorimotor-period spatial knowledge in their series on constructing object relations in space (Table 2-13).

PIAGET'S PREOPERATIONAL SPACE

In their later, specialized book, *The Child's Conception of Space,* Piaget and Inhelder (1967) studied children's understanding of spatial relations in spontaneous and elicited drawings of geometric figures; in linear and circular order of elements; and in knots, points, lines, shadows, geometric sections, geometric transformations, horizontal and vertical coordinates, etc. (Piaget calls spatial-temporal knowledge *infralogical* knowledge as opposed to logical-mathematical knowledge because it deals with whole-part relations rather than class inclusion.) They distinguish three kinds of spatial knowledge, which develop sequentially: (1) topological space, which represents only relationships of proximity, separation, and enclosure, develops during the preoperations period; (2) projective space, which represents relationships of order, point, and continuity, develops during the concrete operations period; and (3) Euclidean space, which represents straight lines, angles, and coordinates, develops during the formal operations period. They use a variety of tasks to trace the development of preconcepts and concepts in each kind of space. We will focus on only one of these tasks, the representation of space in drawings.

Spatial Representation in Drawings. In children's drawings, spatial representation of geometric figures develops as follows:

Stage 0, age 1 to 2 years—Characterized by scribbling.

Stage I, synthetic incapacity, during the symbolic subperiod of the preoperations period from 3 to 4 years—Children represent only the topological properties of geometric figures, for example, drawing closed figures without representing angles when the model is a triangle.

Stage II, intellectual realism, during the intuitive subperiod of preoperations from 4 to 7 years—Children draw pictures that show parts of an object that cannot be seen from a single perspective, for example, pseudorotations. They draw what they know is there rather than what they would see from a single coordinated perspective. Nevertheless, their drawings entail the beginning use of projective and Euclidean elements as straight lines and angles. They are able to correctly draw crosses, triangles, squares, and even rhombuses.

Stage III, visual realism, during concrete operations from 7 to 12 years— Children begin to use perspective and proportions and coordinates. They are able to mentally plan the elements of their drawings.

Spatial Knowledge of Geometric Sections. When children are asked to anticipate the shape of a cross section of a solid geometric figure before it is cut, spatial knowledge is evidenced as follows:

Stage I, up to 4 years—The child cannot distinguish shapes of geometric solids.

Stage II A, at 4 to 6 years—The child cannot anticipate the shape of the cross section.

Stage II B, at 6 to 8 years—The child begins to distinguish the transformation of the intact solid figure by anticipating that a sectional surface will be created by the path of the knife.

Stage III, at 8 years—A sectional surface is correctly anticipated.

REFORMULATIONS OF SPATIAL KNOWLEDGE IN CHILDREN

Case's Studies of Space. In his reformulation of spatial development, Case focuses on figure drawing and block assembly. He notes that, in addition to Piaget, Gesell et al. (1940) and others have studied the development of these abilities. Case (1985) describes the developmental sequence in the relational period from 2 to 5 years:

1. Operational consolidation of elaborated (sensorimotor) coordinations (1½ to 2 years) in which children can draw a line. This requires focusing on the relationship between their pencil and the paper and between the line and the paper edge.

2. Bifocal coordination (2 to 3½ years) in which children can draw a circle. This requires focusing on both the character of the line and the desired end point.

3. Bifocal coordination with elaboration (3½ to 5 years) in which children can draw a figure (e.g., a sun, a tadpole man, and a mandala). This requires incorporating additional elements into the drawing.

Case (1996) notes that when children enter the dimensional period at about 6 years of age, they can place a stick figure in a scene in relation to other objects such as a second stick figure. They can also locate objects in a two-dimensional setting by noting their position in a two-dimensional drawing. This requires coordinating two subordinate structures, an object location structure and an internal object configuration

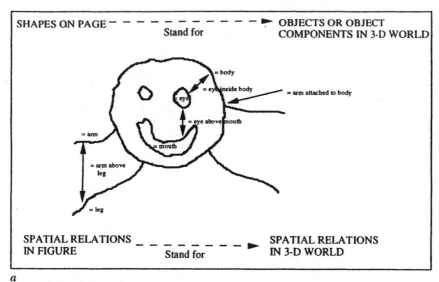

a

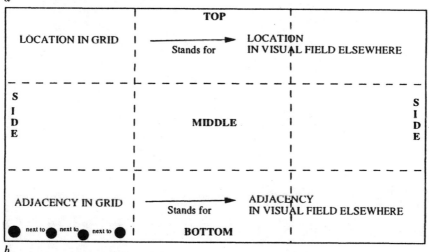

b

FIGURE 2-6
Relational drawing of 4-year-old human child (Case, 1996).
Reprinted with permission.

structure. Figure 2-6 presents Case's model of relational-period drawing.

In his study of block assembly in preoperational children in his relational stage, Case describes the following developmental sequence:

1. operational consolidation (1 to 1½ years), in which children place one block on top of another or next to it

Level 0

A   Tower

B   Flat Train

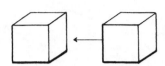

Level 1

A   Train-with-engine

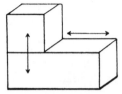

B   Step

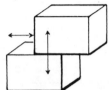

Level 2

A   Fort

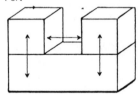

B   Arch

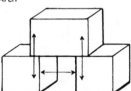

Level 3

A   Window

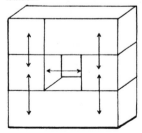

B   Garage with Door

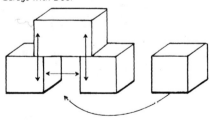

FIGURE 2-7
Levels of block construction in children (Case, 1985).

2. operational coordination (½ to 2 years), in which children can coordinate two operational relations, for example, putting one block on top of one end of another long block ("train with engine"), which coordinates the former two relations

3. bifocal coordinations (2 to 3½ years), in which children become capable of simultaneously focusing on two relationships, for example, in constructing an "arch" or a "fort"

4. elaborated coordinations (3½ to 5 years) in which they become capable of introducing elaborations, for example, transforming the "fort" into a "window" by adding another block on top (see Figure 2-7)

SPATIAL KNOWLEDGE IN MONKEYS AND APES

SENSORIMOTOR SPACE

The spatial series, in contrast to the object concept and the causal series, rarely has been studied (e.g., Parker, 1977; Redshaw, 1978; Poti', 1996). Its neglect by primatologists is somewhat surprising in view of the fact that Köhler studied detour behavior in chimpanzees as many ethologists have in other species. Detours are the sine qua non of the sixth stage in this series.

According to Piaget (1954), the fifth stage in the spatial series is characterized by an understanding of the spatial relations among objects. He describes the most important and typical behavior of this stage as moving and placing objects in relation to other objects; placing them in, on, and under other objects; rolling and sliding them—in other words, experimenting with their relations in space. In their spatial scale, Uzgiris and Hunt list several items as indicative of the fifth stage: namely, putting objects inside other objects, placing objects on top of objects, letting gravity act on objects, and systematically dropping objects and watching them fall. We suggest that the last item, letting gravity act on objects and watching them fall, be chosen as criteria for consistency in comparative studies of spatial knowledge.

In contrast, the sixth stage is characterized by the child's mental representation of invisible displacements of objects (which is also Piaget's criterion for the sixth stage of the object concept series) and by mental representation of his own body in space. (During the sixth stage, the child no longer tries to pull up an object she is standing on, as she did in the fifth stage.) Piaget does describe making *detours* as the typical behavior of the sixth stage, and Uzgiris and Hunt (1975) use detour be-

TABLE 2-14 DEVELOPMENTAL SENSORIMOTOR STAGES IN SPATIAL SERIES IN MONKEYS, APES, AND HUMANS

| Species | Age in months | | | | |
	Stage 2	Stage 3	Stage 4	Stage 5	Stage 6
Humans[a]	2	3 to 5	6 to 9	9 to 15	18
Chimpanzees[b]	3	6	8	9	? By detour; 18 by invisible displacement; 40 by block-stacking criterion[c]
Bonobos	?	?	?	?	?
Gorillas[d]	1.5	3.5	5	8	9 to 10 by invisible displacement; 8 by detour; ? by block-stacking criterion
Orangutans	?	?	?	24	24 by block-stacking criterion[e]
Gibbons[e]	?	?	?	?	?
Baboons[f]	?	?	2	6	?
Macaques[g]	0.5	1	4	6	4.5 to 5 by detour criterion
Cebus[f]	?	?	4	by 15	?

Note. Spatial series is scale 5 in Uzgiris and Hunt (1975). ? = No data available.
[a]Uzgiris and Hunt (1975).
[b]Hallock and Worobey (1984).
[c]Poti' (1996).
[d]Redshaw (1978).
[e]Miles (1990).
[f]Westergaard and Suomi (1994).
[g]Parker (1977).

havior as their criterion for the sixth stage in their spatial series. It is important to note, however, that in contrast to Uzgiris and Hunt, Piaget also describes the child as throwing and displacing objects before using a detour to retrieve them.

We believe that use of Uzgiris and Hunt's detour criterion for sixth stage in the spatial series in monkeys and apes is misleading because it refers to actions of the body as a whole. Most other criteria refer to actions on an object in relation to gravity and/or other objects. We dislike the detour criterion for comparative studies because we think detour behavior may arise from more than one proximate cause. It may, for example, be learned through trial and error. This is indicated by the fact that most mammals display the capacity to make detours. We therefore reinterpret the sixth stage in the spatial series in terms of putting objects in containers or stacking objects. According to these criteria, sixth-stage spatial knowledge develops at about 4 years of age in chimpanzees (see review of Poti's work below).

Table 2-14 gives developmental ages and highest stages for stages 2 to 6 of the sensorimotor spatial series in humans, chimpanzees, gorillas, and macaques. Currently, there are no data for bonobos, orangutans, gibbons, or monkeys other than macaques. The data on macaques are based on observation of spontaneous behavior in a longitudinal study of a single individual (Parker, 1977).

Table 2-14 suggests the following patterns in development of the sensorimotor spatial series: First, humans, chimpanzees, and gorillas traverse the same sequence of stages. Second, like humans, chimpanzees and gorillas traverse all six stages. Third, unlike humans, chimpanzees, and gorillas, macaques probably fail to complete the sixth stage (if we use our substitute criterion of tracing invisible displacements; If we use the detour criteria, however, macaques begin the fifth stage by about 4½ or 5 months). Fourth, humans begin the sixth stage by about 18 months, and chimpanzees at about 12 months, whereas gorillas do so at about 9 to 10 months using our substitute criterion (at 8 months by the detour criteria). In other words, using the detour criterion great apes complete the sensorimotor developmental series in space more rapidly than humans do.

Nesting Cups. Greenfield's (1992) study of hierarchical organization in manual object manipulation provides another measure of spatial development (this idea is discussed further in chapter 4). She identified the following three strategies for combining nested cups: (1) the pairing strategy of placing one cup into another, (2) a pot strategy of placing two cups into a third cup, and (3) a subassembly strategy of placing one cup into another and then placing the two into a third. The study found a developmental sequence in which children began to use the pairing strategy at 11 months, the pot strategy by about 16 months, and the subassembly strategy by about 20 months. This suggests that these activities correspond to fourth- and fifth-stage spatial understanding.

In a replication of this study, Westergaard and Suomi (1994) found that cebus monkeys used all three strategies for manipulating nested cups, though none of them employed the two more complex strategies as a dominant strategy. Seven of the ten cebus in their study used the pairing strategy, five of the ten used the pot strategy, and three of the ten used the subassembly strategy. Based on a similar study of object manipulation in baboons, Westergaard (1993) concluded that baboons used the pairing strategy but not the pot and subassembly strategies.

Langer's Space. Poti' (1996) studied the development of spatial cognition as assessed by block construction in five chimpanzees aged 1 to 4 years. She did a cross-sectional study of five female chimpanzees (Lianne at 1 year, Cindy at 2½ years, Claudette at 3½ years, and Vanessa at 4½ years) and a longitudinal study of one chimpanzee (Sylvia at 2 years, 3 years, and 4 years). None of the subjects had had training in symbol use. The aim of this study was to track the development of spontaneous spatial constructions of young chimpanzees in a free play setting.

The subjects were given a total of 16 objects of four colors and four shapes in sets of six. The objects were presented scattered in random order for 5 minutes of free play. Poti' scored a grouping when 2 or more objects were in contact or within 5 cm proximity after being manipulated by a subject. She examined both the resulting constructions and the constructive processes. In the case of constructions, she tabulated the mean number of objects and the proportion of vertical (towers) and horizontal (trains) and bridge constructions as well as containment and the number of contemporaneous constructions.

The one subject who was followed longitudinally, Sylvia, showed a clear developmental trend in constructions. Beginning at 2 years and increasing significantly at 4½ years, she increased the number of objects in her groups, the number and height of vertical constructions, and the number of contemporaneous groups. She did not produce containments or bridge constructions. Human infants show these trends by the beginning of the second year. They also show horizontal alignments and bridges, which were not constructed by this chimpanzee.[3]

PREOPERATIONAL SPATIAL KNOWLEDGE

According to Piaget and Inhelder (1967), children in the preoperations period focus primarily on topological aspects of space. Consonant with

3. The best-known report of block assembly in great apes, of course, is Köhler's report of chimpanzees stacking boxes to reach a fruit hanging from the ceiling. Based on the fact that only three of the six chimpanzees he tested succeeded and that it took them many attempts to achieve this solution, this was a difficult task, apparently at the limit of their abilities. All three animals stacked two boxes, two stacked three boxes, and one once stacked four boxes. This performance falls within the sensorimotor period.

this, spatial knowledge at the symbolic level in great apes has been assessed in four arenas: block construction, drawings, map and picture reading, and knot tying. In addition, the tube task and the use of the wedge to stabilize the anvil discussed in the previous section on causality also involve spatial cognitions. Great ape performances of these spatial tasks have been analyzed from a variety of perspectives.

From a Casean perspective, the tube task apparently involves elaborated coordinations of topological relations of enclosure between the tube and the goal object, the trap and the goal object, the tool and the tube, and the tool and the goal object. Wedging apparently involves elaborated coordinations of topological relations of *on top of* and *underneath* between the anvil and the wedge, between the goal object and the anvil, and between the tool, the goal object, and the anvil.

Block Assembly. Case himself used block assembly to assess levels of sensorimotor and preoperational thinking in children 1 to 5 years of age. This framework should work well for analyzing comparative data on block manipulations in great apes, some of which have been done using the Bayley scales.

Bayley scales of infant mental development list age norms for the number of blocks used to construct a tower and a train (Bayley, 1969). According to the Bayley scales, the highest achievements are construction of a tower of six cubes (item 143) at 22.4 months average age (16 to 30 range) and construction of a train of eight cubes (item 161) at 30.0 months (22 to 30+ range). This is equivalent to the age norms for the transition from Piaget's sensorimotor period to the symbolic subperiod of preoperations.

Unfortunately the Bayley scales do not include the more complex constructions that Case uses. Miles (1990) studied mental development in Chantek, the signing orangutan, from age 1 year, 11 months, to 7 years, 6 months, using the Bayley scales. According to her table, he passed item 143 (building a tower of six cubes) by 6.4 years, and item 161 (building a train of eight cubes) by 4 years, 8 months.

Drawings and Photographs. Production and comprehension of drawings are another indication of understanding of spatial knowledge at the level of symbolic thought. Piaget and Inhelder (1967) describe the following stages of development of drawings from models during the symbolic subperiod:

Stage 0—scribbling up to 2½ or 3 years

Stage 1A—scribbling with some variation in relation to topological features of the model (i.e., closed vs. open figures)

Stage 1B—representation of topological features such as open or closed figures and crosses, beginning at about 3 years

Stage 2A—representation of curved versus straight lines

Stage 2B—use of angles, beginning at about 4 years

The Bayley Scale on Mental Development (Bayley, 1969) also assays some of these behaviors. According to this scale, spontaneous scribbling (item 112) begins at 14 months on average (10 to 21 range), differentiation of stroke from scribble (item 135) begins at 17.8 months average (13 to 26 range), and imitation of vertical and horizontal strokes (item 147) begins at 24.4 months average (19 to 30+ range). These items are roughly equivalent to Piaget's stages 1 and 2A. In her study of Chantek, Miles (1990) found that he passed the last milestone by 5 years, 6 months.

Morris (1962) reviews data from various studies of drawing by 23 chimpanzees, 2 gorillas, 3 orangutans and 4 cebus monkeys. He also describes data from his own developmental studies of two chimpanzees, Alpha and Congo, based on Rhonda Kellogg's stages. Kellogg describes five stages of pictorial representation: scribbles, diagrams, combines, aggregates, and pictorials. Within the scribbling stages, she describes various kinds of lines, loops, zigzags, and enclosed figures approximating circles. Within the diagram stage, she describes the emergence of simple geometric figures, including the cross, square, circle, and triangle. The latter part of her scribbling stage corresponds to Piaget's stage 2A, and the beginning of her diagram stage corresponds to Piaget's stage 2B.

Congo, a young female chimpanzee at the London Zoo, was studied from the age of 1 to 4 years, during which time she produced 384 pictures (see Figure 2-8). Morris reports that she first began to scribble at 1½ years, progressed through Kellogg's scribbling stages, and began to draw crosses and circles diagnostic of the early diagram stage. He reports that Congo began to make zigzag lines at about 3 years but does not date the onset of circle and cross production. (This rich interpretation has been challenged by Boysen et al. [Boysen, Berntson, & Pentice, 1987] on the basis of controlled studies of scribbling in chimpanzees.)

Morris notes that Congo and other great apes displayed considerable appetite for and interest in drawing and painting. In fact, they would

FIGURE 2-8
Line drawing by a chimpanzee (Congo) (Morris, 1962).

throw temper tantrums if they were interrupted, in sessions that typically lasted 15 to 30 minutes but occasionally up to an hour. Likewise, Miles, Mitchell, & Harper (1996) report that Chantek, the signing orangutan, drew lines and circles on demand when asked to imitate a drawing. Patterson and Linden (1981) also describes drawing by Koko, the signing gorilla, noting that Koko often labels her drawings.

Great apes also display the ability to recognize drawings and photographs of objects. There are numerous reports of picture labeling by symbol-trained apes (B. T. Gardner & Gardner, 1989b; Patterson & Linden, 1981). A recent study of Ai, a symbol-trained chimpanzee, demonstrated that she is able to recognize and correctly label line drawings of individual humans and chimpanzees, including herself (Itakura, 1994b).

TABLE 2-15 SUMMARY OF RELATIONAL (SENSORIMOTOR AND PREOPERATIONAL) SPATIAL KNOWLEDGE IN CHIMPANZEES AND HUMANS

Case's stages	Piagetian equivalent	Tasks	Achievement age (months)	
			Humans[a]	Chimpanzees[b]
Unifocal coordination	Sensorimotor sixth stage/ preoperations symbolic	Stacking one block on another	18 to 24	40
Bifocal coordination	Preoperations symbolic	Scribbling	24 to 40	78
Elaborated coordination	Preoperations intuitive	Drawing a circle and a cross	40 to 60	114

[a] Case (1985).
[b] C. Hayes (1952).

Menzel, Premack, and Woodruff (1978) report that four wild-born 3-year-old chimpanzees (Bert, Sadie, Luvie, and Jessie) were able to find hidden objects in an outdoor field on the basis of representations shown to them on television. They call this ability *map reading*. Two studies have been aimed at testing understanding of three-dimensional scale models in chimpanzees. Premack and Premack (1983) found that their subjects were unable to recognize the parallels between the larger room and the scale model. They had some success, however, when they began with an identical size model and reduced it. When Kuhlmeier and Boysen (1997) did a similar experiment, Sheba, the adult female, performed above chance level in finding hidden food in the real room after seeing its location in the scaled-down room. The juvenile male, Bobby, performed at chance. Similar tests of human children have shown that they can understand such models by the age of 3 years (DeLoache, 1990).

Knot Tying. Another arena of spatial knowledge closely related to that of drawing is knot tying. According to Piaget and Inhelder (1967), children in the preoperational period go through the following stages in learning to tie knots from a model:

Stage 1—winding and inserting without intertwining from about 3 years

Stage 2—tying a simple knot without recognizing the homology between slack and tight knots from about 4 years

Russon and Galdikas (1993) report that an adult female orangutan in the rehabilitant group at Camp Leakey succeeded in hanging a hammock on a tree by winding the rope around a tree trunk, inserting it through a loop, and pulling on it. Although this activity was an imperfect attempt to imitate human hammock hanging, it demonstrates stage 1 abilities comparable to those of 3-year-old human children.

All of the preceding reports in the arenas of block assembly, drawing, symbol reading, and knot tying converge in suggesting that great apes display symbolic levels of understanding of space comparable to those of 3- or 4-year-old human children in the early preoperations period. The one developmental study in this domain suggests that these symbolic abilities develop about 2 or 3 years later in great apes than in human children (Miles, 1990). There are some suggestions that chimpanzees display late preoperations knowledge of space. Table 2-15 summarizes preoperational period physical knowledge in great apes.

Mental Maps. Analysis of the foraging patterns of wild chimpanzees in the Tai Forest reveals that they consistently optimize the distance they travel relative to their nut-cracking sites in gathering hammer stones and nuts. Boesch and Boesch (1984) argue that this pattern implies a mental map and the mental operations of (1) measurement and conservation of distance, (2) comparisons of several distances, and (3) permutations of objects in the map and permutations of reference points. If they are correct, these operations imply an understanding of projective space that is characteristic of human children in concrete operations. Earlier attempts to elicit map reading by chimpanzees using scale models were unsuccessful (Premack & Premack, 1983). Therefore, and in accord with other reports of operational levels of performance, we suspect that simpler mental processes underlie these performances.[4]

CONCLUSIONS REGARDING NEO-PIAGETIAN MODELS OF PHYSICAL COGNITION

What can we conclude from this brief review of various neo-Piagetian proposals? Replication studies of the sensorimotor period, which use essentially the same tasks Piaget used, were mentioned as evidence in

---

4. Piaget's concrete operations tasks in the spatial domain involve construction of spatial relations among objects and/or drawn representations of objects rather than or in addition to movements of the child's body in space. It is unclear whether the task the chimpanzees are performing is equivalent.

favor of the validity of Piaget's model. Nonreplication studies of the sensorimotor period are of two types. The first type, redesigned experiments, focuses on identifying perceptual precursors of later knowledge. Studies by Baillargeon et al. and Starkey et al. fall into this category. In the physical domain, these experiments have been interpreted to mean that infants as young as 3 months expect objects to continue to exist when they are out of sight.

Overall, neo-Piagetian studies of the sensorimotor period present the following picture: First, infants may show very early, if not immediate, understanding of certain relations on a perceptual level in the first few months of life. Second, infants create nascent forms of logical relations in object manipulation on a sensorimotor level beginning at about 6 months. Third, infants reconstruct these nascent sensorimotor forms into symbolic forms beginning at about 2 years.

Most studies of the preoperations period have focused on delineating the least well-known developmental period in Piaget's model. They have aimed at describing children's achievements of this period in positive terms, rather than in negative terms (relative to achievements of concrete operations) as Piaget tended to.

Case's reformulation, which is the broadest, suggests that the knowledge of preoperational (or relational) children is more isolated or modular than the knowledge of operational (or dimensional) children. Case (1996) suggests that the integration of these modules is made possible by maturation of the frontal lobes and other developmental changes in the brain. He emphasizes that integration also depends upon social interaction and playful problem solving. The notion that development proceeds from isolated modules to integrated central structures is especially interesting in light of the notion that the evolution of primate cognition involved a similar progression (Gibson, 1990).

In sum, these studies suggest that construction of physical cognition begins soon after birth and undergoes hierarchical integration at higher levels of organization during each major developmental period. Stated another way, these neo-Piagetian studies reveal recursive cycles of development in virtually every domain beginning in early infancy. Again, this is consistent with Piaget's general model, though the reorganizations he envisioned were more radical in nature than those envisioned by many neo-Piagetians.

Neo-Piagetian research raises new questions about the nature of the neural substrates of cognitive development. One question regards the nature of the neurological mechanisms underlying each level of knowledge—perceptual, sensorimotor, preoperational, etc. How similar or different are they? Another question regards the relationship between these nascent forms of knowledge and subsequent forms of knowledge that develop at higher cognitive levels during development. How directly related, for example, are perceptual and sensorimotor forms of knowledge in terms of their neurological substrates and their developmental courses? Are neural structures modular and isolated during early development and integrated during later development?

SUMMARY OF PHYSICAL KNOWLEDGE IN MONKEYS AND APES

Current knowledge of development of sensorimotor physical knowledge as assessed by Piaget's and Uzgiris and Hunt's measures is summarized in Tables 2-16 to 2-21 on humans, chimpanzees, gorillas, orangutans, and macaque and cebus monkeys. Given that these tables are based on studies of a few individuals of each species, they are likely to be revised in the future.

Tables 2-16 through 2-21 reveal the following patterns. First, species differ in the highest levels achieved. All the great apes but none of the monkey species complete the sixth stage of the sensorimotor period. As we see below, the great apes go beyond the sensorimotor period to develop into the preoperations period as well.

Second, the highest levels achieved vary across sensorimotor series in macaques but not great apes. Macaques develop some but not all of the fifth-stage abilities characteristic of the spatial and the means-ends scales. (They can use a support but cannot use a stick as a tool.)

TABLE 2-16 STAGES OF SENSORIMOTOR-PERIOD DEVELOPMENT OF PHYSICAL COGNITION IN HUMAN INFANTS BASED ON PIAGET'S STAGES AND THE UZGIRIS AND HUNT SCALES

Sensorimotor series	Age in months				
	Stage 2	Stage 3	Stage 4	Stage 5	Stage 6
Object concept	2	4 to 8	7 to 8	7 to 10	11 to 21
Space	2	3 to 5	6 to 9	9 to 15	18
Means-ends	2	3 to 4	8 to 9	10 to 15	19 to 22
Operational causality	2	3	5	12 to 18	18 to 21
Object relations	2	4 to 5	5 to 7	8 to 10	14 to 18
Sensorimotor intelligence	2	3	8	12	18
Imitation	4	7 to 10	11 to 14	14 to 20	18

TABLE 2-17   STAGES OF SENSORIMOTOR-PERIOD DEVELOPMENT OF PHYSICAL COGNITION IN CHIMPANZEES

Sensorimotor series	Age in months				
	Stage 2	Stage 3	Stage 4	Stage 5	Stage 6
Object concept	3	6	9	12	?
Space	3	6	8	9	40 (by block stacking criterion)
Means-ends	3	3	6 to 9	By 22	By 30
Operational causality	?	6	9 to 12	22	?
Object relations	?	?	?	?	?
Sensorimotor intelligence	2 to 5.1	5 to 6	6 to 11½ or 42	11½ to 36	?
Imitation	?	?	?	?	?

Note. ? = No data available.

TABLE 2-18   SENSORIMOTOR-PERIOD DEVELOPMENT OF PHYSICAL COGNITION IN GORILLAS

Sensorimotor series	Age in months				
	Stage 2	Stage 3	Stage 4	Stage 5	Stage 6
Object concept	1.5	3.5	7.5	7 to 9	9 to 10
Space	1.5	3.5	5	8	9 to 10[a] (by detour criterion)
Means-ends	1½	2½	4.5 to 7	36	?
Operational causality	1.5	1.5 to 3	4.5 to 8.4	9.4	?
Object relations	?	?	?	?	?
Sensorimotor intelligence	2.8	3 to 4.5	4.5 to 10	39 (semi-TCRs)	39
Imitation		7 to 24	28 to 37	?	

Note. TCR = Tertiary circular reaction. ? = No data available.
[a]Redshaw (1978).

Third, monkeys show gaps in some series. Macaques fail to display secondary circular reactions in the third stage and tertiary circular reactions in the fifth stage of the sensorimotor intelligence series. Cebus monkeys, in contrast, show most of the abilities characteristic of the means-ends scale (including use of a stick as a tool). They display secondary circular reactions but not tertiary circular reactions.

TABLE 2-19 SENSORIMOTOR-PERIOD DEVELOPMENT OF PHYSICAL COGNITION IN ORANGUTANS

Sensorimotor series	Age in months				
	Stage 2	Stage 3	Stage 4	Stage 5	Stage 6
Object concept	?	?	?	?	?
Space	?	?	?	?	?
Means-ends	?	?	?	?	?
Operational causality	?	?	?	?	?
Object relations	?	?	?	?	?
Sensorimotor intelligence	1½ to 2	2½ to 6	6 to 11½	11½ to 36	24 to 48
Imitation		?	10	15	14 to 17

Note. ? = No data available.

TABLE 2-20 SENSORIMOTOR-PERIOD DEVELOPMENT OF PHYSICAL COGNITION IN MACAQUES

Sensorimotor series	Age in months				
	Stage 2	Stage 3	Stage 4	Stage 5	Stage 6
Object concept	0.75	1 to 1.5	1.75	3.5	None
Space	0.5	1	4	6	4.5 to 5
Means-ends	0.5 to 1	1.3	2.3	By 19[a]	None
Operational causality	?	?	?	?	?
Sensorimotor intelligence	0.5 to 3	1 to 4 No SCR	2½ to 5	No TCR	None
Imitation	?	?	2½ to 6	None	None

Notes. SCR = Secondary circular reaction. TCR = Tertiary circular reaction. ? = No data available.
[a] In lion-tailed macaques (Westergaard, 1988) but not Japanese, stump-tailed, or long-tailed macaques.

Fourth, the rate of development varies across series or scales. The macaques develop faster in the fourth and fifth stages of the object concept series than in any other sensorimotor series. Cebus monkeys also develop slightly more rapidly in this series than in others. Likewise, gorillas develop somewhat faster in the fifth and sixth stages of the object concept and spatial scales (using the detour criterion) than in the means-ends scale and the sensorimotor intelligence series. Chimpanzees, like humans, seem to show greater synchrony among these sensorimotor

TABLE 2-21   SENSORIMOTOR DEVELOPMENTAL STAGES OF PHYSICAL
COGNITION IN CEBUS MONKEYS

	Age in months				
Sensorimotor series	Stage 2	Stage 3	Stage 4	Stage 5	Stage 6
Object concept	1	2 to 3	3 to 7.5	8 to 12	None
Space	?	?	?	?	None
Means-ends	1½ to 2	2 to 2½	3 to 4	9 to 18; 9 = Support; 18 = Stick	None
Operational causality	?	?	?	?	None
Sensorimotor intelligence	1	9 to 12 SCR	12	No TCR	None
Imitation	?	?	?	None	None

Notes. SCR = Secondary circular reaction. TCR = Tertiary circular reaction. ? = No
data available.

scales than gorillas, but this may reflect use of different tasks to diagnose the fifth stage in the means-ends scale. It is interesting to note that
Uzgiris and Hunt (1975) found that human infants begin the fifth and
sixth stages of the object concept series slightly earlier than the other
series.

Fifth, the rate of development in great apes slows down after the
first four stages of the sensorimotor period. Chimpanzees develop at approximately the same rate as human infants during the first four stages,
completing them by about 1 year. Then their developmental rate slows
(except in the object concept series) so that they complete the sixth
stage at about 3 or 4 years in the causality and spatial domains (Poti' &
Spinozzi, 1994).

Sixth, there are similarities among the great apes in overall levels
of achievement. These similarities are accompanied, however, by slight
differences in the rates of development among species. These preliminary data suggest that gorillas develop somewhat more rapidly than
chimpanzees in the Uzgiris and Hunt scales, but at roughly the same rate
in Piaget's sensorimotor intelligence and imitation series (Chevalier-
Skolnikoff, 1983; Parker, in press). There are no strictly developmental
data on orangutans and bonobos based on the Uzgiris and Hunt scales.
Two reports on the development of Piaget's sensorimotor intelligence
series and imitation series suggest that orangutans develop at roughly

the same rate as gorillas and chimpanzees in these series (e.g., Chevalier-Skolnikoff, 1983; Miles, 1990; Miles et al., 1996).

Gaps in these tables indicate lacunae in our knowledge of true developmental patterns in all the sensorimotor-period scales and series of the bonobos and orangutans, the lesser-known great apes. Likewise, they indicate gaps in our knowledge of lesser apes, New World and Old World monkeys, especially baboons.

SUMMARY OF SENSORIMOTOR-PERIOD DEVELOPMENT OF PHYSICAL KNOWLEDGE IN MONKEYS AND APES

The foregoing studies suggest the following patterns. First, macaques, cebus, great apes, and humans traverse the same sequence of stages in the development of physical cognition during the sensorimotor period.

Second, macaques and cebus truncate their development at earlier stages than great apes. Specifically macaques peak at the fourth or fifth stage in causality, but reach the fifth stage in the object concept series and possibly in the spatial series. Cebus achieve the fifth stage in all the series including causality. Rhesus macaques develop very rapidly; cebus develop faster than great apes and slower than macaques.

Third, both great apes and humans complete all the sensorimotor-period stages in physical cognition, whereas macaques and cebus monkeys do not.

Fourth, great apes traverse the developmental stages at a slower rate than human infants do after the first year of life, completing the sixth stage between 3 and 4 years of age as compared to 2 years in humans.

Fifth, great apes traverse developmental stages at different rates in different domains, unlike human infants who traverse all the series at almost the same rate. Specifically, great ape achievement of the fifth and sixth stages in the various sensorimotor series in physical domains differs as follows. The fastest, earliest developing series is in object permanence. This series is completed by about 1 year of age in all the great apes.

If we use stringent criteria, the slowest, latest-developing sensorimotor abilities within the physical domain are spatial and causal knowledge. Using intelligent use of a stick to reach an object as the criterion, fifth-stage causal knowledge is completed at about 3 years of age in chimpanzees and gorillas. Using stacking blocks or putting objects in containers as the criterion, fifth-stage spatial knowledge is completed at about 4 years in chimpanzees and 2 years in orangutans.

Finally, these data also suggest some species differences among the great apes in the rate of development of sensorimotor-period series. Gorillas seem to develop most rapidly, and orangutans most slowly. These differences correlate with differences in the life histories of these species (discussed in chapter 7).

As we indicated in the preceding section, several models have been proposed to describe the development of physical knowledge during the preoperations period. Consequently, we must select a model for comparative developmental studies of specific subdomains. This is necessary because the highest-level abilities of the great apes probably fall under the rubric of symbolic-level thought during the early preoperations period. We have used Case's model to interpret various reports of physical knowledge that might qualify as symbolic.

SUMMARY OF PHYSICAL KNOWLEDGE IN THE PREOPERATIONS PERIOD IN GREAT APES

The foregoing data on the levels of causal and spatial understanding revealed by adult chimpanzees and juvenile and adult orangutans suggest that they surpass the sixth-stage sensorimotor level and achieve at least symbolic levels of ability in these domains. Data on the ages at which they achieve the higher levels of physical knowledge in various domains are sketchy. In some cases these abilities seem to develop by 3 or 4 years, and in others they seem to develop as late a 8 or 9 years. More strictly developmental studies of spatial and causal knowledge of juvenile great apes are badly needed to flesh out the developmental picture.

The foregoing studies suggest that in the physical domains of space and causality, great apes apparently achieve symbolic levels of development spontaneously. In other words, they perform at a functional or optimum level in the domains of space and causality. The sparse developmental data suggest that they develop their highest abilities in these domains as late as 8 or 9 years as compared to about 3 or 5 years in human children.

CAUTIONS REGARDING PHYSICAL KNOWLEDGE IN MONKEYS AND APES

It is important to emphasize that these patterns in the cognitive development of great apes are based on a few studies of a few individuals. They are also based on studies using different instruments and methods to measure development. It is notable that in some studies only one or two individuals achieved the highest stages reported. This fact reflects the marked motivational and intellectual variation among individual great apes and cebus monkeys of the same species. The existence of this variation may also indicate that the behaviors in question are at the outer margins of the species' abilities.

The absence of strictly developmental studies of symbolic abilities in the physical and logical domains emphasizes the need for developmental studies of individuals between the ages of 3 and 10. The sparse data on bonobos, gorillas, and orangutans point to the need for further developmental research on these species. Such studies are prerequisites to identifying differences among the great apes. The absence of developmental data on cognitive development in lesser apes, baboons, langurs, and New World monkeys other than cebus emphasizes the need for studies of these species. Finally, standardization of criteria for various developmental stages will be important in clarifying the picture of cognitive development in monkeys and apes. All of the results reported here should be verified by additional studies.

Some of the events considered in this analysis occurred weeks or even months apart. Perhaps primates keep track of favors given and received, and make assistance to others conditional on the way they have been treated in the past.

DE WAAL, 1996B, P. 157

CHAPTER 3
DEVELOPMENT OF LOGICAL-MATHEMATICAL COGNITION IN CHILDREN, APES, AND MONKEYS

We are accustomed to thinking of apes as tool users and hence as physicists, but less accustomed to thinking of apes as accountants and classifiers. Comparative studies, however, suggest that great apes display logical-mathematical abilities comparable to those of young children. Logical-mathematical knowledge includes logic and classification, number, seriation, and conservation. This chapter reviews comparative studies of these subdomains in human children, monkeys, and apes. Before addressing the comparative data, we will briefly characterize Piagetian and neo-Piagetian ideas about these forms of knowledge.

PIAGET'S APPROACH TO LOGICAL-MATHEMATICAL THINKING

In his early work Piaget focused his inquiry into logical-mathematical knowledge on the concrete and formal operational periods of development. He barely considered the earliest precursors of this form of knowledge in the actions of preoperational children. In later years, however, he turned his attention to earlier expressions of this form of knowledge.

In some of their later work, *Epistemology and Psychology of Functions* (Piaget, Grize, Szeminska, & Bang, 1977) and *Toward a Logic of Meanings* (Piaget & Garcia, 1991), Piaget and his colleagues returned to the issue of the origins of logic in the late sensorimotor and symbolic subperiod of preoperations and began research into the development of "protologic" and "functions" (Langer, 1980, 1986). The purpose of the second book was "to describe some elementary forms of children's responses to simple problems, in order to show the psychogenetic roots of logical relations, leading to operations and their compositions in structures" (Piaget & Garcia, 1991, p. 159). Consistent with earlier formulations, they conclude that, in their most elementary forms, inferences are implications between meanings that are attributed to actions and objects. The meaning of an object at this level is what can be done with it, whereas the meaning of an action is what can be done with it or what it can lead to. They also conclude that the elementary forms of operations (characteristic of concrete and formal operations) can be discerned at the level of actions. At this level, however, they are seen separately and in isolation in the context of specific meanings rather than integrated into a system. It is this latter formulation that has been elaborated by Langer (1980, 1986) and some other neo-Piagetians.

NEO-PIAGETIAN APPROACHES TO LOGICAL-MATHEMATICAL KNOWLEDGE

As in the case of physical knowledge, neo-Piagetian studies of logical-mathematical knowledge fall into three categories: *replications, focal redesigns,* and *systematic reformulations.* The major theme in reformulation studies has been to trace the roots of logical-mathematical thought back to the sensorimotor period (Langer, 1980, 1986). The major theme in redesign research has been to show that Piagetian tasks can be solved by children who are younger than Piaget's subjects were when they solved the related tasks he posed. Work by Spelke, Starkey, Gelman and Gallistel, and others reflects this approach.

This chapter is organized topically by subdomain of logical-mathematical knowledge. Under each topic we discuss Piagetian and neo-Piagetian approaches in human children, monkeys, and apes. Finally, we address possible expressions of logical-mathematical abilities in natural settings.

LOGIC AND CLASSIFICATION IN HUMAN CHILDREN
INHELDER AND PIAGET'S HIERARCHICAL CLASSIFICATION

In their book, *The Early Growth of Logic in the Child,* Inhelder and Piaget (1964) trace the development of *classification* and *seriation* in 3- to 12-year-old children. They distinguish three stages of development of the capacity for *hierarchical classification:*

> Stage I, during the symbolic subperiod of preoperations at 2½ to 4½ years—The child cannot find or maintain a general criterion for clas-

sifying elements. He fails to anticipate, he shifts criteria midway, for example, from shape to color, and consequently fails to produce consistent classes by building one at a time. Piaget and Inhelder called the resulting constructions *graphic collections*.

Stage II, during the transition from preoperations to early concrete operations at 4½ to 7 years—The child is able to produce a consistent classification of objects according to a single criterion (e.g., shape) but does so rather painstakingly through trial and error. Piaget and Inhelder called the resulting constructions *nongraphic collections*.

Stage III, during concrete operations after 7 years—The child takes a systematic approach to classifying objects into classes, constructing several classes at a time. She also understands hierarchical inclusion and exclusion, for example, that there are more beads than there are red or blue beads. She also understands that it is possible to reclassify the same objects according to a second or third criteria (e.g., color or size). This is known as *multiplicative classification*.

LANGER'S SENSORIMOTOR LOGIC AND CLASSIFICATION

Following up on themes from Piaget's later work on logic, Langer (1980, 1986) has demonstrated that infants compose fleeting versions of classes and logical operations on classes as they play with objects. Because these compositions may exist for milliseconds, recognizing them requires microanalysis of videotaped sessions of infants manipulating a standardized set of objects. (Langer calls these fleeting compositions *mobiles* as opposed to the longer-lasting sets of toddlers, which he calls *stabiles*.)

Langer (1996) distinguishes two types of "proto-operations" in infants. *First-order proto-operations*, which dominate the first year, involve construction of single sets of two objects, on which an infant may map simple operations onto the elements. *Second-order proto-operations*, which develop in the second year, involve construction of double sets of several objects each, on which infants perform reciprocal mappings. Langer describes the following developmental sequence.

At 6 months, many infants produce quantitative equivalences within a single set by substituting objects, for example, dropping a doll into a ring, removing it, and dropping another doll in the same ring. By 12 months all infants do this. At 6 months infants also compose sets of objects from different classes (e.g., pairing crosses with rectangles). By 8 to 10 months, their compositions are random.

At 12 months, about half the infants extend substituting within sets to three-object sets. At 15 months they compose sets of identical objects, At this time they also begin to substitute or exchange items between two contemporaneous sets, for example, exchanging a block between two stacks of blocks. At 18 to 36 months, infants construct two contrasting classes of objects. These constructions of and operations on two contemporaneous sets are second-order proto-operations.[1]

Langer's work suggests that similar logical structures recur in more elaborated form at each stage of development; first at the level of actions, then of symbols, then of (stable) object relations, then of concepts. The parallels among the structures at each developmental period are much stronger than initial formulations of Piaget's theory would suggest. As indicated in chapter 1, however, Piaget and his associates were moving toward a parallel formulation (see Piaget & Garcia, 1991)

LOGIC AND CLASSIFICATION IN MONKEYS AND APES

Premack (1976, 1983) was probably the first to study logical-mathematical knowledge in great apes. In analyzing his results, he distinguished two kinds of codes, imaginal codes representing such things as spatial location and abstract or prepositional codes representing relations between relations (Premack, 1983). Abstract codes include, for example, such logical categories as *same versus different, all, one, none,* and *some.* (This distinction shares some elements with Piaget's dichotomy between figural and operative knowledge and his dichotomy between physical and logical-mathematical cognition within operative knowledge.) Premack concluded from his research that chimpanzees and probably other primates naturally use imaginal codes but only chimpanzees are capable of using abstract codes.

In Premack's symbol-training program, Sarah and two other chimpanzees were first trained to understand a string of plastic elements representing such simple propositions as *apple is red, round shape of apple.* Following this, they were trained to replace an interrogative particle ("?") with the correct element. Examples include *red apple ? apple (color of).* Finally, they were trained to recognize such relations between elements as *same versus different* and *all versus none.* As Premack notes, the *same versus different* relation depends upon an analogy.

1. Sometime after 36 months, young children begin to construct three contrasting classes or sets of objects—for example, three sets of cups of contrasting colors. Three-category classifying opens up true hierarchical classification (Langer, 1998).

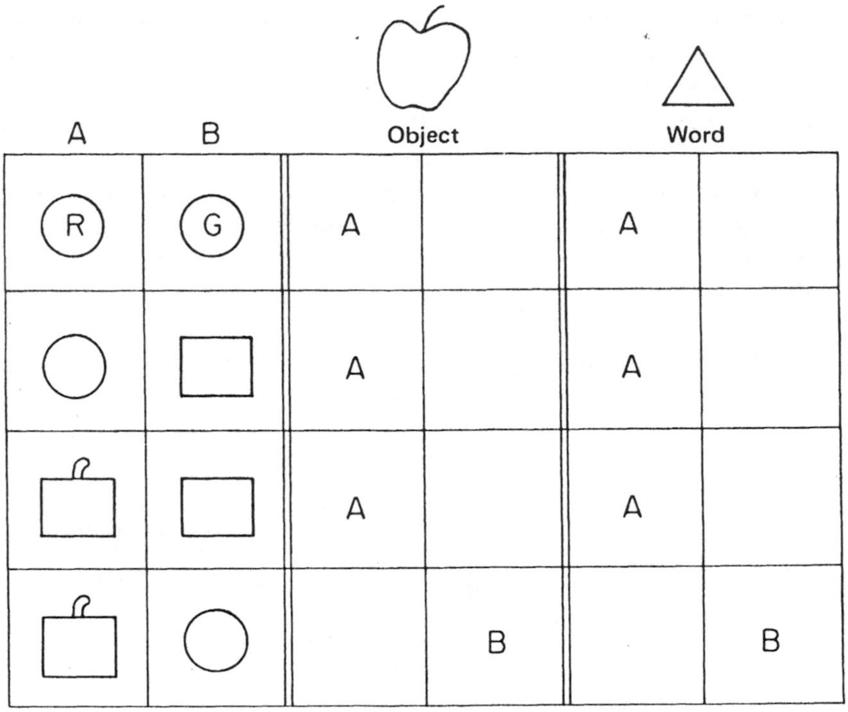

FIGURE 3-1
Multiplicative classification table of feature analysis of the object apple and the symbol for apple by Sarah (Premack, 1976).

Premack (1976) also tested Sarah's ability to classify an apple and its symbol (a triangle) according to the features of color, shape, protuberance: "She chose red over green, round over square, square with a stem-like protuberance over plain square, and for the most part round over square with a stem. . . . She was consistent in her choice of three features and inconsistent on the fourth" (Premack, 1976, p. 170). She repeated the performance exactly when the word for apple (a triangle) was presented. See Figure 3-1 for a depiction of the multiplicative table.[2]

Braggio, et al. (1979) reported similar results in their study of multiple-classification tasks at Yerkes Regional Primate Research Center. Their subjects were six juvenile chimpanzees whose average age was 5½ years.

2. Hayes and Nissen (1971) reported that Vicki, the Hayeses' home-reared chimpanzee, spontaneously sorted her button collection into three contrasting sets, according to color, size, and shape. This indicates that at least some individual chimpanzees are capable of spontaneous multiple classification.

They presented the chimpanzees with a three-dimensional plastic matrix with four cells in which they placed subsets of 3 of 44 plastic objects comprised of combinations of four colors and various shapes, letters, and numbers. In the first experiment the proper response was to place in the fourth cell an object similar in color to the object on one side and similar in shape to the object on the other side. The large number of possible combinations allowed presentation of unique sets in each trial. Over hundreds of tests, four of the chimpanzees achieved above chance levels of correct choice (one individual, Grip, was correct about 67% of the test) and decreased the number of trials to achieve the performance criterion.

Performance on these multiplicative classification tasks seems to parallel that of 5- or 6-year-old children who are in the intuitive stage of the preoperations period. The less than perfect scores suggest the kind of directed groping that precedes the systematic and confident behaviors of concrete operations.

Over a period of years, Premack and his colleagues taught a symbol system to several chimpanzees, first and most notably to Sarah, and later to several others, including Peoney and Elizabeth. He also studied several chimpanzees who had not been trained to understand symbols. These included Bert, Luvie, Sadie, and Jessie.

When Premack administered tests requiring what he calls an *imaginal code*, he found that both the symbol-trained and nonsymbol-trained groups did equally well. When he gave tests that required an *abstract code*, however, he found that only the symbol-trained chimpanzees succeeded. Premack therefore hypothesized that symbol training enhanced chimpanzees' facility with abstract codes. By symbol training, Premack apparently meant lexicon and sentence training (Premack, 1988a, 1988b; Thompson, Boysen, & Oden, 1997). In contrast, we use the term *symbol training* to refer to training in abstract symbols of any kind, lexical and numerical.

Recently, Thompson et al. (1997) have demonstrated that any training in a functional correspondence between any arbitrary token and an abstract relation (lexical, numeric, or other) is sufficient to allow chimpanzees to match relations between relations. These investigators compared the performance of one juvenile (Bobby) and four adult chimpanzees (Sarah, Sheba, Darrell, and Kermit) in a modified analogy problem requiring conceptual matching to sample. Sarah had lexical and language symbol training, the three other adults had token and numerical symbol training, and the juvenile had neither.

The four adults conceptually matched relations among relations without explicit training at levels significantly above chance. These performances were indistinguishable from their performances on physical matching tasks for which they had received rewards. Moreover, they continued to perform at the same level on novel conceptual tasks involving pairs of items whose associative histories in the physical matching tasks would have led to decrements in performance if the choices were based on simple association. Thompson et al. (1997) therefore conclude that these four subjects were using a conceptual response strategy of matching relations among relations.

They also conclude that, contrary to Premack's (1988b) conjecture, these performances depend neither upon language training nor upon persistent "dogged training" over thousands of trials (dogged training does not generalize to novel tasks). Rather, they argue that training in any kind of symbolic tokens (including numbers) allows chimpanzees to re-encode abstract relations exemplified in physically disparate pairs into iconically identical symbols (Thompson et al., 1997).

DEVELOPMENT OF CLASSIFICATION

The only strictly developmental study of classification in apes was done by Spinozzi (1993) using Langer's model. Spinozzi used the same data set on the same subjects described above in Poti's (1996) study of space (see chapter 2), but analyzed it in terms of the structure of classification.

Subjects were given sets of 6 objects selected from among 16 objects of four forms (cups, rings, blocks, and crosses) and four colors (blue, green, red, and yellow). The sets of 6 objects fell into one of three class conditions:

1. two classes of 3 identical objects (additive condition)

2. two classes of 3 objects differing in both color and form (disjoint condition)

3. six objects with either three forms and two colors or vice versa (multiplicative condition)

The resulting constructions were scored according to set size and composition (the sequence of object selection was also scored). Single sets were classified as either unmixed (items with consistent characteristics—similarity or identity) or mixed (with inconsistent characteristics—difference). Classification of objects within a single set is *first-order-classification*. Double or greater sets were classified as either same-same, different-different, or same-different. Classification of objects within one set relative to objects in another set is *second-order-classification.*

This analysis of set size revealed a clear developmental progression in set construction in chimpanzees from single to double or larger sets: until 1 year (15 months), single sets predominate; at 2 to 2½ years (23 to 30 months), double and larger sets increase significantly, though they remain the minority. This pattern compares with the human pattern of transition from single to double sets at about 1 to 1½ years.

FIRST-ORDER CLASSIFICATION

The analysis of within-set composition revealed a developmental progression in at least one chimpanzee from random compositions at 15 months, to grouping by similarity in two-object compositions in the additive and disjoint conditions (but not in the multiplicative condition) at 2 years (23 months), to grouping by similarity in three-object compositions under all conditions at 3 years (37 months), and then to classification by similarity in two-object compositions under all conditions at 3½ years (41 months).

SECOND-ORDER CLASSIFICATION

The analysis of between-set compositions revealed a progression from random relationship between sets at 3 to 3½ years (37 to 41 months) to a substantial number of compositions by similarity or identity between sets under additive and multiplicative conditions by 4 to 4½ years (51 to 54 months). In other words, by 5 years of age, chimpanzees can coordinate relations of similarity between sets of two or three objects. Human infants show this ability by 2 years of age. Macaque and cebus monkeys develop first-order classification by 3 years (34 months) and 4 years (48 months), respectively. Neither monkey species achieved second-order classification (Spinozzi & Natale, 1989).

The analysis also revealed developmental patterns in criteria for object selection within sets. Chimpanzees and human infants showed similar developmental sequences in the construction of both first- and second-order classes. Human infants begin constructing first-order sets at 6 months by composing single categories of different objects. Then, at about 10 months, they switch to random composing. At about 12 months, they switch to composing single categories of identical objects (same). At about 15 months, they compose similar objects into single sets. In second-order sets, they begin by composing two sets, each containing identical sets of different categories (same-different; e.g., all red cylinders in one set and all green crosses in the other) at about 18 months and do so consistently by 36 months.

Regarding compositions in first-order classification in chimpanzees, Spinozzi found that at 12 months classification was random; at 23 and 30 months, they showed classification by identity or similarity in most compositions. Regarding compositions in second-order classifications, she found that up to 37 to 41 months none of her subjects constructed same-different relations between two sets. By 51 to 54 months, however, they did so.

As Spinozzi points out, this study is significant both for its developmental data and for its demonstration that chimpanzees that have not been trained in symbol use have the same capacities as those that have. Similar results were reported by Matsuzawa (1990) in his study of spontaneous classification by an adult female chimpanzee, Ai. However, Ai had been symbol trained.

CLASSIFICATION IN MONKEYS

Macaques and cebus monkeys, in contrast, differ from humans and chimpanzees and from each other in the sequence of compositions they construct. In macaques (studied at 22 and 34 months of age), the trend was from classification predominantly by identity and similarity to systematic classification by similarity. In cebus, the sequence was from classifying partially by identity and similarity (at 16 months) to classifying by difference (at 36 months) to classifying by similarity and identity (at 48 months) (Spinozzi, 1993).

SENSORIMOTOR LOGIC IN PRIMATES ASSESSED BY LANGER'S CRITERIA

Poti' (1997) has used Langer's (1980, 1986) developmental model for protooperations of classification and logical groupings in comparative studies of monkeys and apes. The analysis of protologic used the same data on the same objects by the same subjects as reported above in the study of classification. In this case, the data were analyzed in terms of the composition of the sets and the relationships generated within the elements in a set and between elements in different sets.

The resulting compositions were analyzed using the categories of (1) set size, that is, number of objects in each set; (2) set type, that is, single sets versus binary or multiple sets; (3) sequential transformations within single sets over time (first-order operations); and (4) relations between binary sets (second-order operations).

TABLE 3-1 SOME PARAMETERS OF SET COMPOSITION IN
PROTO-OPERATIONS BASED ON LANGER'S MODEL OF LOGIC

Proto-operation	Single sets—first order	Binary sets—second order

Qualitative equivalence	Exchange operations
	    Replacement
	    Commutivity
	    Substitution
	Reversibility structures
	    Inversion on commutivity
	    Inversion of substitution
		Biunivocal correspondences (1 to 1)
		Exchange operations
		    Replacement
		    Commutivity
		    Substitution
		Reversibility structures

Ordered (quantitative) equivalence operations	Increasing quantity Decreasing quantity	Coordinated series Equivalence and order Adding or subtracting sets

Note. From Poti' (1996).

Sequential transformations within single sets over time include exchange operations (replacement, rearrangement [commutivity], and substitution) and reversibility operations (inversion on rearrangements and substitutions and on series).

Relations between binary sets include equivalence second-order operations between sets of the same size (one-to-one correspondences, exchanges between sets, replacement, rearrangement, and substitutions) and ordered nonequivalence second-order operations between sets of differing sizes. Table 3-1 summarizes some parameters of set composition in proto-operations.

Poti's analysis revealed several patterns. First, like human infants, monkeys and chimpanzees construct single sets of small numbers of items. Second, only chimpanzees and humans construct double sets. Third, only chimpanzees and humans construct temporally overlapping sets, which are prerequisites for double sets. Fourth, like human infants, chimpanzees construct double sets and engage in second-order operations such as exchanging objects between sets. Fifth, unlike human children, chimpanzees fail to develop reversible operations, that is, inverting of exchanges.

As shown in Table 3-2, macaques and cebus are limited to constructing single sets of two or three elements. These are first-order proto-operations similar to those of human infants in the fourth stages of the sensorimotor period. Chimpanzees, in contrast, develop the ability to construct temporally overlapping double sets. These are second-order

TABLE 3-2 DEVELOPMENTAL STAGES OF SENSORIMOTOR PROTOLOGIC IN MONKEYS, GREAT APES, AND HUMANS (AGES IN MONTHS)

	First-order proto-operations on single sets: temporally isolated sequential overlapping			Second-order proto-operations on double sets: temporally overlapping
	Piagetian equivalent			
	Sensorimotor stages			Sensorimotor sixth stage/ early symbolic subperiod
Species	Fourth	Fifth	Sixth	of preoperations
Humans[a]	6 to 8	12 to 15	18	12 to 36
Chimpanzees	?	?	?	51
Bonobos	?	?	?	?
Gorillas	?	?	?	?
Orangutans	?	?	?	?
Gibbons	?	?	?	?
Baboons	?	?	?	?
Macaques[b]	22 to 34	22 to 34	Rare	None
Cebus[b]	16 to 36	36 to 48	Rare	None

Note. ? = No data available.
[a]Langer (1980).
[b]Poti' and Antinucci (1989).

proto-operations similar to those of human children in the fifth and sixth stages of the sensorimotor period. In contrast to human children who develop second-order proto-operations between 1 and 2 years of age, chimpanzees develop these proto-operations when they are 4 years old. Again, it is important to note that these are spontaneous behaviors that represent functional or, at most, optimal levels of performance in Fischer's terms.

Poti' notes that chimpanzees differ significantly from human infants in the way they manipulate objects: "The general and peculiar characteristics of all chimpanzees' operations is that operations strictly depend on the momentary features of the ongoing process of construction" (Poti', 1997, p. 57). In other words, chimpanzees continue to produce mobile constructions, whereas human infants increasingly produce stable constructions, which allow them to detach the object configurations from their own actions. Poti' argues that it is the externalization of actions and detachment of interobject relations that allows human infants to reflect and elaborate on observable relations among objects.

NUMERICAL KNOWLEDGE IN CHILDREN

PIAGET'S STAGES

In his book *The Child's Conception of Number*, Piaget (1965a) follows the development of concepts of number, which he sees as the integration of classes and series. Conservation of discontinuous quantity, that is, of discrete, countable objects, is the analog of conservation of quantity of continuous, noncountable objects. He distinguishes stages of development of one-to-one correspondence that parallel those in the development of conservation of continuous quantities:

Stage I, during the intuitive stage at 4 to 5 years—The child recognizes one-to-one correspondence between two sets of elements when they are spatially arrayed next to one another but is unable to see that the correspondence exists when they are spatially rearranged.

Stage II, during the transition to concrete operations at 5 to 6 years—The child waivers and shows uncertainty regarding the conservation of correspondences when the objects in one set are spatially transformed.

Stage III, during concrete operations after 7 years—The child is convinced that the quantity is conserved and demonstrates it by counting the elements or placing them back in one-to-one correspondence with elements in the other set.

Piaget also traces development of the concepts of *ordination* (the number of elements at a given point in the sequential order of a set of elements) and *cardination* (the number of elements in a set represented by a numeral). He describes three stages, based on a task in which the child is asked to tell how many stairs a doll has to climb to reach a given point:

Stage I, during the intuitive subperiod at 4 or 5 years—The child fails to understand that she has to figure out the number of steps in serial order.

Stage II, during the transition at 6 or 7 years—The child understands that she has to figure out the number of steps in serial order, but counts the whole series above and below the designated step and has trouble translating the step in the series (ordinal number) with the numeral represented by that step (cardinal number).

Stage III, during concrete operations after 7 years—The child immediately counts to the right step (ordinal number) and pronounces the numeral represented by that step (cardinal number). He understands that the total number of steps is the sum of the step the doll occupies plus the remaining steps in the series.

As we indicated above, Piaget's description of the preoperations period is couched primarily in terms of young children's disabilities relative to the abilities they display in the concrete operations period. As in the case of gaps in the formulation of the sensorimotor period, Piaget and his colleagues had begun to address this limitation in his model at the end of his life (Piaget & Garcia, 1991).

Following this lead, Robbie Case has developed an integrated model for mental development in children 4 to 10 years of age. This period embraces the end of his relational period and the bulk of his dimensional period. He proposes that they display "central conceptual structures" that cut across physical, logical, and social domains (Case, 1996). We briefly describe his formulation of numerical and spatial knowledge during this period.

CASE'S STUDIES OF NUMBER

In his reformulation of numerical knowledge in the preoperations period, which he calls the *relational period*, Case (1985) describes the developmental sequence in children from 1 to 5 years:

1. Operational consolidation of elaborated sensorimotor coordinations in which children can sequentially touch a series of items when prompted.

2. Bifocal coordinations (2 to 3 years) in which children can sequentially touch a series of items while simultaneously using number words to label two to five items. This requires coordination of the movements with objects and the words with the movements.

3. Elaborated coordinations (3½ to 5 years) in which children can count a random spread of items or a subset of items. This requires elaboration of the preceding coordinations.

In a subsequent discussion, Case (1996) elaborates on his model. He notes that during the relational period children can use number tags and can make nonnumerical judgments of quantity. He emphasizes, however, that they are unable to integrate these two abilities until they enter the *dimensional period* at about 6 years of age. Until then they answer the question "Which is more/ which is bigger?" at chance levels despite their ability to count to five and judge which of two arrays is larger.

NEOINNATIST STUDIES OF SENSORIMOTOR NUMBER

INFANT NUMBER STUDIES

Starkey, Spelke, and Gelman (1990) devised a series of habituation experiments to test the numerical knowledge of infants. They also employed the two- test design and relied on longer looking time to reveal surprise. They reported that infants between 6 and 9 months of age were able to recognize the correspondence between two visual arrays of the same number of objects versus two visual arrays of different numbers of objects. They also reported that infants of this age were able to recognize the correspondence between visual arrays of a given number of items and auditory displays of the same number of items versus noncorrespondence between number of items in visual and auditory displays. In both cases these abilities were limited to arrays of two or three items.

On the basis of these results Starkey et al. conclude that infants are capable of establishing one-to-one correspondences of small numbers of items. They also conclude that children have some sort of representation of numerosity across modalities that does not depend upon language or counting or cultural learning.

Starkey (1992) also devised a series of experiments to test the understanding of numerical reasoning and numerical abstraction in children 1 to 4 years of age. One set of experiments involved a procedure of placing balls into a box and then asking the child to remove the same number that had been inserted. He found that children as young as 18 months could perform this numerical reasoning task when the number of items was two. Children 3 years and older could do so when the number of items was three. Another of his experiments involved adding or subtracting items after the child had inserted them, then asking the child to remove the number of items that resulted.

He found that even 24-month-old children correctly adjusted their behavior according to whether items had been added or removed. When the number of items remaining was one or two, all the children 36 months and older gave correct answers above chance levels, thereby displaying numerical reasoning abilities earlier than Gelman and Gallistel reported.

PREOPERATIONAL NUMBER STUDIES

Gelman and Gallistel (1978) designed a series of experiments that provide evidence that children as young as 2½ years understand some aspects of number that Piaget attributed to 6- or 7-year-old children and that children as young as 4 years understand all the aspects of number that he attributed to concrete operational children. The experiments these investigators used were quite different from those Piaget used.

Gelman and Gallistel presented their tasks as magic games. First, children were asked to judge whether an array of n items still contains n items after the experimenter alters the length, color, and identity of items in the array. Then they are asked to undo the changes. Gelman and Gallistel found that preschool children judged whether two sets of objects are numerically equal on the basis of their cardinal number when counted. They also found that the children often constructed two sets with the same number of items when asked to undo the changes. They distinguished between numerical reasoning and numerical abstraction.

Gelman and Gallistel describe three developmental stages in understanding of number:

Stage I, numerical knowledge in 2 year olds—Characterized by representations of specific numerosities, that is, by the ability to count small arrays sequentially (though the specific sequence of numbers may be idiosyncratic, they are consistently applied, e.g., 1, 3, 4, 6). They can compare the cardinal number of two arrays but cannot conserve the number of items in an array that is displaced relative to another array using one-to-one correspondence.

Stage II, semialgebraic knowledge in 3 year olds—Characterized by the ability to reason about effects of transformations on the number of items in a single array but not the effects of transformations of the relationship between the number of elements in two arrays.

Stage III, algebraic knowledge in 4 year olds—Characterized by understanding of relations among numerosities, that is, by the ability to manipulate systems of numerical relations rather than specific numbers. Hence they are able to conserve the number of items in an array that has been transformed relative to another array by using one-to-one correspondences.

Gelman and Gallistel note that their experiment differs from Piaget's experiment in that in his experiment the two arrays of objects were placed next to each other, whereas the two arrays were separated in their own experiments. They suggest that the perceptual array creates two problems for the preschool child. First, it invites a judgment of equivalence in a form that exceeds the child's reasoning principles. Second, it impedes the application of accessible reasoning based on cardinal value achieved by counting each row. In other words, the nature of the conservation task requires algebraic reasoning.

Starkey (1992) noted that, contrary to Gelman and Gallistel's claim, many of the children who correctly performed these tasks did not yet know how to count. They also noted that even those children who knew how to count did not do so during the tasks. He therefore concludes that early numerical reasoning abilities cannot rely on counting as these authors claim.

Case (1996) in his reformulation of the development of number (see next section) notes that although young children may be able to count a limited number of items, they do not know how to use numbers to judge quantities.

NUMBER IN APES

In one of the earliest publications on the subject, Romanes (1898) reported that he taught a chimpanzee to respond to verbal requests for one to six straws. She did so with few errors, but was inaccurate when asked to give eight, nine or ten straws. He also reported that she bent over straws to double them rather than picking up additional ones, suggesting some rudimentary notion of multiplication. His pupil, Morgan (1895), disagreed with Romanes's interpretation, arguing that the performance could be explained by simpler processes. In the United States, Yerkes (Yerkes, 1916; Yerkes & Yerkes, 1929) championed Romanes's view, whereas Watson championed Morgan's view (Rilling, 1993).

In his review of studies of counting in primates and other animals, Rilling (1993) notes that these two opposing perspectives have been reflected in research on number from that day till this. They are reflected in continuing disputes over definitions of numerical competence (Davis & Perusse, 1988). Although a steady stream of papers on number in animals were published by proponents of both views between 1900 and 1990, until the seventies only a few of the studies focused on primates. (A notable exception was Ferster's training study [Ferster, 1958] in which he taught two chimpanzees to match binary number symbols to arrays of one to seven items.)

Premack (1976, 1983) tested the symbol-trained chimpanzee Sarah and nonsymbol-trained chimpanzees on analogies between items that had been divided into various proportions. First, they were asked to match to a sample of, for example, one-fourth apple from two alternatives, for example, one-fourth apple and three-fourths apple. All passed this test. Next, they were asked to match to a sample of a glass cylinder that was one-half full to a sample of one-half apple or three-fourths apple. Only Sarah passed this test. Premack argues that the latter test involved an abstract code representing proportions. This task seems to require bifocal dimensional coordinations characteristic of 6-year-old children in the late intuitive substage of preoperations.

Matsuzawa (1985b) taught his language-trained chimpanzee, Ai, to label arrays of one to six objects with six symbols on a computer keyboard. Likewise, Rumbaugh and Washburn (1993) taught their symbol-trained chimpanzees, Lana, Sherman, and Austin, to place either arabic numerals or black dots on boxes in accord with the value of a target number.

In a recent study, Boysen and Berntson (1990) followed the development of numerical abilities in chimpanzees. They report that their subjects, Kermit, Darrell, and Sheba, were able to sequentially tag several items by 2½ to 3½ years (30 to 40 months) of age (Table 3-3). This performance is typical of human children 1 to 1½ years in Case's operational consolidation stage (sixth-stage sensorimotor period). They were able to do sequential tagging by 4½ years (54 months) of age. This is typical of human children 1½ to 2 years old in the operational coordination stage (sixth-stage sensorimotor period). They were able to do sequential tagging and labeling by 6 to 7 years (72 to 84 months) and counting by 6½ years (78 months). These abilities are typical of human children 2 to 3½ years of age in the bifocal coordination stage (symbolic subperiod of preoperations), and 3½ to 5 years of age in the elaborated coordinations stage (intuitive subperiod), respectively.

As Table 3-3 indicates, with training, chimpanzees seem to be able to achieve numerical knowledge similar to that of human children in the intuitive subperiod of preoperations. This, of course, reflects a scaffolded rather than a functional level of performance (in Fischer's terms). It is notable that, as they develop, they increasingly fall behind human children. At the end of the sensorimotor period, chimpanzees are about 3 or 4 years old as compared to 2 years in children; in the middle of the symbolic subperiod, chimpanzees are 6 or 7 years old as compared to 3 to 4 years in human children; and in the intuitive subperiod, chimpanzees are 8 or 9 years old as compared to 4 or 5 years in human children.

TABLE 3-3 DEVELOPMENTAL STAGES OF NUMBER IN MONKEYS, GREAT APES, AND HUMANS BASED ON CASE'S RELATIONAL STAGE (AGES IN MONTHS)

	Case's stages			
	Operational consolidation: making 1-to-1 correspondences	*Unifocal coordination: sequential tagging*	*Bifocal coordination: coordinating, tagging, and labeling*	*Elaborated coordination: counting subsets of an array*
	Piagetian equivalent			
Species	*Sensorimotor sixth stage*	*Sensorimotor sixth stage/ preoperations symbolic*	*Preoperations symbolic*	*Preoperations intuitive*
Humans[a]	12 to 18	18 to 24	24 to 40	(40 to 60)
Chimpanzees[b]	30 to 40	54	(72 to 84)	(78)
Bonobos	?	?	?	?
Gorillas	?	?	?	?
Orangutans	?	?	?	?
Gibbons	?	?	?	?

Note. Case's relational stage is roughly equivalent to Piaget's sixth sensorimotor stage and symbolic substage of preoperations). ? = No data available.
[a]Case (1985).
[b]Boysen and Bernston (1990).

SERIATION IN HUMAN CHILDREN

Brainerd (1974), among others, designed a transitivity task that apparently showed that children were able to solve the (operational) task at an earlier age than Piaget had reported. Their reformulation of Piaget's task, however, changed a crucial element. Piaget (1968) designed the experiment so that three sticks A, B, and C of increasing length were presented in pairs A with B and B with C so that the child could compare the length of each pair but could not compare the three simultaneously. This required that when the child was finally asked whether A is smaller than C while A is hidden she had to infer the answer.

In Brainerd's redesign, however, the three sticks, A, B, and C were laid out together in serial order so that the child could see that the stick to the right was always larger and hence could answer the question correctly from spatial information rather than inference. Likewise in Bryant and Trabasso's (1971) redesign, pairs of sticks were presented so that the stick to the right was always larger. (This redesign has been used in studies of seriation in monkeys and apes.)

A critical test of seriation (Chapman & Lindenberger, 1988) presented children with the two alternative versions of the task. The children performed better on the redesigned task, answering the questions in a manner that indicated that they were using preoperational functions based on spatial cues rather than operational logic (Lourenco & Machado, 1996). In contrast, Piaget's procedures (Inhelder & Piaget, 1964; Piaget, 1968) avoided any correlation between the length of sticks and their spatial position (Lourenco & Machado, 1996).

SERIATION IN APES AND MONKEYS

Gillan (1981), one of Premack's colleagues, designed a study of seriation in chimpanzees based on the transitivity involved in seriation tasks. This task involved presentation of a series of four pairs of colored plastic containers designated as A – B+, B – C+, C – D+, D – E+ wherein A, B, C, D, and E represented different colors and – and + indicated the absence or presence of food. The use of a single food item in one of each pair indicated which of the pair was higher in the series. First, the three 5- to 6-year-old chimpanzee subjects, Jessie, Luvie, and Sadie, were trained to choose the container with the food to a criterial level. Then, they were presented with adjacent pairs and nonadjacent pairs to test their choices. When presented with the nonadjacent pairs BD, Jessie chose D 7 out of 12 times, Luvie chose D 5 out of 12 times, and Sadie chose D 12 out of 12 times. Subsequent versions of the experiment were run to eliminate such possible effects as color preferences or linear biases involved in serial presentation of the pairs. Performances were depressed under the nonlinear presentation.

Boysen, et al. (1993) studied seriation in their three older juvenile chimpanzees, Kermit (8 years), Darrell (8½ years), and Sheba (6½), replicating and extending Gillan's method. In the replication, all three individuals were trained to select the larger of two adjacent object pairs. Following training, all three selected D, the larger of the nonadjacent pairs, with novel test stimuli at frequencies significantly above chance.

A second, parallel experiment using arabic numerals in place of seriated objects was designed to examine possible inferential processes in two individuals who had shown a facility for using number symbols. They were trained to select the larger of two adjacent numerals. Only one individual, Sheba, chose the larger nonadjacent numeral on novel trials at a significant frequency. A third experiment reversed the direction of choice, training the subjects to select the smaller of two adjacent numerals. After training, Sheba correctly selected the smaller nonadjacent numeral in novel tests more than 90% of the time. A fourth experiment replicated the range of arabic numerals used in the second experiment after the subjects had received additional training in counting. After training, all three subjects selected the larger of the nonadjacent numerals in novel tests; two of them did so with 100% accuracy. Boysen et al. argue that these results provide evidence that these chimpanzees understand transitivity (seriation) and ordinality.

There is an anecdotal report of spontaneous seriation in rehabilitant orangutans. Princess, the symbol-trained rehabilitant orangutan at Camp Leakey in Kalimantan, "once 'categorized' sticks by length, neatly ordering them from shortest to longest as she played with them on the ground" (Galdikas, 1995, p. 363). According to Inhelder and Piaget (1964), seriation, when accomplished in a systematic manner rather than through trial and error, is diagnostic of concrete operations. Although Princess may have been imitating human models, it is notable that imitative performances rarely exceed the natural performances (Piaget, 1962). In contrast, it seems likely that the performances of chimpanzees depended on preoperational strategies.

There is one report of seriation in squirrel monkeys (McGonigle & Chalmers, 1977). The experimental protocol was based on one designed by Bryant and Trabasso (1971). It was like that used later by Gillan and replicated by Boysen et al. Although the monkeys performed at a level similar to that of the 4-year-old children in the model study, McGonigle and Chalmers suggest that the monkeys do so through a binary decision process rather than transitive reasoning.

CONSERVATION IN HUMAN CHILDREN

In *The Child's Construction of Quantities,* Piaget and Inhelder (1974) studied the development of the child's understanding of conservation of properties of objects undergoing various kinds of transformations. They examined children's explanations for various transformations, such as changes in shape of clay, dissolution of sugar in water, and expansion of seeds during growth. They identified the following stages of development of the conservation of quantity under deformation of a piece of clay, or so-called conservation of continuous quantities:

Stage I, during the intuitive subperiod of preoperations from 4 to 6 years—The child fails to grasp the fact that the same amount of clay is present after a ball is flattened as there was before it was flattened. This is true even though she has previously agreed that that ball was the same as a matching ball of clay that has not been deformed. She is overwhelmed by perceptual factors, focusing on one dimension only, thickness and thinness, or height, but not both. Piaget characterized these as one-way functions.

Stage IIA, during the transition to concrete operations from about 6 to 7 years—The child oscillates between her previous response and the mature response that the amount remains the same despite alterations. This intermediate response reflects a conflict between perception and conception.

Stage IIB, during concrete operations after about 7 years—The child immediately responds that the amount of substance is the same despite deformations and explains that this must be so because the operation can be reversed or because the increased surface area is compensated by the decreased height of the clay. (Piaget notes that understanding of the conservation of weight and volume is achieved later because it involves metric quantifications.)

CONSERVATION IN MONKEYS AND APES

Several investigators have tested monkeys and apes for their ability to understand that the amount of a substance is conserved when it undergoes transformations. These studies have been done on both continuous quantities of such substances as liquids and clay and discontinuous quantities such as buttons and candies. (The latter studies are reported under the rubric of "numerical understanding.")

Woodruff, Premack, and Kennel (1978) tested the symbol-trained chimpanzee Sarah for conservation of liquid and solid quantities and number when she was 14 years old. Sarah answered queries regarding the substances using the symbols for *same* and *different*. The procedure involved placing two items in front of Sarah, demonstrating a transformation of one of them, and then leaving Sarah to select either *same* or *different* tokens from a covered dish.

The control involved showing Sarah the two items without demonstrating the transformation. During the experiments but not the controls, Sarah performed significantly above chance on conservation of continuous quantities (liquid and clay) but failed on conservation of discontinuous quantities. As discussed in the next section, other investigators have found that chimpanzees can conserve number, that is, that they understand one-to-one correspondences.

More recently, Call and Rochat (1996) tested four orangutans aged 6 to 16 years and ten children aged 6 to 8 years on conservation of continuous quantities. In a series of experiments, they asked subjects of both species to point to the larger quantity. Three of their tests asked the subjects to choose the larger amount following visible transformations through pouring the liquid in two containers into two differently shaped containers. This dual procedure was apparently designed to allow nonsymbol-using subjects to compare two copresent items by pointing to one, rather than comparing a before and after using *same/different*.

The experiment differed from those of Piaget, Inhelder, and Szeminska (1960) in that a set of two transformations was produced in each case, one of a larger amount of fluid, the other of a smaller amount of fluid. We suggest that, in Case's terms, this corresponds to a bivocal coordination of dimensions rather than a univocal coordination of dimensions.

In the first transformation experiment, all the orangutans successfully discriminated the larger amount even under transformations, as did five out of six of the children. In the second transformation experiment, which involved a greater contrast in the container shape, three of the four orangutans successfully chose the larger amount; three of the six of the children failed to do so. In the third transformation experiment in which the fluid in one cup was poured into six small cups, none of the orangutans successfully chose the larger amount, but four of the six children did.

The authors conclude that although the orangutans can discriminate the quantities under minor transformations, they are unable to do so under major transformations. More important, they fail to understand that conservation is a logical necessity, as human children do when they reach concrete operations. They argue that Woodruff et al.'s (1978) results are inconclusive because they failed to test the limits of their subjects' understanding of transformations.

In our judgment, these two experiments are incommensurate. Call and Rochat's protocol is more complex than that of Woodruff and Premack and, indeed, than that of Piaget and Inhelder. Their version of the conservation task requires the subject to compare two transformations rather than one. This may be the reason that many of the 6- to 8-year-old children were unable to conserve.

In another early experiment on conservation, Thomas and Peay (1976) tested four adult squirrel monkeys to see if they understood the conservation of length. First, the monkeys were trained to respond to the right-hand food well when two lengths were the same and to the left-hand food well when they were different. Then they were tested as to the sameness or difference between two blocks after one of the blocks was moved laterally by about half its length. Two of the four monkeys met the criteria for recognizing equivalency after transformation, although it was impossible to tell whether they had attended to the transformation itself.

According to Piaget, Inhelder, and Szeminska (1960), human children first manifest understanding of conservation of quantities and of length when they are entering concrete operations at about 7 years of age. According to Gelman and Gallistel (1978), children show understanding of conservation of quantities at 4 years of age (in a scaffolded situation). In either case, chimpanzees seem to show late preoperational performances in this realm. Of course, these are scaffolded rather than functional levels of performance in Fischer's terms (Fischer et al., 1993). Clarification awaits further research.

SUMMARY AND CONCLUSIONS
LOGICAL-MATHEMATICAL KNOWLEDGE IN CHILDREN

The first type of neo-Piagetian studies using redesigned tasks in the numerical domain have been interpreted to mean that infants as young as 6 months of age are able to perceive the correspondence between two visual arrays of the same number of objects. These results may reflect early perceptual biases in favor of numerical knowledge. Alternatively, they may reflect flawed experimental design (Bogartz, Shinskey, & Speaker, in press). In either case, their focus on perception rather than action places them outside the range of phenomena that Piaget considered in the sensorimotor period (Rivera, Wakeley, & Langer, in press).

The second type of neo-Piagetian studies, reformulated models, focus on discovery of sensorimotoric (as opposed to perceptual) precursors of later knowledge. Langer's studies of the development of logical cognition fall into this category. Analysis of spontaneous patterns of object manipulation reveals that infants as young as 6 months display recognizable precursors of logical thinking in their patterns of object manipulation. This analysis also reveals that children as young as 2 years of age display precursors of numerical thinking in their manipulation of objects.

These conclusions on logical cognition contrast with Piaget's initial conclusions that children begin to understand classification and number during the intuitive subperiod. They are, however, consistent with his later groping movement toward the concept of proto-operations (Piaget & Garcia, 1991).

Overall, neo-Piagetian studies of the sensorimotor period present the following picture: First, infants may show very early, if not immediate, understanding of certain relations on a perceptual level in the first few months of life. Second, infants create nascent forms of logical relations in object manipulation on a sensorimotor level beginning at about 6 months. Third, reconstruction of these nascent sensorimotor forms into symbolic forms of logical relations begins at about 2 years.

LOGICAL-MATHEMATICAL COGNITION IN MONKEYS
AND GREAT APES

Studies of the development of logical knowledge during the sensorimotor period in monkeys and apes reveals that macaques and cebus monkeys peak at a level equivalent to the fourth stage, whereas chimpanzees peak at a level equivalent to the fifth and sixth stages. Specifically, like human infants, chimpanzees can classify objects into two sets on the basis of similarity and difference, whereas monkeys can construct single sets. Likewise, chimpanzees can construct binary sets and perform exchange operations between them. The performances of chimpanzees peak at the equivalent of the sixth stage, whereas those of monkeys peak at the equivalent of the fourth stage of the sensorimotor period. Chimpanzees achieve their highest level at about 4 years as compared to about 2 years in human children.

At least some adult symbol-trained great apes may display abilities characteristic of the intuitive substage of preoperations in the subdomains of number, seriation, and conservation. In most cases this occurred after extensive training. It is important to recall that these are primarily scaffolded levels rather than functional levels of performance. Nevertheless, such a level of performance is noticeably higher than scaffolded performances in most other subdomains. It is notable that these performances are achieved only in individuals who are 6 or 7 years old or older. Table 3-4 summarizes preoperational logical-mathematical knowledge of primates.

Several reports suggest that some great apes have achieved the intuitive stage of preoperational levels of logical-mathematical knowledge: (1) Woodruff et al.'s (1978) report of conservation and Premack's (1976) and Braggio et al.'s (1979) reports of multiplicative classification in the symbol-trained chimpanzee, Sarah; (2) Gillan's (1981) and Boysen et al.'s

TABLE 3-4 SUMMARY OF SENSORIMOTOR AND PREOPERATIONAL ABILITIES IN THE PHYSICAL AND LOGICAL-MATHEMATICAL SUBDOMAINS OF THE NONSOCIAL DOMAIN IN CHIMPANZEES

	Achievement age in months		
Level	Sixth sensorimotor	Symbolic	Intuitive
Physical knowledge			
Space	51	?	None
Causality	36?	76	Transitional
Logical-mathematical knowledge			
Protologic	51	?	NA
Seriation	?	?	(60+)
Classification	51	?	(60+)
Conservation	?	?	(By adulthood)
Number	(12 to 24)	(24 to 40)	(40 to 60)

Note. ( ) denotes symbol-trained great apes and trained performances. ? = No data available.

(1993) studies of seriation and transitivity in chimpanzees; and (3) Call and Rochat's (1996) study of conservation in orangutans. It remains to be seen whether all four great apes display similar logical-mathematical capacities.

SUMMARY OF LOGICAL-MATHEMATICAL DEVELOPMENT IN MONKEYS, APES, AND HUMANS

The foregoing studies suggest the following patterns. First, macaques, cebus, great apes, and humans traverse the same sequence of stages in the development of logical as well as physical cognition during the sensorimotor period.

Second, both great apes and humans complete all the sensorimotor-period stages in logical as well as physical cognition, but macaques and cebus monkeys do not.

Third, great apes traverse the developmental stages at a slower rate than human infants do after the first year of life, completing the sixth stage between 3 and 4 years of age as compared to 2 years in humans.

Fourth, development within logical-mathematical domains is slower-later than in most physical domains. Both classification and logical groupings achieve their highest development, probably equivalent to sixth sensorimotor stage, at about 4½ years in chimpanzees. Development in the physical domain of space, however, apparently runs at a parallel rate in chimpanzees

Poti' (1996) has suggested that development in both logical-mathematical and spatial domains is apparently limited by patterns of object manipulation typical of chimpanzees as contrasted with those of human infants. Most notable is the continuing tendency to focus on their actions with objects rather than the relations between objects that their actions have created. This is both caused by and reflected in their tendency to handle two or more objects at a time with one hand or foot rather than using a separate action sequentially on each object. It is also caused by and reflected in their tendency to continue to make mobile rather than stabile constructions with two or more objects and to place objects on their bodies rather than on a substrate. All of these tendencies deprive chimpanzees of the opportunity to construct more elaborate relationships among objects and to reflect on those constructions (Poti', 1996, 1997).

Fifth, chimpanzees who have not been trained in symbol use achieve levels of logical thought equivalent to those of 4-year-old children in the subdomain of classification.

Sixth, chimpanzees who have been trained in symbol use achieve levels of logical thought similar to those of 4-year-old children in the subdomains of number and seriation.

The foregoing studies suggest that great apes may achieve the level of intuitive subperiod of preoperations in the logical-mathematical domains of seriation, classification, conservation, and number. They develop numerical abilities at a slower rate than human children, at about 6 or 7 years of age as compared to about 4 or 5 years in children. The ages at which they may develop these abilities are not known.

The oldest males (Yeroen and Luit) do not allow personal preferences to influence their interventions. My interpretation is that Yeroen and Luit are extremely aware of the functional effect of what they do. Their interventions are in accordance with a policy directed at increasing their power. The flexibility with which they make and break coalitions gives one the impression of policy reversals, rational decisions and opportunism. There is no room in this policy for sympathy and antipathy.

DE WAAL, 1983, P. 196

CHAPTER 4
DEVELOPMENT OF SOCIAL COGNITION IN CHILDREN, APES, AND MONKEYS

Like the models of physical and logical cognition reviewed in the preceding chapters, many models of social cognition in children had their genesis in Piaget's theory. Perhaps his most influential work in the social domain is his study of the development of imitation and pretend play. His work on imitation has also been the subject of the greatest criticism and revision by neoinnatists. This chapter begins with a brief review of Piagetian and neo-Piagetian studies of imitation and pretend play, which have influenced students of language acquisition and of primate cognition. Following that, it reviews studies of theory of mind and self-awareness in human children and great apes.

IMITATION IN HUMAN INFANTS AND CHILDREN

Piaget's (1962) book *Play, Dreams and Imitation in Childhood* has greatly influenced developmental and comparative psychologists interested in language and social cognition. It covers the development of imitation and play from birth to 3 or 4 years of age.

PIAGET'S STAGES OF SENSORIMOTOR IMITATION

Like other sensorimotor series, the imitation series is characterized by six sequential stages of development. During the first and second stages, from birth to 2 months, the infant engages in sporadic imitation of schemes from her own repertoire such as crying or vocalizations. During the third stage, the infant begins systematically imitating vocalizations and visible actions from her own repertoire; this imitation seems to be a form of the *secondary circular reactions* aimed at prolonging the spectacle of the model's behaviors. She is not yet capable of imitating actions from her own repertoire that are invisible to her or of imitating new schemes. This limitation seems to result from her difficulty in differentiating elements into independent components.

During the fourth stage, beginning at about 8 months, she becomes capable of imitating schemes from her own repertoire that are invisible to her, for example, facial expressions. This involves differentiation of schemes into components as well as a matching of her kinesthetic schemes with the schemes she sees the model performing.

During the fifth stage, beginning at about 12 months, she begins imitating novel vocal, gestural, and facial schemes as *tertiary circular reactions* allow her to "go beyond mere application and accommodation of existing schemata to accommodation through systematic and controlled trial and error" (Piaget, 1962, p. 52).

Finally, during the sixth stage, beginning at about 18 months, the child shows *deferred imitation* of novel schemes, performing them for the first time when the model is no longer present. She also becomes capable of immediate imitation of more complex schemes and begins to imitate the actions of objects. Piaget traces the origins of symbolic *representation* to interiorized imitation of object relations. This sixth and final stage of the sensorimotor period begins the transition into the symbolic subperiod of the preoperations period.

Piaget also briefly describes stages of development of practice play during the sensorimotor period: the secondary and tertiary circular reactions and the intervening coordination of secondary schemata are "ritualized" and repeated for their own sake, for the pure joy of assimilation rather than for some adaptive purpose. Indeed Piaget characterizes imitation as almost pure *accommodation* and play as almost pure *assimilation* and adaptive behavior as a balance between the two (Piaget, 1962).

Piaget's sensorimotor stages of development in the gestural and vocal imitation series have been replicated by Uzgiris and Hunt (1975) in their large-scale study of the sensorimotor period. Table 4-1 summarizes their sequence in the gestural modality.

NEOINNATIST STUDIES OF INFANT IMITATION

In contrast to this replication of Piaget's model of imitation are neoinnatist studies that show that imitation develops earlier than Piaget had

TABLE 4-1 SENSORIMOTOR GESTURAL IMITATION

Step	Scheme	Month of age
1	Gestural response to familiar gesture	4
2	Makes familiar gesture on seeing it performed	7
3	Inability to imitate novel body movement, e.g., banging two blocks	8
4	Beginning ability to imitate novel body movements, e.g., banging two blocks after groping	9
5	Immediate imitation of novel body movements	10
6	Imitates several novel gestures he can see himself perform	11
7	Tries to imitate facial gestures but does not succeed	14
8	Imitates at least one facial gesture	14 to 17
9	Imitates more than one facial gesture	14 to 20

Note. Adapted from Uzgiris and Hunt (1975).

indicated. Meltzoff and his colleagues, for example, have done a series of studies revealing imitation in neonates and older infants. The neonatal studies reveal that infants can imitate facial movements far earlier than Piaget (1962) had reported. These studies contradicted behaviorist accounts that explained imitation as trained behavior. Meltzoff and Moore (1983, 1989) propose that neonatal and early infant imitation is mediated by an *active intermodal mapping* (AIM) of a mental representation of the infant's behavior onto a model of the model's behavior.[1]

Meltzoff's neoinnatist studies on older infants reveal early expressions of the capacity for imitation of novel actions and *deferred imitation* of novel actions as early as 9 months. Piaget described these abilities as emerging during the fifth and sixth stages at 12 and 18

1. In 1977, Meltzoff and Moore reported that infants 12 to 21 days of age are able to successfully imitate lip protrusion, mouth opening, tongue protrusion and sequential finger movements. Moreover, they were able to do so from memory (as constrained by use of a pacifier to prevent immediate imitation). This study also indicated that these infants began with imperfect imitations and then homed in on the proper actions without subsequent exposure to the model. A subsequent study by the same investigators showed that neonates 0.7 to 71 hours old had the same abilities (Meltzoff & Moore, 1983). In a third study (Meltzoff & Moore, 1989), they showed that neonatal infants are also able to imitate head-turning movements. These studies have been replicated by many but not by all other investigators. Meltzoff and Moore (1983, p. 300) suggest that their experiments may have succeeded where others failed because they eliminated nonattending infants in a pretest and used a burst-pause procedure, which kept the infant's interest during the presentations. It is noteworthy that neonatal facial imitation seems to go underground for several months before it emerges again (Bower, 1979).

months, respectively, that is, about 3 months earlier. Meltzoff therefore questions Piaget's argument that mental representation first arises at 18 months during the sixth stage through interiorization of imitation (Meltzoff, 1985). Mounoud and Vinter (1981) and Mandler (1988) make similar points.[2]

It is important to note, however, that the kinds of actions imitated by 9-month-old infants are few in number and much simpler than those described by Piaget. It is also important to note that active *intermodal matching* may develop in complexity from infancy to 18 months. Contrary to Meltzoff's interpretation, neonatal imitation is relatively narrow and stereotyped as compared to sensorimotor imitation. It seems likely, therefore, that the kind of representation Piaget was describing is qualitatively different from that described by Meltzoff in neonates and even 9-month-old infants. This conclusion is reinforced by the observation that pretend play, first referential gestures and words, and drawing and other expressions of symbolic capacities first emerge during the sixth stage of the sensorimotor period.

IMITATION IN APES AND MONKEYS

IMITATION IN GREAT APES

Because of our comparative developmental focus, we concentrate on Piagetian and neo-Piagetian models of the development of imitation. These human models are particularly relevant because they place imitation in the context of sensorimotor and preoperational cognitive development. Imitation was the first ability in the social domain to be approached from a comparative Piagetian perspective (e.g., Chevalier-Skolnikoff, 1976, 1977; Parker, 1977; Miles et al., 1996). Aside from a few explicitly Piagetian studies, anecdotal descriptions in books on home-reared chimpanzees and gorillas, and ape symbol studies, there are few data on the development of imitation in great apes. Unfortunately, most

---

2. In subsequent studies of precocious imitation, Meltzoff (1985) has demonstrated that infants as young as 9 to 14 months old are able to imitate novel actions and, moreover, to imitate them after 24 hours without subsequent modeling. In one of these studies, the modeled actions included taking apart and reassembling a specially designed dumbbell shaped toy. These studies included an imitation condition, a baseline control involving simple playing with the toy, and an activity control, moving the toy in a circle for the same time that the action was modeled in the experimental condition. Meltzoff has also shown that these infants recognize when they are being imitated, as compared to being contingently responded to in some other way, and that they prefer being imitated (Meltzoff, 1990).

of these are cross-species studies in which great apes imitate human actions rather than actions of members of their own species.

In the following sections, we briefly review results of studies of great ape imitation, both developmental and nondevelopmental, interpreting them in a Piagetian frame. We have divided the discussion according to the modality of imitation. We begin with gestural imitation and proceed to imitation of tool use.

IMITATION OF GESTURES AND MOVEMENTS BY APES

Cathy and Keith Hayes home raised a young female chimpanzee, Viki, virtually from birth to 3 years of age (C. Hayes, 1952). In her book, *The Ape in Our House*, Cathy Hayes reported various incidents of imitation and pretend play beginning when Viki was about 16 months old. These included such household tasks as dusting, washing dishes, and brushing her hair. When she was between the ages of 17 and 34 months, Viki was trained to imitate modeled actions on the command "Do this" and was tested on her ability to imitate 70 items, of which she eventually mastered 55. The initial set of tasks included sounds (*Mama* and Bronx cheer), mouth and hand movements, eye blinking (which she helped with her hands), and activating toys. When she was 36 months old, the Hayeses demonstrated solving such problems as stick and tunnel, stick and string, string and candle (burning the string to break it), and ball throwing. They also tested human children on the same tasks and concluded that Viki performed about as well as 3-year-old children (K. J. Hayes & Hayes, 1951). Unfortunately, the Hayeses did not include clear data on the ages at which Vicki accomplished each imitative feat.

Three contemporary studies of great ape imitation have used the Hayeses' technique (which according to Miles et al. [1996] was pioneered by Furness in 1916). Custance and Bard (1994) and Custance, Whiten, and Bard (1995) used the "Do as I do" paradigm to test imitation. Custance and Bard tested two 4-year-old chimpanzees, Scot and Katrina, who had been nursery reared at Yerkes Regional Primate Research Center. After a 3-month training period, they were tested with familiar and novel actions, all of which were imitated. These included facial, head, and whole body movements, touching both visible and invisible parts of the self, making symmetrical and asymmetrical hand movements. (Table 4-2 presents the results of this experiment.)

The third recent study to use the "Do as I do" paradigm is a study of gestural imitation in the symbol-immersed orangutan Chantek, when he was 6 to 8 years old (Miles et al., 1996). The authors categorized the

TABLE 4-2 GESTURAL IMITATION IN CHIMPANZEES

Category	Activity			Category	Activity	
Touch	Shoulder	**		Single	Open hand	**
in	Elbow	**		hand	Wiggle fingers	**
sight	Stomach	***			Wave stiffly	**
	Thigh	*			Arch fingers	**
	Foot	**		Facial	Protrude lips	*
Touch	Back of head	***			Lip smack	***
out of	Top of head	**			Teeth chatter	**
sight	Nose	***			Puff cheeks	*
	Ear	*		Face-	Mouth pop	**
Sym-	Clap	**		head	Lip wobble	**
metrical	All-digit touch	***			Pull mouth sides	*
hand	Interlink fingers	***			Look up	**
	Roll fists	**			Look right	*
	Peekaboo	**		Whole	Jump	**
Asym-	Clap back of hand	***		body	Flap arms	**
metrical	Clap two digits	*			Hug self	***
hand	Grab thumb	***			Foot to foot	**

Notes. From Whiten and Custance (1996). Table includes all actions identified at least once: *** = Identified for both subjects by at least one coder. ** = Identified for one subject by at least one coder. * = Identified for at least one subject, on coder's second guess. Descriptions given here are intended to convey the range of actions presented.

modeled behaviors according to modalities: kinesthetic-visual, auditory-auditory, and visual-visual, as well as facial, bodily, and bodily-craniofacial. The number of demonstrations for each behavior is tabulated. Chantek's responses are coded as either full imitation, partial imitation, or nonimitation. Fifty-six percent of responses were full imitations, 34%, were partial imitations, and 9% were nonimitations. His responses are also broken down by percentage that were mirror images of the modeled behavior (the majority).

Chantek's partial imitations were more informative than his full imitations about the processes involved in imitation. He had difficulty with four actions: lip-popping noises, foot-stomping noises, jumping, and eye blinking. In each of these cases, he responded to repeated modeling by trial-and-error attempts at closer approximations of the behaviors. In the case of the foot-stomping noises, his first attempts (clapping his hands, slapping his foot) apparently focused on producing the sound, while subsequent attempts involved his foot. In the case of jumping (it was apparently impossible for him to lift his weight off the ground), he lifted and stomped his foot, then lifted each foot alternately, then dropped both legs from a sitting position, etc. Finally, in the case of eye blinking, he pushed his lids up and down with his fingers (as the chimpanzee, Viki, had done in the Hayeses' study).

These repeated responses (*rehearsals*) honing in on the modeled action reveal that Chantek was not sure which aspects of the modeled behavior were to be imitated. The microdevelopmental course of Chantek's imitations also illustrate the interactional nature of imitation, showing the crucial role of continuing feedback from the demonstrator in helping to achieve this honing-in process (Miles et al., 1996).

Additional data on gestural imitation in chimpanzees come from the Gardners' study of American Sign Language (ASL) acquisition in four young chimpanzees. Although the most extensive data on gestural imitation in great apes comes from studies of sign-symbol acquisition, these data have been neglected by most students of imitation. Indeed, the Gardners themselves downplay imitation in their descriptions. This may be a reaction to Terrace, Petitito, Saunders, and Bever's (1979) dismissal of imitation as an indication of lack of comprehension of the meaning of the signs.

B. T. Gardner, Gardner, and Nichols (1989) provide detailed descriptions of the form of 430 ASL signs produced by their four young chimpanzees. The description is based on the PCM system, that is, *place*, the making of the sign on the body or elsewhere; *configuration*, the shaping of the hands; and *movement*, the movement of the hand in space including its orientation, contactor, and direction. The standard or citation model from the *Dictionary of American Sign Language* (DASL) for each ASL sign is given next to the variants produced by the young chimpanzees.

Although some ASL signs were acquired through molding of the hands by the teacher, most of them were acquired through imitation of models. Therefore, these data provide a detailed description of the degree of matching of each component of gestural imitation in each individual for the PCM components. The order of acquisition of the signs for each chimpanzee is also listed. The degree of matching on all three components is striking. Of the three, place is the least accurate. Young chimpanzees, like young children learning ASL, often sign on the body of their audience or on the object rather than on their own body. Like young children, they also had difficulty with some complex finger configurations. Similarly, they produced simpler movements. (ASL signing is further discussed in chapter 5.)

In their longitudinal study of a peer group of 5- to 8-year-old chimpanzees, Tomasello, Gust, and Frost (1989) identified several processes by which gestures changed developmentally. Most gestures were conventionalized through interactions with conspecifics. Some gestures were learned through direct imitation of a model ("second-person imitation"), but none seemed to be learned through observation of gestures used between other conspecifics ("third-person imitation").

Overall, these studies suggest that chimpanzees and orangutans, and by implication, gorillas, who have also learned ASL, are capable of imitating novel gestures to a moderate degree of accuracy. It is interesting to note, however, that studies of these same species, in some cases the same individuals, show that they are less adept at imitating novel actions on objects.

IMITATION OF TOOL USE AND OBJECT MANIPULATION BY APES

In addition to testing for gestural imitation, Whiten, Custance, and colleagues (Whiten & Custance, 1996; Whiten, Custance, Gomez, Teixidor, & Bard, 1996) also tested for imitation of movements of parts of an artificial fruit that could be opened according to two methods for each of several versions. Each set of paired alternative methods (twists versus pokes of a bar in one case and turns versus pulls, in the other) was a control for the other. They compared eight chimpanzees aged 3 to 8 years to an unspecified number of children aged 2½ to 4½ years.

The chimpanzees were less successful in their imitation of movements of parts of the artificial fruit than in their imitation of gestures. Specifically, they were better at imitating pokes and turns than at imitating pulls and turns. They did show some imitation of some specific movement patterns, for example, twisting. Generally, the chimpanzees performed less well than the children, both in overall kinds of imitated actions and in finesse of actions. These results lead the investigators to propose a continuum of copying fidelity from virtually exact copying to emulation (in which the observer copies the results of the models' actions).

Likewise, in a series of studies, Tomasello and his colleagues report lack of imitation of tool use and other forms of object manipulation in both chimpanzees (Nagell, Olguin, & Tomasello, 1993) and orangutans (Call & Tomasello, 1995). In one experiment, a model demonstrated use of a rake to rake in objects for chimpanzees and children. In one condition the model rotated the rake to the tines down position and, in the other condition, in the flat side down position. Chimpanzees raked in the goal object without regard to the position of the rake. Children rotated the rake to the demonstrated position. A similar experiment with twelve juvenile and adult orangutans produced the same result. Tomasello and his colleagues interpreted these results to mean that these great apes were emulating the goal of the model rather than imitating his movements. They argue that this means that they are unable to differentiate the means from the goals. They also argue that imitation is confined to human-reared "enculturated" apes.

In a third study designed to control for emulation, the model demonstrated two sequential actions on a manipulandum (rotate plus pull; rotate plus push, etc.) that resulted in delivery of a reward. In this case the mechanism by which the reward was delivered was invisible. A prior experiment determined that 10 of the 12 subjects were capable of at least three of the individual actions. As in the previous experiments, the orangutans failed to replicate the two sequential movements within the time frame necessary to qualify as imitation.

Four orangutans were successful when the correct action was "rotate plus pull," but not in any other condition. This reflected the fact that each orangutan had a favorite action, four were pullers, four were pushers, and four were rotators. Three-year-old children performed slightly better than the orangutans, but 4-year-old children were highly successful. The 3-year-old children showed a different pattern from the orangutans. They tended to perform the two actions in reverse order to their demonstration. Finally, in another version of the experiment, another orangutan was trained to serve as the model. The results were the same. Call and Tomasello (1995) believe that this experiment definitively demonstrates that the great apes are unable to imitate problem-solving actions with objects. Both the experimental design and their interpretation of the results are based on Thorpe's definition of imitation (see below).

Out of the laboratory, Russon and Galdikas (1993, 1995) describe imitation of object manipulation and tool use in free-ranging rehabilitant orangutans in Kalimantan, Borneo. As in the previous studies, humans —staff and workers at Camp Leakey—were the models. Russon (1996) argues that the conditions at Camp Leakey provide built-in controls that amount to a natural experiment in that the imitated actions are arbitrary, exceptional, and atypical from the orangutans' perspective. Moreover, the rapidity and productivity of their actions argue against experiential learning processes. Russon (1996) describes 59 examples of imitation in 11 rehabilitant orangutans who had been released from cap-

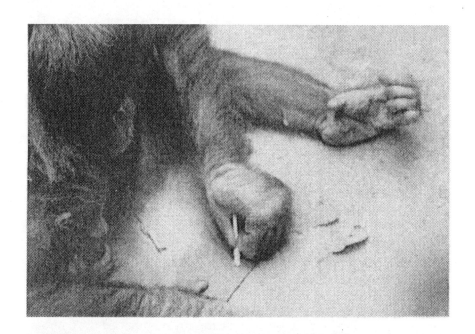

FIGURE 4-1
Rehabilitant orangutan imitating: (*a*) sharpening point, (*b*) siphoning gas
(photographs by Anne Russon).

tivity at various ages (see Figure 4-1). (As Call and Tomasello, 1995, note, some of these routines may have been learned in captivity.)

Virtually all of the 59 behaviors analyzed in this study involved use of objects, either in some sort of self-grooming (3 cases) or in some kind of tool use or implementation (54 cases). This is in contrast to the gestural behaviors that were studied by previously cited investigators. Most of these behaviors were partial or incomplete replications of such house-keeping and camp routines as cooking, sweeping, and carpentry. The imitated behaviors are analyzed in terms of the level of matching to the model and the patterns of substitution, chaining, and errors. This study also reveals a high level of rehearsal and repetition of actions.

There is one detailed report of instrumental object imitation in wild great apes. Boesch (1991b; 1993) describes in detail an incident involving demonstration teaching by a mother and imitation by her offspring. A young female chimpanzee, Nina, who had struggled unsuccessfully for 8 minutes to crack open a nut, gave her tool to her mother, Ricci. She watched her mother demonstrating the orientation of the hammer and then, adopting her mother's grip, she carefully maintained the hammer in the demonstrated position as she successfully cracked four nuts in succession. Later, even after she experienced a string of failures, she was able to succeed again.

PROGRAM-LEVEL VERSUS ACTION-LEVEL IMITATION

Most discussions of imitation in great apes have focused on the imitator's matching of specific actions of the model. Recently Byrne and Russon (in press) have argued for an alternative approach to "action-level" imitation, which they call *program-level imitation:* "copying the structural organization of a complex process (including the sequence of stages, subroutine structure, and bimanual coordination), by observation of the behaviour of another individual while furnishing the exact details of actions by individual learning."

They argue that this *hierarchical level* of imitative ability has been favored for learning of novel instrumental action because it is more efficient than exact action-level imitation in acquiring complex sequences of actions whose fine details can be efficiently acquired through trial-and-error learning. They suggest that controversies over great apes' capacity for imitation stem from the focus on action-level as opposed to program-level imitation (Byrne & Russon, in press).

TABLE 4-3 SENSORIMOTOR STAGES OF IMITATION (MANUAL, GESTURAL, AND FACIAL) IN MONKEYS AND APES

Species	Age in months			
	Stage 3	Stage 4	Stage 5	Stage 6
Humans[a]	4	7 to 10	11 to 14	14 to 20
Chimpanzees[b]	?	?	15	?
Bonobos	?	?	?	?
Gorillas[c]	?	7 to 24	27 to 37	?
Orangutans[d]	?	10	15	35
Gibbons	?	?	?	?
Baboons	?	?	? ·	?
Macaques[e]	?	2½ to 6	None	None
Cebus[f]	?	?	None	None

Note. Imitation is scale 3 in Uzgiris and Hunt (1975). ? = No data available.
[a]Uzgiris and Hunt (1975).   [d]Miles, Mitchell, and Harper (1996).
[b]C. Hayes (1952).   [e]Parker (1977)/Patterson and Cohn (1994).
[c]Parker (in press).   [f]Visalberghi and Fragaszy (1996).

CONCLUSIONS ON IMITATION IN GREAT APES

Even after a century or more of discussion and research on imitation in apes and monkeys, strictly developmental studies of imitation are almost nonexistent. Chevalier-Skolnikoff (1977, 1983) reported that the sequence of development of imitation in chimpanzees and gorillas paralleled the sequence in human children. Miles et al. (1996) present some rare data on the development of imitation. Chantek's first reported imitation (opening and closing a door) occurred when he was 10 months old, shortly after Project Chantek began. His first pretense (feeding a toy animal) occurred at 20 months. His first delayed imitation of a sign occurred at 21 months. Developmental data on chimpanzee imitation come from B. T. Gardner et al. (1989). Developmental data on orangutans (Miles et al., 1996) and gorillas (Patterson & Cohn, 1994; Parker, in press) suggest a similar pattern in these species.

The limited data on the development of imitation in great apes suggest that they traverse Piaget's (1962) six sensorimotor-period stages of imitation in the gestural and manual modes at approximately the same rate as human infants, except perhaps in tool use, which develops much later in apes than in human infants. These data also suggest that imitation may peak at a later age in great apes than in human infants, perhaps between 4 and 6 years of age as compared to 2 to 4 years of age. Pretend play may occur even later. Table 4-3 summarizes development of imitation in monkeys, great apes, and humans.

DEFINITIONAL AND METHODOLOGICAL CONTROVERSIES OVER APE IMITATION

Imitation has been the focus of both developmental and comparative psychologists since before the turn of the century (see historical reviews by Galef, 1988; Mitchell, 1987; and Whiten & Ham, 1992). Many investigators have attempted to define imitation and to distinguish it from other social processes that influence the frequency and form of behaviors in animals. A century later, imitation continues to be the focus of debate. Controversy in developmental and comparative psychology continues to focus on the same basic issues: (1) definitions, (2) mechanisms, (3) development, (4) evolutionary history, (5) adaptive significance, (6) experimental methodologies, and (7) taxonomic distributions.

In the realm of definitions, Whiten and Ham (1992) have summarized and classified a variety of mimetic processes in animals, including contagion, stimulus enhancement, observational learning, goal emulation, and imitation. Although most investigators agree on the importance of distinguishing among various kinds and degrees of copying, they use different terms to refer to the various mechanisms. Whiten and Ham use the phrase *mimetic processes* as a generic term to embrace all kinds of processes leading to behavioral mimicry of A by B. Mitchell (1994) uses the term *simulation* in the same generic sense. Parker (1996a) uses *imitation* as a generic term.

Whereas early investigators attributed imitative abilities to monkeys, most modern primatologists argue that monkeys fail to display imitation as distinguished from stimulus enhancement or goal emulation and other mimetic processes (e.g., Hauser, 1988; Visalberghi & Fragaszy, 1990; Whiten & Ham, 1992; Tomasello, Kruger, & Ratner, 1993). Current debate centers on whether or to what degree great apes display imitative abilities as distinguished from other mimetic processes. This debate involves many of the basic issues enumerated above, that is, definitions, methodologies, and mechanisms. Several factors contribute to differing conclusions about great apes' capacity for imitation. These include differing definitions and criteria and differing conceptions of the nature of communication and the contexts in which it occurs.

Comparative psychologists favor Thorpe's definition of true imitation as "copying of a novel or otherwise improbable act or utterance, or some act for which there is no instinctual tendency" (Thorpe, 1956, p. 135). They often interpret this definition to require immediate, completely accurate imitation upon one or a few presentations of the novel act by the experimenter (Call & Tomasello, 1995). From the perspective of imitation in human children, this is an unrealistic standard.

Comparative developmentalists, in contrast, favor looser definitions of imitation. They see imitative behavior as a complex phenomenon with a virtually infinite set of components. In other words, from a Piagetian perspective, imitation is constructed through a process of goal-directed trial-and-error matching of components to the model's behavior (Piaget, 1962). As such, it takes time and repetitions. If virtually perfect matching of all the aspects of the models' actions on the first trial is the criterion for imitation, then very few examples will be discovered in great apes or in human infants and young children (Moerk, 1978).

Moerk (1978) points out that language imitation in humans is a "fuzzy set" of phenomena that varies along the following dimensions: (1) levels of processing, (2) temporal dimensions, and (3) degrees of similarities between model and production. In his microanalysis of imitation of language in young human children (sample taken from Adam, from 27 to 35 months of age), Moerk (1978), for example, identified ten subsets of imitation. Indeed, he argues that the idealized model of perfect matching on the first attempt is impossible neurologically and developmentally. Moerk emphasizes that imitation is inherently interactional and as such involves repeated cycles of trials, followed by feedback from the model, followed by improved trials. It is an information-processing device that uses information from conspecifics. As such, it involves repeated attempts to approximate elements of a model and/or to recombine and substitute elements to elicit more information from the model.

Equally significant, human children, unlike great apes, are immersed in a didactic social environment. Caretakers frequently model behaviors for their children, often segmenting, labeling, and repeating them. The fact that great ape mothers rarely engage in such tutorial actions is an important species difference that contributes to the relatively low level of imitation in great apes (Russon, in press)

In addition to using idealized criteria for imitation, some investigators have built into their research design the assumption that at any age and in any context great apes are equally likely to imitate actions of any model. All these assumptions are questionable (Russon, 1996). There is considerable evidence that human infants and children select which models to imitate and which actions to imitate. There is also evidence that these factors change developmentally. Specifically, there is evidence that children imitate models to whom they have affectional ties, models they perceive as like themselves, (e.g., in gender), and models with prestige or relevant expertise. There is also evidence that they selectively model actions (or terms) that they are just beginning to understand or partially understand (e.g., Moerk, 1978). Clearly both of these variables change developmentally.

Finally, in the case of tool use, imitation of actions is constrained by spatial and causal variables imposed by the materials and laws of physics. "To effect desired outcomes, manipulations of tool-object relationships must be flexibly fine-tuned to the ecological contingencies of the problem" (Russon, in press). This means that imitation of tool use is more likely than imitation of gestures to involve pragmatic adjustments that diverge from those of the model. Byrne and Russon (in press) suggest that action-level imitation is inefficient for learning novel instrumental behaviors and therefore program-level imitation is the major factor in the acquisition of such skills in human children as well as great apes.

In their review of the history of imitation research, Whiten and Ham (1992) laud Dawson and Foss's (1965) "two-action" experimental design for distinguishing imitation from local enhancement, that is, modeling two different actions on the same manipulandum. They and Whiten and Custance (1996) argue that the methodology of imitation experiments cries out for new refinements. Custance et al. (1995) suggest distinguishing and tracking shape, extent, speed, laterality, orientation, and sequence of actions. (It is interesting to note in this context that studies of sign language acquisition provide just this sort of detailed description of the topology of imitated actions [see B. T. Gardner et al., 1989].)

Russon's (1996) comments on model and action selectivity in great ape imitation suggest that these factors should be considered in experimental design. Likewise, if imitation is inherently interactional with continuing improvement resulting from selective repetition by models of missing elements, then repeated modeling should be a variable (see, e.g., Miles et al., 1996). Finally, if rehearsals gradually hone in on modeled actions, they too should be a variable (see Russon, 1996).

We conclude that controversy over imitation in great apes will continue until the gap between definitions and criteria for imitation as conceived by experimentalists and developmentalists is narrowed. If research designs are to take into account construction, motivational factors, selective repetition by models and rehearsals by imitators, and pragmatic adjustments to problems, then at the very least, we need expanded time frames for experiments and more flexible definitions of imitation. In the meantime, we must do our best to interpret existing studies of imitation and pretend play in primates in light of these issues.

IMITATION IN MONKEYS

In her longitudinal study of sensorimotor-period development in an infant stump-tailed macaque, Parker (1977) concluded that the infant occasionally displayed fourth-stage imitation of some actions from his own repertoire when he saw those actions performed by his mother or his playmate. He failed to achieve fifth-stage imitation of novel actions. Reviews of social learning in monkeys suggest that this pattern applies to all monkeys.

In their reviews of the literature on social learning in various Old World and New World monkeys, Visalberghi and Fragaszy (1990; 1996) and Whiten and Ham (1992) conclude that there is no good evidence for imitation of novel actions in monkeys. They carefully analyzed such purported examples of imitation in monkeys as sweet potato washing in Japanese macaques (Itani & Nishimura, 1973) and Acacia pod dipping in vervets (Hauser, 1988)

They conclude that monkeys are simply performing species-specific actions in new contexts or that individuals have acquired the same knowledge independently through experiential learning aided by stimulus enhancement. In his review of imitation in animals, Mitchell (1987) cites one possible example of imitation or pretense in a rhesus macaque. Hauser (1988) cites another possible instance of imitation in a wild vervet. Parker and Poti' (1990) and Visalberghi and Fragaszy (1996) observed that cebus monkeys failed to learn tool use tasks from observation of tool-using peers. This is particularly significant given the tool using abilities of this species.

SUMMARY AND CONCLUSIONS REGARDING IMITATION IN MONKEYS AND GREAT APES

The parameters of imitation in great apes are gradually emerging. As has been known for many years, great apes have poor voluntary control over their vocalizations and therefore fail to produce vocal imitations. All the great apes are capable of imitating many novel facial, body, and hand gestures. They are also capable of imitating some novel actions on or with objects. Their failure to replicate a sequence of rotate and push or pull movements on the same object, for example, suggests that they have difficulty imitating some actions on objects. Further research may identify which actions are most difficult. Success at object manipulation may require many rehearsals and perhaps demonstration of particular components of action. Research on naturalistic modeling and teaching with selective demonstrations of missing components may also help explain the discrepancies between experimental and naturalistic reports.

More complete understanding of the relative complexity of various imitative acts may be gained through application of Case's (Case & Khanna, 1981) neo-Piagetian analysis. This analysis would clarify the number and kinds of coordinations involved in particular actions. This analysis suggests that imitation of actions on objects mediated by other objects (e.g., tools) involves a larger number of coordinations, such as coordinations between the action and the tool and between the tool and the goal object, as well as coordinations between the model's actions and the imitator's actions.

PRETEND PLAY AND ROLE PLAYING IN HUMAN CHILDREN

Piaget describes pretend play as the preoperational extension of imitation. It develops coordinately with language and lays the groundwork for games with rules and moral judgment. Piaget noted that the transition from practical to symbolic play begins when the child creates a symbol by repeating a sensorimotor schema "outside its usual context and in the absence of its usual objective," for example, lying down pretending to sleep during the day (accompanied by laughter) (Piaget, 1962, p. 119). But "the symbol is not yet freed as an instrument of thought. It is the behaviour, or the sensorimotor schema, which is the symbol, and not this or that individual object or image" (Piaget, 1962, p. 120).

PIAGET'S SYMBOLIC PLAY STAGES

Piaget distinguished two stages of symbolic development in pretend play in his children. Stage 1, the symbolic subperiod of preoperations from about 1 to 4 years, was characterized by the following subtypes:

— Type IA, *projection of symbolic schema onto new objects*, for example, having a doll sleep

— Type IB, *projection of imitative schema onto new objects*, for example, using a variety of objects as a telephone

— Type IIA, *simple identification of one object with another*, for example, moving a box along making the noise of a motor car (Piaget, 1962, p. 123)

— Type IIB, *simple identification of the child's body with that of other people or with things,* for example, crawling on all fours mewing like a kitty (Piaget, 1962, pp. 124–25)

— Type IIIA, *simple combinations* with the construction of scenes, for example, washing and ironing the doll's clothes (Piaget, 1962, p. 139)

— Type IIIB, *compensatory combinations* involving correcting rather than reproducing reality, for example, pretending to pour water from a vessel, which was forbidden (Piaget, 1962, p. 131)

— Type IIIC, *liquidation combinations* involving transforming unpleasant realities, for example, pretending that another child (rather than the self) hurt her lip (Piaget, 1962, p. 132)

— Type IIID, *anticipatory symbolic combinations* involving playful invocation of anticipated consequences of actions, for example, telling about an imaginary playmate who slipped on a stone and fell after being warned of such a possibility (Piaget, 1962, p. 134)

Stage 2, the intuitive subperiod of preoperations from about 4 to 7 years, is characterized by increasing orderliness, exactness of imitation, and use of collective symbols. It is also characterized by increasing use of realistic props. One of Piaget's children during this period, for example, enacted whole cycles of scenes dealing with families, houses, weddings, and school. Eventually she created an entire imaginary village. Stage 3, the concrete operations period from 7 to 12 years, is characterized by the decline of symbolic play (pretend play) and the gradual rise of games with rules and moral judgment (Piaget, 1965b).

NEO-PIAGETIAN APPROACHES TO IMITATION AND PRETEND PLAY IN HUMAN CHILDREN

Since the publication of *Play, Dreams and Imitation in Childhood* (Piaget, 1962), many investigators have described the development of imitation and pretend play in young children. In a major review of the literature, Bretherton (1984) traces the origins of event representation in pretend play in 1½- and 2-year-old children, attributing the seeds of this model to Piaget. She argues that pretend play schemes represent events. In other words, they are *scripts* that include role representations, action representations, and object representations, props. In her view, "Joint make-believe constitutes . . . collaborative event representation on two levels, planning and acting" (Bretherton, 1984, p. 7).

Bretherton traces the development of roles through 11 stages: (1) self-representation, repeating one's own scheme out of context (e.g., pretending to sleep); (2) representing another's behaviors; (3) representing others as passive recipients of the child's actions; (4) engaging in parallel roles; (5) using replicas as active recipients of actions; (6) using a replica as an agent; (7) assuming another person's role; (8) using a replica as an active partner; (9) engaging in collaborative role play; (10) using replicas to play several interacting roles; and (11) engaging in collaborative play with several interacting roles. She also traces the development of object representation through the stages of (1) single schemes to (2) combined schemes to (3) multiordered schemes.

Bretherton notes work by Nelson and Gruendel (1986) showing that children as young as 3 years have scripts for such common daily routines as going to McDonald's or making cookies. Children this age can recount the sequence of events. Moreover, children show unanimity in their recounting of such sequences. These and other studies suggest that pretend play calls on the ability to understand causality.

FISCHER AND WATSON'S MODEL FOR DEVELOPMENT OF SOCIAL ROLES IN PRETEND PLAY

M. W. Watson and Fischer (1977, 1980), M. W. Watson (1981, 1984, 1990), Fischer, Pipp, and Bullock (1984), and Fischer et al. (1993) have also proposed a model for the development of social role playing in pretend play. Their model traverses 14 steps through Fischer's four skill levels: sensorimotor systems, single representations, representational mappings, and representational systems (which roughly parallel Piaget's four periods). The significance of Fischer's model is that it presents development of social roles in levels and steps that parallel the steps and levels Fischer describes in the development of nonsocial cognition. Hence, it provides an explicit bridge between social and nonsocial cognitive development. (Case's, 1985, 1996, model of social interaction and role-taking provides another parallel bridge.)

Overall, Watson and Fischer and colleagues did a series of studies of role development in 258 children aged 1 to 13 years. Role development was assessed in the context of family roles and doctor-patient roles. In spite of differences in roles and tasks in these two contexts, both sequences developed in close synchrony. These studies all supported the validity of both the scalable sequence and the underlying developmental model. See Table 4-4 for a description of the steps in this scheme (M. W. Watson, 1984). They are also consistent with Piaget's early work

TABLE 4-4 SEQUENCE OF SOCIAL ROLE DEVELOPMENT

Level		Description	Example
Sensorimotor	1	Self as agent: child pretends to carry out one or more behaviors	Child pretends to drink from a cup
	2	Passive other agent: child pretends to make an object carry out one or more behaviors	Child makes a doll drink from a cup
	3	Passive substitute agent: child pretends to make an object act as a passive agent	Child makes a block drink from a cup
Single representations	4	Active other agent: agent can perform one or more behaviors	Child has a doll drink from a cup as if it were carrying out the action itself
	5	Active substitute agent: child makes object behave as an active agent	Child has a block drink from a cup as if it were carrying out the actions itself
	6	Behavioral role: an agent performs several behaviors fitting a social role	Child has a doll, as a doctor, use a thermometer
	7	Shifting behavioral roles: one agent performs one behavioral role and then a second agent performs a different behavioral role	Child has a doll, as a doctor, use a thermometer and then has another doll, as a patient, say it is sick and go to bed
Representational mappings	8	Social role: one agent behaving according to one role relates to a second agent behaving according to a complementary role	Child has a doctor doll examine a patient doll and respond appropriately to the patient's complaints
	9	Shifting social roles with one common agent: two agents perform a social role, and then one performs a different social role with a third agent	Child has a doctor doll examine a patient doll and respond appropriately to the patient's complaints and then has the patient doll interact appropriately with a nurse doll
	10	Social role with three agents: one agent in one role relates simultaneously to two other agents in complementary roles	Child has a doctor doll relate to a patient doll and be aided by a nurse doll
	11	Shifting roles for the same agents: two agents perform a social role and then perform a different social role	Child has a doctor doll relate to a patient doll and be aided by a nurse doll and then has the doctor doll act as father
Representational systems	12	Role intersection: two agent-complement role relations are coordinated so that one agent has two roles simultaneously relating to both complementary roles	Child has a doctor doll examine a patient doll and act as a father to the patient, who responds as both daughter and patient

Note. From J. S. Watson (1984).

on children's developing understanding of the concepts of social roles (Piaget, 1929).

Each level in the sequence is characterized by a different kind of role relationship. Sensorimotor systems, for example, involve a single child pretending to carry out one or more behaviors. Representational mappings, for example, involve two children playing complementary social roles. Representational systems, for example, involve a child's use of "role intersections," that is, the understanding that a person can play two complementary roles relative to two different individuals (e.g., be simultaneously a father to one and a son to another).

Each step in the sequence involves increasingly complex kinds of agency and increasing numbers of agents. The earliest sensorimotor steps distinguish between passive and active agency in terms of whether the infant treats the object as a mere recipient of her actions (passive) or as if it had a will of its own (active) (M. W. Watson & Fischer, 1977). Like Bretheron, Fischer and Watson have elaborated Piaget's stages of development of pretend play in terms of the roles children assume at different developmental stages. They based their model on data from free play and elicited play with the same objects in the same setting.

In one of their studies, 40 children aged 1½ to 4¼ years were tested . under these two conditions. In the elicited assessment condition, the experimenter told the child that he wanted to the child to use the toys to act out modeled stories. Each story was modeled for 30 to 50 seconds. If the child did nothing, the experimenter repeated the modeling. Five stories of differing levels of complexity were modeled in scrambled order. In the free play condition following the elicited assessment, the experimenter suggested that the child act out some stories of her own. Sixty-eight children were tested with an additional condition in which they were asked to enact the best story they could (M. W. Watson & Fischer, 1980).

In the elicited assessment condition of this study, children followed the predicted first eight-step sequence described in the Table 4-4. Beginning with step 5 (first-level representational mapping) in the free play condition, however, approximately half the children did not display their highest step. They suggest that by 3 or 4 years of age children's imitation may outstrip their spontaneous abilities. This result is consistent with two aspects of Fischer's skill theory: first, that children 4 years old and older show functional levels of performance in spontaneous activities and optimal levels of performance in scaffolded activities; and, second, that the gap between spontaneous and supported performances is greater at higher levels within a tier and increasingly greater across development (Fischer et al., 1984).

The Fischer and Watson model begins with sensorimotor system step 1 in pretense, the "self as agent." M. W. Watson (1990) subsequently alludes to a preceding step 1, "other as instrument of the self," which emerges at about 12 months. This step corresponds roughly to stage 4 in Piaget's sensorimotor causality series. During this stage the infant sees another as an agent who can be stimulated to do something, such as activate a toy (Piaget, 1954). Old step 1 in their systems corresponds to stage 6 in Piaget's (1962) imitation series, which is the bridge into pretend play.

The early development of these steps is significant because, at the time Fischer and Watson published their research, many investigators doubted that social roles begin when children are as young as 1 year of age. These earlier studies, which had used a more rigorous definition of role playing, had placed the emergence of role playing at 7 or 8 years of age (M. W. Watson & Fischer, 1980). This research extends the developmental sequence backward to the first year of life.

In summary, Fischer and Watson's model of role development is theoretically significant because it provides a systematic framework for assessing social and nonsocial cognitive development in parallel terms. Topically, it is important because it focuses attention on role development in early childhood. Methodologically, it offers systematic procedures for diagnosing behaviors characteristic of many steps and levels. Fischer's collaborative skill theory is important because it systematically treats cognitive development in a social context. Specifically, it offers the concept of developmental range based on assessment of the influence of social support on performance. In other words, this model builds on Piaget's studies of pretend play but overcomes many of the weaknesses identified in Piagetian theory.

According to Fischer's model, each child shows a developmental range of skills rather than a static skill level within any domain. The lower end of her range is defined by a child's functional level, and the upper end of her developmental range is set by her scaffolded level. Indeed, Fischer and colleagues say that skill theory "predicts that context affects the developmental level or stage of a person's competence even when the effects of experience and domain are controlled for" (Fischer et al., 1993).

Research on the developmental range of skills in storytelling by children under spontaneous and supported conditions revealed some systematic effects. The gaps between skill levels in the two conditions appeared between levels 4 and 5, at the boundary between the level or tier of single representations and that of representational mappings. In other words, the gap appeared at the boundaries between tiers, reinforcing the notion that tiers reflect significant developmental discontinuities (Fischer et al., 1984). It is also noteworthy that the gap between upper and lower ranges increased as a function of age. This reflects the increasing importance of collaboration and teaching during later development (Fischer et al., 1993). The increasing benefit from tutelage may explain the performance gap between schooled and unschooled children cross culturally.

NEO-PIAGETIAN VIEWS OF THE RELATIONSHIP BETWEEN IMITATION AND PRETEND PLAY AND REPRESENTATION

As indicated earlier, Piaget (1962) argues that during the sixth stage of the sensorimotor period deferred imitation arises from interiorization of imitation and that this provides the bridge to semiotic or representational functions of symbolic play, drawing, and language. This interpretation has been challenged by many neo-Piagetians. On one hand, neoinnatists have argued that representation occurs much earlier than the sixth stage (Meltzoff & Gopnik, 1993, see above). On the other hand, some constructivists have argued that representation develops as late as the 6th year of age (Lillard, 1993a, 1993b; Perner, 1991). Still other constructivists agree with Piaget that representation emerges during the transition from sensorimotor to preoperational intelligence but argue that Piaget's account of that emergence is flawed.

Lillard (1993a, 1993b) says that mental representation presupposes an understanding that one thing can represent another and that pretend play in young children does not entail such an understanding. She shows that object substitution in pretend play is based on an understanding of similarity that involves "acting as if" one object is another rather than an understanding that it represents the other. She also shows that character enactment and sociodramatic pretend play in children 1½ to 4 or 5 years of age is based on routines or simple scripts rather than true representation. Therefore, she dismisses as an illusion the apparent precocity of understanding in symbolic play as compared to performance in false belief tasks.

Perner (1991) agrees that true mental representation arises after symbolic play. He argues that pretend play depends upon multiple representations, one of the real situation, the other of a past situation. In other words, pretend involves representation of different situations or contexts. (See section on theory of mind for a discussion of Perner's model of mental development.)

Muller, Sokol, and Overton (in press) agree with Piaget that pretend play involves mental representation. They argue, however, that Piaget's account of the origin of representation is circular because it depends upon the very function it seeks to explain. They also argue that Piaget's account is contradictory in implying that representation arises from figurative (perceptual) rather than operative (action-based) roots (Muller et al., in press). Following Langer and others (Langer, 1986), they argue that representation arises through reflection on the external arrangements of objects children have constructed through *second-order proto-operations* during the second year of life (see chapter 3 for a description of second-order proto-operations). They suggest that these objectified stabilized external productions serve as means for detaching the products of actions from the actions themselves. As noted in chapter 3, chimpanzees achieve the level of second-order proto-operations, but they fail to produce the stabilized external productions that are characteristic of human infants.

PRETEND PLAY, ROLE PLAYING, AND TEACHING IN GREAT APES

As with imitation, investigators disagree on the issue of symbolic capacities of great apes. Some argue that they go beyond the sensorimotor period to the early symbolic subperiod of preoperations in the domain of pretend play. Most of the reports of pretend play have come from home-raised and/or symbol-immersed apes. One of the unresolved issues regards the criteria for diagnosing actions as true pretend play. Different investigators have used different criteria.

The best-known example of pretend play is Cathy Hayes's (1952) report that the young chimpanzee Viki played an elaborated and continuing pretend game with an imaginary string on an imaginary pull toy when she was about 16 months old. More recently, Jensvold and Fouts (1993) report six incidents of pretend play by the signing chimpanzees Washoe, Dar, and Moja on 15 hours of videotape. These included eating pretend food, addressing a toy bear, taking a picture with a toy camera, and using a wooden block as a hat.

Miles et al. (1996) report that the signing orangutan Chantek pretended to feed a toy animal at 20 months of age. Likewise, Patterson and Cohn (1994) describe the signing gorilla Koko engaging in pretend feeding of her doll and molding the dolls' hands to make the signs for *drink mouth*. Savage-Rumbaugh and MacDonald (1988) report that the symbol-immersed bonobo Kanzi plays a variety of pretend games, including pretend eating, hiding pretend objects, and giving pretend objects to others. These reports classify eating pretend food and acting on a passive agent as pretend play.

In contrast to these reports, the first experimental study of pretend play in young chimpanzees (Mignault, 1985) concluded that the chimpanzee subjects did not reach the transition between sixth-stage sensorimotor and symbolic activity described by Sinclair (1970). This study was based on Inhelder et al.'s neo-Piagetian study of the development of symbolic play in human children. Inhelder et al. classified the behavior of 10- to 30-month-old infants into: (1) activities without objects, (2) activities with objects, which included (a) interpretable activities including functional and make-believe activities and (b) noninterpretable activities including those involving use of one object (on self or not) and two objects. The development of interpretable activities has the following course: (1) conventional use of objects at about 12 months during the sixth stage, (2) conventional use of objects on the self, such as brushing own hair, at about 15 months, (3) use of objects as passive partners at about 18 months and later use of objects acting as agents and at the transition to symbolic play marked by (4) use of substitute objects to pretend at about 22 months.

Mignault used these criteria to analyze object play in four young chimpanzees in three experiments. The first experiment occurred when Maya, Merlin, Sophie, and Spock were 1 to 2 years old. A 30-minute videotaped sample of their activities with 15 objects was analyzed. The second experiment, done 2 years later when the chimpanzees were 3 and 4 years old (34, 38, and 50 months), analyzed the chimpanzees' activities with these objects following a demonstration modeling the following actions: (1) sweeping with a broom, pretending to fill a bucket with water, and pretending to wash the floor with a sponge; (2) diapering a doll after pretending to rub the diaper with oil from a bottle; (3) pretending to fill a glass with juice and filling a plate with food followed by pretending to feed a doll. The behaviors of the four young chimpanzees were analyzed in terms of categories taken from Inhelder's study of pretend play in human children of the same ages. The chimpanzees showed a similar profile to that of the children, that is, a very low level of interpretable behavior.

Mignault's third experiment was done when the chimpanzees were

TABLE 4-5 DEVELOPMENTAL STAGES OF PRETEND PLAY IN GREAT APES AND HUMANS ACCORDING TO INHELDER'S MODEL (AGE IN MONTHS)

Species	Stage 1: conventional use of objects	Stage 2: conventional object use on self	Stage 3: use of objects on passive partners	Stage 4: use of substitute objects
		Piagetian equivalent		
	Sensorimotor fifth stage	Sensorimotor fifth stage	Sensorimotor sixth stage	Preoperations symbolic
Humans	12	15	18	22
Chimpanzees	?	?	by 31 to 41	?
Bonobos	?	?	?	?
Gorillas	?	?	about 36	?
Orangutans	10	?	20	?
Gibbons	?	?	?	?

Note. ? = No data available.

almost 4 and 5 years old (44 to 61 months). In the spontaneous play, the chimpanzees showed much lower rates of interpretable behavior than human children aged 2½ to 3½ years (31 to 41 months), which was the oldest comparison group available. Even following modeling, the noninterpretable activities were predominant in the chimpanzees. Although the interpretable behaviors had increased slightly, they were still significantly lower than in human children. The noninterpretable behaviors with two objects were also much lower in frequency than in the human children. Mignault notes that the chimpanzees displayed deferred imitation of some conventionalized activities, including pretend eating and brushing a doll, but few behaviors involved two objects or the use of a substitute object, as required by Inhelder's criterion (Table 4-5).

There are at least two reports of pretend play in wild chimpanzees that meet Inhelder's criterion. Interestingly, both involve use of a log as a doll by wild juvenile chimpanzees. The first report comes from a study in the Kibale Forest in Uganda:

Last year, I watched a lonely boy chimpanzee, eight-year-old Kakama, playing for four hours with a log. He carried it on his back, on his belly, in his groin, on his shoulders. He took it with him every time he moved. He carried it up four trees, and down again. He lay in his nest and held it above him like a mother with her baby. And he made a special nest that he didn't use himself except to put the log in. Three months later, he did it again, watched by my field assistants in the Kibale Forest. They recovered the log, and pinned to it a description of the behavior. Their report was headed "Kakama's toy baby." (Wrangham, 1995, p. 5 [used with permission of the publisher]; see also Wrangham & Peterson, 1996)

The second report of pretend play in wild chimpanzees, which is strikingly similar, comes from Bossou in Guinea:

Jire (the mother) was observed to carry the sick infant, and moved from one tree to the other. Ja (8 year old female juvenile) was following the mother. Ja stopped at an enormous tree, Aningeria. She shook a dead branch of the Aningeria tree and got a rod about 50 cm long and 10 cm in diameter. Ja took the rod on her shoulder and followed the mother. Ja moved the rod from her shoulder to the place under the upper arm to hold it. She stopped at a large horizontal branch to take a rest. She placed the rod on the branch, just balancing it. She slapped the rod with one hand several times, as if softly slapping the back of an infant. Ja moved further into the trees . . . and was lost from sight. Ja seemed to handle the rod as if it were a doll. After all, the native Manon people in the village Bossou also have dolls like this. (Matsuzawa, 1995; see also Matsuzawa, 1997)

These reports, combined with Boesch's report on imitation in tool use in Tai Forest chimpanzees cited above, suggest that imitation and pretend play occur in wild chimpanzees. If so, these phenomena cannot be explained as an artifact of "enculturation" by humans. On the other hand, imitation and pretend play are certainly much less frequent and less robust in great apes than in humans. One reason for this may be that great ape mothers provide little reciprocal imitation and other forms of gestural and vocal scaffolding as compared to human mothers (Parker, 1993).

Experimental studies on the comprehension of role reversal in apes and monkeys suggest that, with appropriate scaffolding, chimpanzees can cooperate in level 5 pretend play. Povinelli, Nelson, and Boysen (1992) trained four chimpanzees—Sheba, 7 years, Kermit, 9 years, Darrell 9½ years, and Sarah, 28 years—to perform one of two complementary roles in a so-called communication apparatus that requires cooperation between an informant and an operator. The operator cannot see in which cup the food is placed, and the informant can see where the food is placed but cannot operate the handles to get the food. The operator takes his cue from pointing by the informant and pulls the appropriate handle, which delivers food to both the informant and the operator. In each role, the chimpanzees were paired with humans. After they had achieved 85% correct responses for three consecutive days, the chimpanzees were placed in the opposite role. Their performance in their new role was compared with their performance in the old role.

Three of the four subjects immediately performed the opposite role at the same level they had performed their old role. The investigators interpret this result as indicating that chimpanzees "learned role taking by learning the intentional significance of pointing from the perspective of the opposite role" (Povinelli, Nelson, and Boysen, 1992, p. 638). They also take this as evidence for cognitive empathy. Povinelli, Parks, and Novak (1992) performed the same experiment on four adult rhesus monkeys. They found little evidence that the monkeys understood their partner's social role.

Incipient role reversal and perspective taking are also necessary for demonstration teaching. The demonstrator gauges the pupil's skill level, identifies elements that need improvement, isolates or parses out those elements, and represents them to the pupil. Typically, she slows down, exaggerates, and repeats these segments after she has captured her pupil's attention. She then watches his imitation of her demonstration and sometimes repeats her demonstration (Wood, Bruner, & Ross, 1976). Although simpler forms of teaching such as opportunity teaching and coaching are not uncommon among birds and mammals, demonstration teaching is rare (Caro & Hauser, 1992).

As indicated in the section on imitation, Boesch (1991b, 1993) has presented written and videotaped descriptions of demonstration teaching by two chimpanzee mothers in the Tai Forest. In the case described previously, Ricci carefully demonstrated to her juvenile offspring the orientation and movement of a wooden club necessary for nut cracking. She then placed the tool in the proper position in her hand and watched as she imitated her movements. She succeeded in cracking the nut.

Likewise, Fouts, Fouts, and Van Cantfort, (1989) have described demonstration teaching by Washoe of ASL signs to her adopted son, Loulis. Washoe used both manual demonstration and shaping of Loulis's hands, techniques used by human parents to teach signs to their children. Similarly, Miles et al. (1996) describe teaching by signed prompting *Do this* by Chantek, the signing orangutan. It is important to note that teaching by demonstration depends upon the pupil's capacity to imitate the teacher.

We are unaware of studies of the development of teaching in human children. Anecdotal evidence from descriptions of pretend play suggest that pretend teaching emerges early in the symbolic subperiod. Children as young as 3 years adopt special linguistic registers when they are playing mommy or teacher (Cook-Gumprez, 1992).

As noted above, M. W. Watson and Fischer (1980) found the following sequence of role concept development in human children: level 1, representing self as agent emerges at about 12 months; level 2, representing another as agent, at about 24 months; level 3, representing an active substitute agent (Inhelder's criterion), at about 30 months; level 4, representing several related behaviors, at about 36 months; and level 5, representing one social role in relation to another, at about 48 months. (As Lillard, 1993a, points out, however, social role play may be based on enactment of routines or simple scripts.)

The foregoing reports of pretend play suggest that great apes can spontaneously achieve a level of pretense involving an active substitute agent. The report of Chantek signing to his caretaker suggests that some apes can even attain role playing. This would make their peak abilities in this domain comparable to those of 4-year-old children. The age at which they first attain the highest level is unknown, but it may be as late as 8 to 10 years.[3]

It is interesting to note in this context that, unlike monkey mothers, great ape mothers typically play with their offspring during the first 2 or 3 years of the infant's life (Parker, 1984, in press; Parker & Milbrath, 1994). They play a variety of physical games involving dangling, tickling, and gentle wrestling. They do not engage, however, in the kinds of mutual vocal and facial imitation games that are typical of play between human mothers and infants (Parker, 1993). It seems likely that mother-infant play in apes is associated with maternal teaching of tool-using skills, as mother-infant play in humans is associated with language acquisition (Stern, 1977).

PERSPECTIVE TAKING AND THEORY OF MIND IN HUMAN CHILDREN

Piaget's early research into the development of the child's notions about thought prefigured modern research in the arena now known as *theory of mind* (Wellman, 1990; Perner, 1991). In *The Child's Conception of the World*, Piaget (1929) distinguished three stages of development from

---

3. This late age is suggested by the fact that demonstration teaching in wild chimpanzees, which implies some rudimentary role playing, has been seen only in adults (Boesch, 1991b, 1993).

"realism" to objectivity. In the first stage, at about 6 years, the child confuses thoughts with things and identifies thinking with the mouth. In the second stage, at about 8 years, the child still materializes thought as air or other insubstantial matter and identifies thinking with a voice inside the head or the neck. In the third stage, at about 11 years, the child transcends these confusions and recognizes thought as immaterial and coming from the brain or mind. Piaget's criteria for distinguishing thought from things include (1) situating thoughts as intangible entities in the head; (2) distinguishing thoughts from things, that is, distinguishing reference from referent; and (3) situating dreams in the head.

PERNER'S THEORY OF MIND MODEL

The phrase *theory of mind,* which identifies a burgeoning area of research in social cognition (Astington, Harris & Olson, 1988; Baron-Cohen, 1995; Perner, 1991; Wellman, 1990; Whiten, 1991), comes from an experimental article on chimpanzee cognition written by Premack and Woodruff (1978). Early investigations in social perspective taking by the neo-Piagetian John Flavell and his students (Flavell, Botkin, Fry, Wright, & Jarvis, 1968), however, preceded the coining of this phrase. The theory of mind model differs from other models of social cognition in that it focuses on the child's understanding of the mental states rather than just the behaviors of others.

Success at the so-called false belief task is one of the hallmarks of the child's developing understanding of other minds. In the original version of this task, children were shown doll play enactments of the following vignette: a mother and her son Maxi are unpacking shopping bags. Maxi places chocolates in the green cupboard and plans to get them to eat when he returns from school. While he is gone, his mother uses some of the chocolate for her cake and returns the box to the blue cupboard. She leaves and Maxi returns. The child is asked where Maxi will look for the chocolates.

Most children 3 years old say that Maxi will look in the blue cupboard, whereas most older children say that he will look in the green cupboard. The researcher's interpretation is that the younger children respond in terms of what they know as observers, whereas older children respond in terms of what Maxi knows or doesn't know. Though this interpretation has been challenged, various iterations of this experiment have yielded the same result (Perner, 1991).

We focus on Perner's theory of mind model because it offers the

TABLE 4-6 DEVELOPMENTAL STAGES OF THEORY OF MIND IN MONKEYS, GREAT APES, AND HUMANS IN MONTHS

Species	Primary representations	Secondary representations	Meta-representations
		Piagetian equivalent	
	Sensorimotor period, stages 1 to 5	Sensorimotor stage 6 and symbolic subperiod	Intuitive subperiod
Humans	12	30 to 36	48
Chimpanzees	18	32 to 49	No
Bonobos	?	?	No
Gorillas	18	31 to 48	No
Orangutans	?	48	No
Gibbons	?	No	No
Baboons	?	No	No
Macaques	?	No	No
Vervets	?	No	No
Cebus	?	No	No

Note. Adapted from Perner (1991). ? = No data available.

most comprehensive reformulation of Piaget's model for children's understanding of mental states (addressing the role of representation in imitation, pretend play, empathy, intentionality, deception, and mirror self-recognition) and because it has influenced comparative research on primate cognition.

According to Perner's "common sense psychology" model, young children develop an understanding of the mind through three sequential levels he calls primary, secondary, and metarepresentations (see Table 4-6). He defines *representation* as "something that stands for something else," that is, as the representational medium rather than its content (Perner, 1991, pp. 16 & 280). Representational medium and content are connected by a *representational relation*. This definition covers such diverse phenomena as pictures, models, photographs, and mental states.

Primary Representations. Primary representations, characteristic of the first year of life, are single updatable mental models that represent real situations only. They are not modality specific and are not limited to current perceptions. They underpin early simple forms of deferred imitation and the understanding of object permanence, but not invisible object displacements.

Secondary Representations. Emerging in the second year of life, secondary representations are multiple mental models that can represent present and past situations. These secondary representations underpin reconstruction of invisible displacements and the onset of pretend play characteristic of Piaget's sixth sensorimotor stage and early preoperations period. Unlike Piaget's symbolic schemata, secondary representations are still presymbolic. According to Perner, these secondary representations also underpin mirror self-recognition, which involves recognizing two models of the self, and pictures of self and others, which involve recognizing self and others in two situations or contexts. Likewise, multiple models underpin the ability to understand correspondences between a little three-dimensional model and a real room. These models allow children to understand the difference between thinking and reality (Wellman, 1990), but only as two different situations. Finally, these models underpin early language use. This is 'the reason Perner calls children at this level, *situation theorists* and says that they have a "mentalistic theory of behavior."

Metarepresentational Models. Emerging at about 4 years of age, metarepresentational models are mental models of models, that is, entities that represent other entities. These metarepresentations entail an understanding of this relationship of representation. They entail an understanding that the representation is an interpretation and, consequently, that there can be alternative interpretations and misrepresentations. It is this understanding that underpins the realization that people can have false beliefs, and differing perspectives, and that appearances can differ from reality and can deceive people. Metarepresentations entail an understanding of the difference between knowing and guessing and an attempt to understand how people know what they know. This is the reason Perner calls children at this level *representation theorists* and says that they have a "representational theory of mind." This level corresponds in age to the intuitive subperiod of preoperations, although it involves more precocious understandings than Piaget attributed to that subperiod. Figure 4-2 is a cartoon illustration of the difference between situation and representational theorists' understanding of an event.

Distinguishing the Three Levels of Understanding. Perner suggests criteria for distinguishing three levels of understanding of the mind: at the primary level, mind is understood as inner experience; at the secondary level, as inner experience linked with situational models; at the meta-

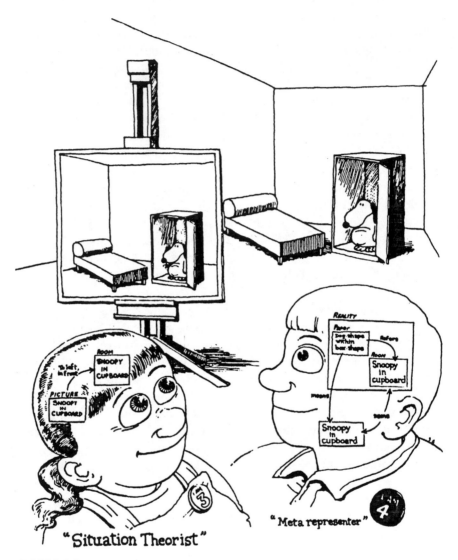

FIGURE 4-2
Different mental models of a picture and the depicted.
(from Perner, 1991, p. 84).

level, as possible nonexistence, aspectuality (feature abstraction), and misrepresentation.

To illustrate the differences between the secondary representational and metarepresentational levels of thought, Perner contrasts the understandings of knowledge by younger and older children. Younger children think knowledge is desirable to have because it helps people achieve goals. They believe that certain actions such as seeing, being told, and so forth, lead to knowledge, but they are unsure of the connection or mechanism leading to knowledge.

Older children, in contrast, understand that knowledge is a representation of some fact and that such knowledge is necessary for achieving goals. They also understand that there are necessary conditions for correctness and that certain experiences lead to knowledge acquisition. They understand, for example, that it is necessary to see something to know that it has happened.

In relation to the question of understanding of knowledge, Perner also contrasts younger and older children's understanding of the relationship between seeing and knowing. Studies of showing pictures to another reveal the following progression:

At 1½ years children hold the picture horizontally so that they can see it at the same time as the viewer; by 2 years, children hold the picture upright facing the viewer, suggesting that they understand that the viewer has her own inner visual experience. Likewise 1½-year-old children uncover the eyes of the viewer who covers them when shown the picture but do not seem to notice if the viewer closes her eyes. Three-year-old children understand the request to place an object such as a doll in front of themselves but are less able to understand the request to place an object in front of another object such as a dog that has a front. Only after 4 years are children able to understand the request to place an object in front of another object such as a glass that has no front.

SENSORIMOTOR-PERIOD ANTECEDENTS TO THEORY OF MIND

Most theory of mind research has focused on children in Piaget's preoperations period, between the ages of 2 and 6 years. A few investigators have tried to understand its origins in earlier stages of development. Leslie (1988), for example, has suggested that theory of mind develops out of pretend play at about 18 to 24 months. He argues that pretend play is an early form of theory of mind in which the young child already distinguishes between fantasy and reality. In support of his theory, Leslie notes that autistic children, who generally fail theory of mind tests, show little pretend play.

Meltzoff and Gopnik (1993) push the origins of theory of mind back even further into the sensorimotor period. They argue that theory of mind emerges out of imitation and that, since imitation begins within hours of birth, the foundations for theory of mind are present at birth. Their argument takes the form:

1. Imitation, but particularly facial and gestural imitation, involves a cross-modal mapping between the behavior of a visually apprehended model and the invisible behavior of the self.

2. This mapping allows the infant to understand that his body is like the body of his model.

3. Since imitating the model's facial expressions involves experiencing the same emotional states he is experiencing, it allows the infant to understand that his emotions are like the emotions of his model.

4. Mutual imitation provides the infant with a social mirror of its own behavior.

5. Since only social objects engage in mutual imitation, mutual imitation allows the infant to understand the difference between social and nonsocial objects.

6. This mapping involves a primitive form of perspective taking since the infant sees his own actions from a different perspective than he sees the actions of the model.

7. Therefore, imitation of the model's manipulation of objects by older infants (9 to 14 months of age) entails mutual focus of attention on the same objects and actions.

According to this model, imitation provides the basis for distinguishing social from nonsocial objects, for distinguishing self from nonself, and for experiencing empathy. Empathy provides a basis for identification with the emotions of others characteristic of 2½ year olds: "In other words, imitation of behavior provides the bridge that allows the internal mental state of another to 'cross over' to and become one's own experienced mental state" (Meltzoff & Gopnik, 1993, p. 358). Meltzoff and Gopnik point to experimental evidence that autistic children also lack the ability to imitate facial expressions and gestures, particularly those associated with emotions.

PERSPECTIVE TAKING AND THEORY OF MIND IN GREAT APES AND MONKEYS

The term *theory of mind* was coined by Premack and Woodruff (1978) in their landmark study of chimpanzee mentality. By a theory of mind, they meant that "the individual imputes mental states to himself and to others" (Premack and Woodruff, 1978, p. 515). By mental states, they meant wanting, thinking, and believing. They also meant distinguishing between pretense and reality and between guessing and knowing.

In this study, they presented the 14-year-old symbol-trained chimpanzee Sarah with a series of problems designed to test her ability to impute mental states to others. In the first test, they showed a series of videotapes of a human trying to get bananas that were out of his reach in four different situations. They stopped the videotape and presented Sarah with two photographs depicting alternative "solutions" to the problem. She chose the appropriate picture in each case but one (which required the person to remove rocks from a box before dragging it under the fruit). As they pointed out, however, there are several alternative explanations for Sarah's performance, including learned associations and empathy.

In order to rule out nonmentalistic explanations for Sarah's success, Premack and Woodruff devised additional problems. One series depicted two different trainers, one of whom Sarah liked and one of whom she disliked, with each trainer trying to solve the same problem. Sarah was then given a choice between a good alternative and a bad alternative for each problem. She generally chose the good alternative for her favorite trainer and the bad alternative for the other trainer. Some of the bad alternatives, however, were irrelevant to the problem.

They concluded that Sarah was not choosing the alternative on the basis of empathy. They tried a test to see if Sarah could distinguish between knowing and guessing, but the results were inconclusive. They also tried to test to see if four other chimpanzees could discriminate between a lying and a truthful trainer by deceiving the liar and helping the truthful trainer. One chimpanzee was able to lie in comprehension and production. The authors concluded that "not even the chimpanzee will fail tests that require him to impute *wants, purposes,* or *affective attitudes* to another individual, but he may fail when required to impute states of knowledge. . . . The ape may be incapable of differentiating between *guess,* and *know, doubt* and *believe,* and so forth" (Premack and Woodruff, 1978, p. 526).

Subsequently, Woodruff and Premack (1979) reported that some chimpanzees can discriminate between a lying and a truthful trainer and can deceive the liar. Four infant chimpanzees, Bert, Sadie, Luvie, and Jessie, who were aged 22 to 28 months old at the beginning of the series, participated in a series of experiments over a 3-year period. In the first phase, the chimpanzees, who were behind a barrier, were rewarded for indicating to a naive trainer which container was baited. Or, rather, trainers tried to determine from the chimpanzees which container had food. There were two kinds of trainers, cooperative trainers who shared the food with the subjects and competitive trainers who withheld the food. In the first phase, which lasted 5 months, both kinds of trainers were able to determine from the subjects which containers were baited.

The second phase began 11 months after the beginning of the first phase and lasted for 14 months. In this phase, the subjects were tested for both production and comprehension of cues. They now played both roles, alternately producing and comprehending cues regarding which containers were baited. In this phase, the cooperative trainer always oriented toward the baited container, and the competitive trainer always oriented toward the unbaited container. The third phase, a brief retest, began 10 months after the end of the second phase.

By the end of the second phase the chimpanzees were 4 to 4½ years (47 to 53 months) old. In the comprehension condition, three of the four chimpanzees consistently avoided the container indicated by the competitive trainer. Likewise, in the production condition, two chimpanzees consistently withheld cues from the competitive trainers, and two gave consistently misleading cues to the competitive trainer. These results persisted in the third phase.

In a later article on the same subject, Premack (1988a) retrenched a bit and shifted his experimental focus to seeing, wanting, and expecting. He notes that 3-year-old children fail to understand that another's knowledge depends upon his perception, his seeing or hearing. He then invokes his rule of thumb that chimpanzees fail at tasks that 3½-year-old children fail. He describes various tests of chimpanzees' understanding of seeing. In one experiment Woodruff put four juvenile apes with a trainer who had the key to a locked food container. The chimpanzees led the trainer to the container by beckoning. When the trainer was blindfolded, three of them tried to drag him, while only one removed the blindfold.

In another test, the same juvenile chimpanzees were exposed to two trainers: They could see that one trainer could see which container was baited. They could also see that the other trainer could not see. Three of the four chimpanzees chose the trainer who could see which container was baited. One of these individuals, however, failed to heed this trainers' advice. Finally, in a test similar to the false belief test, Sarah failed to show awareness of the trainer's lack of knowledge. Premack concludes that "the only states of mind the chimpanzee may attribute—if it attributes any at all—may be the simple ones—seeing, wanting, expecting" (Premack, 1988a, p. 175).

Povinelli and his associates tested chimpanzees on a related problem. They tested four chimpanzees, Sheba, 7 years old; Kermit, 9 years old; Darrell, 9½ years old; and Sarah, Premack's subject in earlier experiments, now 28 years old. In their experiment, they used a so-called communication apparatus that requires cooperation between an informant and an operator. The operator cannot see in which cup the food is placed, and the informant cannot operate the handles to get the food. In the crucial test, the operator has the choice of believing one of two informants. The chimpanzee can see that the first informant, the knower, has seen where the food is placed. He can also see that the second informant, the guesser, could not have seen where the food is placed (either because he was out of the room, in the first case, or had his back turned, in the second case). The knower always pointed to the correct location and the guesser always pointed to the incorrect location.

All four subjects in this experiment showed a significant preference for the knower's information in both conditions, transferring their knowledge immediately. The authors tentatively hypothesize that "chimpanzees, like young children, believe that those who see an object or event have a different understanding of that object or event than others who do not see it" (Povinelli, Nelson, & Boysen, 1990, p. 208). They also note that children develop this understanding sometime during their third year.

More recently, based on extensive experiments with 5- and 6-year-old chimpanzees, Povinelli and Eddy (1996) conclude that their subjects "displayed no clear evidence of appreciating the mental connection engendered by visual perception" (Povinelli & Eddy, 1996, p. 122). This conclusion was based on their subjects' apparent failure to appreciate that various forms of eye covering (buckets, blindfolds, closed eyes, etc.) might prevent actors from knowing where a test object had been hidden. The investigators suggest that older subjects might be more successful.

Premack and Woodruff's memorable phrase *theory of mind* was retrospectively applied to a growing body of research on understanding of minds by human children. This research, in turn, fed back into studies of theory of mind of monkeys and apes (Whiten, 1991). Related research in mind reading and deception in monkeys and apes (Mitchell & Thompson, 1986; Whiten & Byrne, 1988; Whiten & Ham, 1992) fits in this category as well. This research connects to the popular theory—the Machiavellian Intelligence hypothesis—that primate intelligence arose as an adaptation for social manipulation, which we discuss in chapter 10 on the adaptive significance of primate cognition.

Intentionality and Deception.   Premack (1988a) later posed his question about theory of mind in chimpanzees in terms of the philosophical concept of intentionality: "Is the ape an intentional system?" Premack's question preceded an article by the philosopher Daniel Dennett entitled "The Intentional Stance in Theory and Practice." In it, Dennett (1988) distinguishes four levels of intentionality in intentional systems.[4] He introduces these distinctions in his analysis of an anomalous deceptive behavior of a vervet monkey who gave a leopard alarm cry that distracted a rival from a desired object and allowed the distractor to get it, reported by Cheney and Seyfarth.

In an explicitly nonexperimental approach to deception, Whiten and Byrne used Dennett's classification of *intentionality* to categorize a corpus of reports of deception in monkeys and apes. Their full corpus comprised 253 records of primate behaviors from a large sample of field-workers that met their standards for deception: "Acts from the normal repertoire of the agent, deployed such that another individual is likely to misinterpret what the acts signify, to the advantage of the agent" (Byrne & Whiten, 1992). On the basis of their second round of reports, they re-

---

4. According to Dennett, a zero-order intentional system has no desires or beliefs; the communicator acts reflexively with stimulus-bound responses. In contrast, a first-order intentional system "has beliefs and desires but no beliefs about beliefs and desires." A second-order intentional system "has beliefs and desires . . . about beliefs and desires—-both those of others and its own" (Dennett, 1988, pp. 84–85). A third-order intentional system can want the responder to have beliefs about the actor's beliefs. Finally, a fourth-order intentional system can want to create the belief that it understands what belief the communicator has. Dennett concludes his discussion of intentionality with the following prediction: "It will turn out on further exploration that vervet monkeys (and chimps and dolphins, and all other higher nonhuman animals) exhibit mixed and confusing symptoms of higher-order intentionality" (Dennett, 1988, p. 196).

classified reports into the functional categories of: (1) concealment, by silence, hiding, object hiding, inhibiting interest in, or ignoring an object; (2) distraction, by calling, looking, threatening, leading, etc.; (3) attraction, by calling, looking, leading, etc; (4) creation of an image, neutral, affilitative, or threatening; (5) deflection to a third party; (6) use of a social tool; and (7) counterdeception (Byrne and Whiten, 1992).

Elaborating on Byrne and Whiten's categories, Mitchell (in press) has identified 66 kinds of deceit in great apes. The majority of these have been described in all four species, 54 in chimpanzees, 41 in gorillas, 35 in bonobos, and 34 in orangutans (yielding no significant differences in frequencies across species).

Hiding and Concealment. Reports of deception via concealment or hiding are common in descriptions of great ape behavior. These can be classified as hiding the self, hiding a behavior, hiding an object, hiding physiological evidence of arousal, and, in theory, hiding evidence. Mitchell (1991) reports repeated episodes of hiding of the self and of the self's actions by two adult male gorillas who were trying to gain access to an infant without alerting her mother. Tanner and Byrne (1993) report that a young adult female gorilla repeatedly hid her play face from her companion with one or both hands. De Waal (1983) reported that an adult male chimpanzee hid his erection from the dominant male with his hand. Savage-Rumbaugh and MacDonald (1988) reported that the bonobo Kanzi frequently hid himself and objects. See Table 4-7.

In order to qualify as tactical deception, an act had to be rated as first order—level 1— or higher by Byrne and Whiten (1992). If the outcome could be explained as incidental to or reciprocal to the act, it did not qualify. The qualifying records were classified as first or second order— level 1 or level 2. They were also classified by genus and corrected according to the number of studies of that genus. This analysis revealed the largest number of records of deception in chimpanzees and baboons. There were no records of tactical deception in prosimians. There were many records of tactical deception in monkeys, but these were all at level 1, involving the visual perspective of another monkey.

Only the great apes showed higher-level intentionality involving attribution of intentionality to another. Byrne and Whiten (1992) cite the following example of purported counterdeception, higher-order intentionality in great apes: Two chimpanzees are able to see a desirable object that was rather inobvious. The lower-ranking chimpanzee inhibits its impulse to look at it (apparently to keep the higher-ranking chim-

TABLE 4-7 DEVELOPMENTAL STAGES IN DECEPTION IN MONKEYS, GREAT APES, AND HUMANS AS MEASURED BY HIDING

	Age in months			
	Hiding own body and/or actions	*Hiding objects*	*Hiding or exaggerating signs of own affect*	*Hiding or modifying physical signs of lying*
	Piagetian equivalent			
Species	*Symbolic subperiod*	*Symbolic subperiod*	*Symbolic subperiod*	*Intuitive subperiod?*
Humans[*]	?	30 to 40	?	40 to 60
Chimpanzees[a]	By 56	By 56	? (Adults)	Not reported
Bonobos[a]	36?	60?	?	Not reported
Gorillas[b]	15	?	?	Not reported
Orangutans	?	?	?	Not reported
Gibbons	?	None	None	None
Baboons	?	None	None	None

Note. Adapted from Wellman (1990). ? = No data available.
[a] Savage-Rumbaugh and MacDonald (1988).
[b] Parker, in press.

panzee from noticing it or noticing that he sees it). The higher-ranking chimpanzee leaves, as if suspecting nothing, but hides behind a tree and peaks out.

Clearly, linguistic communication among humans entails higher levels of intentionality. Indeed some philosophers define communication in terms of intentionality. Anecdotal evidence from studies of great apes suggests that their spontaneous communication involves higher levels of intentionality. Gomez (1990), for example, describes the development of intentional communication of the request to open the door by a young captive gorilla. She does this by leading the human to the door, placing his hand on the knob, looking in the eyes of the human and then at the latch. This ability emerged at about 18 months. Likewise, Tanner and Byrne (1996) describe the intentional use of iconic gestures to communicate requests among zoo gorillas. Similar intentional use of iconic gestures in spontaneous communication among bonobos has been reported by Savage-Rumbaugh et al. (Savage-Rumbaugh, Wilkerson, & Bakeman, 1977).

DEVELOPMENTAL PERSPECTIVES ON THEORY OF MIND IN APES

Premack and Woodruff, Cheney and Seyfarth, Povinelli and his colleagues, and Whiten and Byrne all refer to the literature on the development of theory of mind in human children. Their studies are only pseudodevelopmental because they identify terminal-level abilities of their nonhuman subjects. So far there have been no explicit studies of the development of theory of mind in monkeys and apes. There are, however, developmental data on the use of emotion terms from ape symbol studies that provide hints about the development of theory of mind. We use Perner's framework to interpret the development of theory of mind in great apes.

If we examine the foregoing reports in terms of Perner's developmental levels, we see that great apes apparently approach his second level of multirepresentational thought typical of 3-year-old human children. This is probably equivalent to Dennett's second-order intentionality. The age at which great apes first achieve this ability remains to be discovered. Anecdotal evidence from ape symbol studies suggests that they do so before 3 years of age.

But what of the earlier sensorimotor-period stages or levels of intentionality? According to Piaget (1954), children in the third stage of the causality series display an incipient intentionality in the form of the secondary circular reaction. Children in the fourth stage display an incipient ability to dissociate means from ends in their coordination of secondary schemes. Children in the fifth stage display an ability to discover how to use a means to an end through trial-and-error groping in their tertiary circular reactions. They also display the ability to match their behavior to that of a model through imitation. They display intentionality and understanding that their actions are like the actions of others but without understanding the role of their own feelings or those of others. Hence, we conclude that sensorimotor stages 3 through 5 correspond roughly to Dennett's first-order intentionality, that is, desires without beliefs about desires. Sensorimotor stages 1 and 2 correspond to zero-order intentionality. Available evidence on monkeys suggests that they probably display no more than first-order intentionality.[5]

5. Various developmental psychologists have noted that attribution of intentionality and theory of mind both depend upon the prior achievement of joint attention between mother and infant (Baron-Cohen, 1995; Butterworth, 1991). Butterworth (1991) describes three stages of development of visual attention in human infants: In the ecological stage, at 6 months babies look to the same side of the room as their mothers

If we add Butterworth's 12-month joint attention landmark to Perner's landmarks, we create an earlier stage we might call "joint attention psychology" (which probably coincides with Piaget's fifth sensorimotor stage). An even earlier stage that we might call "simple attention psychology" (fourth-stage sensorimotor period) might apply to monkeys (Cheney & Seyfarth, 1991).

CONCLUSIONS ON THEORY OF MIND IN GREAT APES

Great apes apparently achieve early stages of development of theory of mind, but stop short of false belief. The scanty developmental data suggest that, in Perner's terms, great apes probably develop joint attention psychology by about 18 months (Gomez, 1991; Itakura, 1994a) and multi-representational thought by about 40 to 50 months (B. T. Gardner et al., 1989; Miles, 1990; Patterson & Cohn, 1994). Given that most of the evidence for these achievements in apes comes from scaffolded contexts, the functional level of development is unclear. Reports of strategic behavior in captivity and in the wild (e.g., de Waal, 1983; Goodall, 1986), however, suggest that these may represent functional levels of achievement.

DECEPTION AND THEORY OF MIND IN MONKEYS?

What is the evidence concerning theory of mind in monkeys? Fewer studies have been devoted to the study of this ability in monkeys than in apes. In an attempt to study theory of mind in monkeys, Povinelli, Parks, and Novak (1992) tested four adult rhesus monkeys, Tuck, Stud, Fuzzy, and Sundari, on the same task they used to study chimpanzees. In contrast to the chimpanzees, the macaques failed to discriminate between the knower and the guesser even after receiving six hundred to eight hundred trials. As the authors of this paper note, the contrasting results of this experiment on chimpanzees and macaques is consistent with differences in capacity for mirror self-recognition in these two

---

but cannot localize the target and cannot localize a target behind themselves. In the geometric stage, at 12 months, babies begin to localize stationary targets but still fail to look behind themselves. This is also the age at which they begin to understand referential pointing at objects. In the next stage at 18 months, babies can accurately localize the target but only search for a target behind themselves when nothing is in their field of view. Butterworth (1991) concludes that at about 12 months, human infants become capable of establishing joint attention with their mothers. This ability, which coincides with the fifth stage of the sensorimotor period, is important in both symbol development and theory of mind development.

species. Cheney and Seyfarth (1991) report similar findings regarding the inability of Japanese macaques to take into account their infants' knowledge or ignorance.

Cheney and Seyfarth (1991) devised a test for theory of mind in Japanese macaques. For this purpose, they devised three alternative conditions based on the fact that macaque infants adopt their mother's rank. In the first condition, the dominant observer could see and be seen by her dominant offspring and a subordinate adult through a glass partition. In the second condition, the dominant "observer" was obscured behind an opaque partition. In the third condition, the dominant observer could be seen by the dominant offspring and the subordinate adult but could not see them because the partition was a one-way mirror facing her direction. This experiment was designed to distinguish the effects of an observer's presence from the effects of the observer's knowledge of key events.

Under the first—clear—condition, when the dominant mother could see, the subordinate adult engaged in more friendly and less aggressive behavior toward the infant before and after the mother's release. In the second—opaque—condition, when the dominant mother could not see or be seen, the subordinate adult engaged in more threatening behavior toward the infant before the mother's release and less after her release. In the third—one-way mirror—condition, when the two interactants could see the dominant mother, but she could not see them, the subordinate adult engaged in more threatening behavior than in the glass condition, but less than in the opaque condition, before the mother's release, and less after her release. Both the infant and the subordinate adult spent more time near the partition under the one-way mirror condition.

Cheney and Seyfarth note that the results are mixed, some arguing for theory of mind, others against it. They conclude that the most conservative interpretation is that the subjects were monitoring the observer's behavior and apparent attentiveness rather than recognizing her mental state. They note that it is "by no means a trivial feat to adjust one's own behaviour according to the other individuals' orientation and direction of gaze. The ability certainly demands that monkeys recognize that attentiveness can strongly affect *actions*. It remains to be determined, however, whether monkeys also recognize that attentiveness can affect knowledge" (Cheney & Seyfarth, 1991, p. 193).

Mitchell and Anderson (1997) reported deceptive pointing in cebus monkeys who were tested using the same paradigm designed by Woodruff and Premack (1979) for chimpanzees. Three wedge-capped capuchins, Boy, aged 19 years, her son Coluche, 13 years, and Churchill, 26 years, were trained to point (using the whole hand) to the bowl that had been baited (by the experimenter) when the naive trainer came to the scene. The monkeys were confronted with two conditions. In one condition, the distinctively dressed "good" or cooperative trainer gave them the food when they correctly pointed to the baited container. In the other condition, the "bad" or uncooperative trainer kept the food himself when they correctly pointed to the baited container. By the end of the training, all three cebus showed 90% to 100% accuracy of pointing to the baited bowl with the cooperative trainer, and one, Coluche, pointed with almost 100% accuracy to the unbaited bowl with the competitor.

Mitchell and Anderson argue that this suggests that cebus monkeys, like chimpanzees and 4-year-old human children are capable of deceptive pointing. On the other hand, their results could be explained as the outcome of associative learning. The latter interpretation is more consistent with other reports of cebus cognition, particularly with the absence of imitation and pretend play, which are developmental precursors, and probably prerequisites of theory of mind.

According to Perner's criterion, macaques and cebus show level 1 primary representation and no evidence for theory of mind. This is consistent with their failure to complete the sensorimotor period in physical and logical-mathematical domains. In contrast, chimpanzees show the rudiments of theory of mind typical of level 2 secondary representations characterized by multirepresentational thought.

SELF-AWARENESS AND OTHER AWARENESS IN HUMAN CHILDREN

Piaget did not study the development of self-awareness in infancy and early childhood, but others have appealed to his developmental stages in various domains to explain the emergence of self-awareness. Case (1991), for example, has presented the only neo-Piagetian stage model for the development of self-awareness during infancy and early childhood. His model is based on his own stage model.

In the earliest, sensori-orienting stage from 1 to 4 months, the infant assembles a model of the mother's facial and bodily features, and he experiences affective states in his interactions with her. At this stage the infant may have an implicit sense of the subjective self.

In substage 1 of the sensorimotor stage, from 4 to 8 months, the infant becomes capable of unifocal interactions. He develops a subjective sense of agency in interactions with social and nonsocial objects. He also develops a sense of the objective self coincident with increasing control over his own hands and other body parts. In substage 2 of the sensorimotor stage, from 8 to 12 months, the infant becomes capable of bifocal transactions. He develops a sense of indirect agency; for example, he begins to share his interest in objects with his mother by looking at her to see if she is watching. In substage 3 of the sensorimotor stage, from 12 to 18 months, the infant develops a more explicit sense of indirect agency, for example, explicitly sharing objects with his mother. He also develops the ability to match the behaviors of two of more body parts. This is when he begins to recognize his image in mirrors.

Then, in substage 1 of the interrelational stage, from 1½ to 2 years, the infant begins to differentiate his own actions on an object from his mother's actions on an object and to create a relationship between their actions. This shows up in reciprocal pretend play routines, such as sweeping and holding a dustpan. In substage 2 of the interrelational stage, from 2 to 3½ years, and substage 3 from 3½ to 5 years, the child expands and coordinates the elements in an interaction. He becomes capable of playing complementary roles of two or more members of a social system (Case, 1991).

According to J. S. Watson (1994), the earliest form of self-knowledge, self-detection, occurs when infants are able to detect the difference between images they are creating with their own leg movements and the images created by delayed playback of their own movements. This ability emerges between 3 and 5 months (J. S. Watson, 1994).

The next developmental landmark, mirror self-recognition (MSR), is attributed to infants when they respond to their mirror image by touching a mark that has been placed on their face without their knowledge. MSR develops between 15 and 24 months of age in association with contingent facial movements, self-labeling, and self-conscious behavior (Lewis & Brooks-Gunn, 1979). MSR is followed at about 24 months by pictorial self-recognition and use of personal pronouns (Lewis & Brooks-Gunn, 1979).

These early indicators of self-awareness are followed at about 36 months by self-evaluative emotions of shame, guilt, hubris, and pride. These emotions apparently reflect internalization of social rules (Lewis, Sullivan, Stanger, & Weiss, 1989). Shadow self-recognition arises as about the same age (Cameron & Gallup, 1988). An early form of theory of the minds of self and others also appears (Perner, 1991). These indicators are

TABLE 4-8 DEVELOPMENTAL MANIFESTATIONS OF SELF-AWARENESS IN HUMANS IN RELATION TO PIAGET'S STAGES AND PERIODS OF DEVELOPMENT

Age (and stage)	Subjective self	Objective self
2 to 3 months (Stage 2 of the sensorimotor period)	Coordination of primary schema	
3 to 5 months (Stage 3 of the sensorimotor period)	Self-detection	
9 to 12 months (Stages 4 and 5 of the sensorimotor period)	Contingent body movements in mirror Use of mirror to locate objects	
15 to 24 months (Stages 5 and 6 of the sensorimotor period)	Self-conscious behavior (embarrassment)[a] Contingent facial movements in mirror Mark-directed behavior in mirror	Verbal self-labeling Pointing at self in mirror
2 to 3 years (symbolic subperiod of preoperations)	Possessiveness (symbolic)	Personal pronoun labeling Pictorial self-recognition
3 to 6 years (preoperations)	Shadow self-recognition Self-evaluative behavior (shame and pride) Theory of mind	Self-adornment Authoritarian morality Role reversal
6 to 11 years (concrete operations)		Consensual morality
11 years (formal operations)		Principled morality

Note. From Parker, Mitchell, and Boccia (1994).
[a]Reported in Lewis (1994); reported as occasionally occurring in 12-month-old infants in Amsterdam (1972) and Lewis et al. (1989).

followed at about 4 years of age by the beginning of an autobiographical self (Snow, 1990). Table 4-8 shows possible associations between these indicators and Piagetian periods.

As Baldwin (1906) proposed long ago, the development of self-knowledge seems to depend critically on imitation. Between about 9 and 14 months, mutual imitation of mother and infant provides the infant with a "social mirror" that reflects his own behaviors back to him

(Meltzoff, 1990). Indeed, Gopnik and Meltzoff (1994) argue that imitation is a form of self-knowledge, which is manifested in its most primitive form in neonatal imitation at birth. Various studies, as well as comparative data, suggest that MSR probably depends upon fourth- or fifth-stage sensorimotor causality and imitation rather than fifth- or sixth-stage object concept (Parker et al., 1994). Pretend role play, which grows out of imitation, provides the young child with a more elaborated social mirror. This feedback probably underpins the development of more complex forms of self-knowledge such as self-evaluative emotions. Thus, subsequent development of self-awareness depends upon role playing, as Mead (1970) proposed nearly six decades ago.

On the basis of various studies and anecdotal data, psychologists have suggested that certain cognitive abilities are prerequisites for MSR. Some developmental psychologists have argued that object permanence is a necessary, if not sufficient, cause for MSR (Bertanthal and Fischer, 1978; Lewis & Brooks-Gunn, 1979; Parker, 1991; Mitchell, 1993). It also seems likely that understanding of space and operational causality are prerequisite for MSR.

Others have proposed that imitation is a necessary cause for MSR (e.g., Parker, 1991; Mitchell, 1993). Mitchell (1993) argues, following Guillaume (1971), that *kinesthetic-visual matching* of movements of the self to visually apprehended movements of the model is the mechanism underlying MSR as well as gestural imitation. The idea that MSR depends upon imitation is also supported by the taxonomic distribution of various sensorimotor abilities. Although all primates achieve fifth-stage object permanence and at least one genus of monkeys—cebus— displays intelligent tool use, only the great apes show evidence of MSR and show the ability to imitate facial and manual movements. This suggests that fifth-stage object permanence and tool use are insufficient to explain MSR. If sixth-stage representation is necessary for MSR (Custance & Bard, 1994; Gergely, 1994), then it is more difficult to distinguish among the alternative sensorimotor series since representational abilities occur in the sixth stage of all the series.

Gopnik and Meltzoff (1994) argue that incipient self-awareness is present at birth. They believe that the intermodal matching in neonatal imitation implies some early form of self-awareness. At least one study has shown developmental correlations between imitation and MSR in human infants (Hart & Fegley, 1994). The idea that self-awareness depends upon imitation, of course, goes back to James Mark Baldwin (1897), Charles Cooley (1983), and George Herbert Mead (1970). Only developmental studies of the course of development of MSR in relation to the various sensorimotor-period scales will settle this issue.

Before we leave the topic, we should note that MSR is an index of some early developing form of self-awareness rather than an evolved adaptation. Work by Meltzoff (1990) suggests that MSR is an artifact of mutual imitation between infants and caretakers, or the "social mirror" that each provides the other during imitative interactions. If the "social mirror" is the origin of MSR and related forms of self-awareness, then imitation is the key to this domain of social cognition.

SELF-AWARENESS IN GREAT APES

Interest in self-awareness in great apes was stimulated by Gallup's (1970) invention of the mark test for studying MSR in monkeys and apes. This technique involves (1) exposing individuals to a mirror, (2) marking their face and/or brow in locations invisible to them while they are unaware of being marked, (3) re-exposing them to their mirror image, and (4) comparing the frequency of touching of parts of the face before and after they were marked. Increased frequency of visually directed touching of the marked area is taken as diagnostic of MSR. Gallup also noted other signs of MSR in chimpanzees, including visual inspection of otherwise invisible body parts. (See Heyes, 1995, for suggested refinements on this methodology.)

In a series of studies using this technique, Gallup and others reported that chimpanzees and orangutans passed the test, while gorillas and macaques did not (Gallup, 1977; Ledbetter & Basen, 1982). Subsequent studies have revealed that some gorillas pass the mark test and/or show self-directed behaviors in response to their mirror image (Patterson & Cohn, 1994; Swartz & Evans, 1994). Likewise, one study suggests that bonobos are capable of MSR (Hyatt & Hopkins, 1994). Further studies of other monkey species, including baboons and cebus monkeys, have demonstrated that they also fail to pass the mark test (Anderson, 1994).

Subsequent studies also suggest that MSR is a somewhat elusive phenomenon in chimpanzees as well as gorillas. These studies reveal that MSR and self-directed behaviors are not always coincident and that contingency behaviors do not correlate with MSR. Swartz and Evans (1991) performed the first study to reveal the elusiveness of MSR in chimpanzees. Of 11 chimpanzees they tested with the mark test, only 1 passed. Three of the chimpanzees who failed the mark test, however, displayed other self-directed behaviors while looking at their images. Later, none of three gorillas they tested passed the mark test. Two gorillas, however, showed a few self-directed behaviors and no social behaviors. The third showed social-directed behaviors to her mirror image (Swartz & Evans, 1994).

Subsequent research by Povinelli et al., (1993) has extended and elaborated these results in a series of studies on a much larger sample of 105 chimpanzees aged 10 months to 40 years. In one experiment on 30 subjects who had been pretested for other evidence of MSR, 10 passed the mark test and 20 failed. Although the ratio of pass to fail of 1:2 in this experiment is less extreme than the ratio of 1:10 in the Swartz and Evans experiment, the elusiveness of the phenomenon is robustly confirmed. The reason for this intraspecific variation is unclear. If MSR is at the upper boundary of the cognitive abilities of great apes, then more variation would be expected at this level than in cognitively simpler tasks.

This same experiment also revealed that only 8 of the 17 chimpanzees who had been pretested and classified as self-recognizing by virtue of self-exploratory behavior in front of the mirror passed the mirror test. In other words, 9 of the chimpanzees who used the mirror to examine otherwise invisible parts of their body did not pass. Conversely, 1 of the 7 chimpanzees who had been pretested as negative for self-recognition by the same criterion passed the mark test. So did 1 of the 6 subjects who had been pretested as ambiguous.

The demonstrated decoupling between success or failure in the mark test and apparent MSR measured by exploration of otherwise invisible parts of the body using the mirror is unexplained. Povinelli et al. (1993) suggest that individuals who earlier had demonstrated self-examination may have forgotten what they knew between the pretest and the experiment. We suggest that some great apes may be shy or embarrassed by the mark and avoid touching it. There is anecdotal evidence for this in the case of the signing gorilla Michael, who failed the mark test but turned away from the mirror and surreptitiously wiped the mark off his brow (Patterson & Cohn, 1994). Similar responses have been seen in human children (Lewis & Brooks-Gunn, 1979; Lewis, personal communication).

The series of studies by Povinelli et al. (1993) has also revealed that contingent movements of the face and the body—which typically co-occur with success at the mark test and with self-exploratory behaviors in front of the mirror—also occur in the absence of these two measures of MSR. Hence they are unreliable indicators of MSR (contra Parker, 1991). These contingency-testing behaviors, which seem to rely on fourth- and fifth-stage imitative abilities, may represent precursors to MSR both developmentally and microgenetically. The fact that only great apes and human infants display these behaviors supports this hypothesis (Parker, 1991).[6]

Studies of the development of MSR and other indices of self-awareness in great apes are in their infancy. Linn et al. (Linn, Bard, & Anderson, 1992) reported that chimpanzees as young as 2½ display MSR, while Povinelli et al. (1993) reported that chimpanzees typically do so between 4½ and 8 years of age (one of their subjects did so at 3½ years, however). One solution to this dilemma may lie in the complex trajectory of MSR development. According to some investigators, the earliest manifestations of MSR in human infants are unstable; that is, these manifestations either alternate with or coexist with other manifestations, suggesting that the infants are looking for another creature behind the mirror. Only after several months does MSR stabilize (Zazzo, 1982). Miles (1994) reports a similar developmental course in the signing orangutan Chantek, who displayed a stabilized understanding of MSR at about 3½ years.

Anecdotal data from studies of symbol-immersed apes indicates that, like human children, they begin to self-label and to show self-conscious behaviors at about the time they achieve MSR (e.g., Miles, 1994; Patterson & Cohn, 1994). A recent study reveals that some years after they pass the mark test, sometime between 3 and 8 years, chimpanzees also develop the capacity to recognize their own shadows (Boysen, Bryan, & Shreyer, 1994).

Given use of parallel methodologies in human and nonhuman primates, MSR is a natural candidate for comparative studies. Research on MSR in human infants reveals that the proportion of infants displaying MSR gradually increases from age 15 months to 24 months, when it is universal. This research also reveals that verbal self-labeling and signs of coyness and embarrassment develop in the same period. Following this, beginning at about 24 months, children begin to recognize their own pictures (Lewis & Brooks-Gunn, 1979). Beginning at about 36 months of age, children begin to display self-evaluative emotions of shame and

---

6. Several macaques have been trained to tolerate exposure to their own mirror image without threatening it. When subsequently given the mark test they have made mark-directed movements while glancing at their own image (Boccia, 1994; R. L. Thompson and Boatright-Horowitz, 1994). Their responses to their mirror images, however, differ both qualitatively and quantitatively from those of great apes. Whereas monkeys glance and hurriedly swipe at the mark, great apes look intently at their image and engage in collateral investigations of otherwise invisible body parts. These differences have led most investigators to conclude that monkeys lack MSR (Anderson, 1994).

TABLE 4-9 LEWIS'S CONCEPTS OF DEVELOPMENTAL STAGES OF SELF-AWARENESS IN MONKEYS, GREAT APES, AND HUMANS (IN MONTHS)

Species	Contingency testing	Mirror self-recognition and verbal self-labeling	Self-evaluative terms
		Piagetian correlates	
	Sensorimotor fourth or fifth stage	Sensorimotor sixth stage	Symbolic
Humans[a]	15 to 18	24	36
Chimpanzees	24?	40?	?
Bonobos	?	?	?
Gorillas[b]	24?	40?	51
Orangutans[c]	24	36	48
Gibbons[d]	No?	No?	No?
Baboons	No	No	No
Macaques	No	No	No
Baboons	No	No	No
Cebus	No	No	No

Note. ? = No data available.
[a]Lewis and Brooks-Gunn (1979).   [c]Miles (1990).
[b]Patterson and Cohn (1994).   [d]Patterson (personal communication).

pride (Lewis et al., 1989). They also begin to recognize their own shadows (Cameron & Gallup, 1988). Because psychologists have yet to describe clear stages of development of self-awareness from early infancy through early childhood, comparative psychologists have certain landmarks but lack clear-cut stages that can be used in comparative studies. A developmental study correlating various aspects of self-awareness with Piagetian stages would be ideal for our purposes. (See Table 4-9.)

SUMMARY AND CONCLUSIONS
SOCIAL COGNITION IN MONKEYS AND APES

Current knowledge regarding social cognition and its development in great apes and monkeys is summarized in Table 4-10, which reveals the following patterns in the development of social cognition in great apes. (Symbolic communication is treated separately in chapter 5.)

First, in social cognition, as in nonsocial cognition, great apes traverse the same stages or sequences of development as human infants and

TABLE 4-10 SUMMARY OF HIGHEST-LEVEL SOCIAL COGNITION IN GREAT APES AND MONKEYS

Species	Imitation	Pretend play	Self-awareness	Mentality
Chimpanzees	Sixth stage SM by ? months	Symbolic SP by 41 months	Symbolic SP by 40 months	Symbolic SP by 32 to 49 months
Bonobos	Sixth stage SM by ? months	Symbolic SP by ? months	Symbolic SP by 56 months	Symbolic SP by ? months
Gorillas	Sixth stage SM by ? months	Symbolic SP by 36 months	Symbolic SP by ? months	Symbolic SP by 39 to 51 months
Orangutans	Sixth SM by 35 months	Symbolic SP by 20 months	Symbolic SP by ? months	Symbolic SP by 48 months
Macaques	Fourth stage SM?	None	None	Fifth stage SM?
Vervets	Fourth stage SM?	None	None	Fifth stage SM?
Cebus	Fourth stage SM?	None	None	Fifth stage SM?

Notes. SM = Sensorimotor. SP = Subperiod (of preoperations period).

children do, up to their terminal level in the symbolic subperiod. Second, they apparently traverse these stages only slightly later than human children do. It is interesting to note that great apes achieve their terminal levels by between 3 and 8 years of age in various social domains, as compared to terminal achievements at 6 to 8 years in some physical domains. As we have mentioned, human children develop more or less synchronously in all domains.

Second, all the great apes achieve sixth sensorimotor stage in the imitation series (deferred imitation of novel actions) in the manual/gestural and facial modalities. They seem to do so by about 3 years of age. Third, all the great apes achieve early symbolic subperiod preoperational abilities to pretend play as measured by pretend actions on a passive partner by about 3 years of age. A few great apes have been reported to use substitute objects in pretend play by 8 years of age, thereby fulfilling Inhelder's more rigorous criterion for pretend play.

Fourth, at least some individuals of each great ape species have displayed MSR. They typically achieve stable MSR by about 4 years of age. Stable MSR probably represents sixth-stage sensorimotor abilities. Symbol-trained great apes also manifest symbolic-level self-awareness as indicated by use of self-evaluative terms. They seem to do so by about 3 or 4 years of age.

Fifth, at least some chimpanzees show evidence of symbolic subperiod level of theory of mind as measured by such intentional acts of deception as hiding objects. They may even display intuitive levels of theory of mind as manifested by misleading uncooperative interactants.

In contrast to great apes, macaque and vervet monkeys show no evidence of imitation, pretend play, MSR, or theory of mind. Given that all the foregoing abilities (excepting imitation) depend upon symbolic abilities, this is consistent with their failure to achieve sixth-stage mental representation and symbolic thought. Although monkeys show deceptive strategies, these seem to be limited to the sensorimotor level. Cebus monkeys, however, may be an exceptions to this.

We should note that these conclusions regarding great ape achievements in the social domain differ significantly from Tomasello and Call's conclusion that "there are no compelling differences in the social cognition of monkeys and great apes" (Tomasello and Call, 1994, p. 273). Several factors account for their conclusion. First, Tomasello and Call question reports of imitation and teaching in great apes. Second, in accordance with Tomasello's enculturation hypothesis, they attribute most of the higher-level performances of great apes to training by humans:

> The fact that apes raised in human-like cultural environments learn to
> do some very human-like things, such as communicating with symbols
> and imitatively learning new actions on objects, argues for some poten-
> tialities in apes that may be realized in certain kinds of environments
> (and perhaps monkeys too, if they were given the opportunity). And it
> is even possible that being raised in a cultural environment . . . is nec-
> essary not only for apes but also for human children to develop human-
> like social cognitive and social learning skills as well. (Tomasello and
> Call, 1994, p. 300)

In contrast, we accept the reports of imitation and pretend play in great apes. We agree with Tomasello and Call regarding the importance of social rearing for cognitive development in human and ape children. We suggest, however, that the relevant question is to what extent great apes provide cognitively stimulating environments for their own offspring and group members in natural environments. We suspect that great apes "enculturate" their own offspring in the wild. To answer this question, we need further research into conspecific imitation and teaching in naturalistic settings. We will address this issue again in chapter 10 on the adaptive significance of primate cognition.

CONCLUSIONS ABOUT SOCIAL COGNITION

The neo-Piagetian models of social cognition we have reviewed strongly suggest that social imitation and pretend play are prime movers in the development of social cognition. First of all, they suggest that imitation is one of the prime means by which human infants learn to distinguish between the social and nonsocial worlds. Second, they suggest that imitation is also a prime mover in the early development of empathy, self-awareness, and theory of mind. Finally, these studies suggest that pretend play, which develops out of imitation, is the prime mover of event representation through language and social role playing.

We believe that *event representation* via imitation, pretend play, language, drawings, and theory of mind is the essence of social cognition in apes and humans. Deception can be understood as a strategic violation of a script shared by two or more participants, the sophistication depending on the complexity of the script (Mitchell, in press). These various forms of event representation are apparently developmentally cumulative and synergistic. Event representation in imitation is prerequisite to that in pretend play. Event representation in pretend play may be prerequisite for case grammar, and simple grammatical event representation feeds back into and supports more complex role playing in pretend play. Likewise, reciprocal role playing is probably a prerequisite for understanding the desires and beliefs of self and others, and theory of mind feeds back into and supports more complex role relationships.

Nature is a miser. She clothes her children in hand-me-downs, builds new
machinery in makeshift fashion from sundry old parts, and saves genetic
expenditures whenever she can by relying on high-probability world
events to insure and stabilize outcomes. . . . We are looking at one of
Nature's most interesting achievements, the construction of the capacity
for symbols. . . . If we trace this marvel to its beginning in human infancy,
we will see that this particular work of art is a collage, put together out of
a series of old parts that developed quite independently.

BATES, 1976, P. 1

CHAPTER 5
DEVELOPMENT OF LANGUAGE IN
YOUNG CHILDREN AND APES

This chapter compares the development of symbolic communication in
great apes with that of human infants and children. We first briefly review some approaches to language acquisition in late infancy and early
childhood that have incorporated Piagetian concepts. Our focus is on
approaches that place the development of symbolic abilities within the
context of the development of other cognitive abilities, particularly on
studies and models of language acquisition that are relevant to comparative studies of symbol use in the great apes.

LANGUAGE DEVELOPMENT IN CHILDREN

Piaget (1962) argued that symbolic play, drawing, and language were parallel manifestations of the emergence of the capacity for symbol formation. He placed these at the onset of the preoperations period at about
2 years. Piaget's (1929) study of language development in young children has had little currency among psycholinguists. His sensorimotor
and symbolic stages, however, have been widely used as landmarks in
studies of the emergence of language.

These psycholinguistic studies have revealed much more specific associations between sensorimotor intellectual abilities and various aspects of language than Piaget's theory implied. They have implied local
homology or specificity rather than generality (Bates, Thal, & Marchman, 1991). On the other hand, Case (1985, 1996) has argued for a general
relationship, showing that the sequence of operational structures that

emerge during language acquisition parallels those in other cognitive domains. The differences between these two approaches may reflect the level of analysis.

DEVELOPMENT OF SYMBOLIC COMMUNICATION

The terms *signal, vocalization, gesture, symbol, sign,* and *language* have been variously defined by various linguists and psychologists. We therefore begin our discussion of language development with definitions of these terms, which we have taken from Volterra (1987). *Sounds* and *movements* refer to behaviors produced without communicative intention. *Vocalizations* and *gestures* refer to signals, that is, context-bound intentional communications. In linguistics, *symbols* refer to words and signs that are decontextualized and combinable. A *word* or a *sign* has to fulfill any two of the following criteria: (1) reference to a class of entities, (2) reference to absent entities, (3) reference to the same entity in different contexts, and (4) use in combination with other symbols. *Language* refers to symbol combination. (In mathematics, symbols refer to numbers and other abstract relations that are used in combination. In the broader sense *symbol training* can refer to lexical, numerical, or other iconic or arbitrary symbols [Thompson et al., 1997].)

A pioneering study of language acquisition has shown that certain sensorimotor series are more closely correlated with the emergence of words and signs than others. In a series of studies, Bates and her colleagues (Bates, Benigni, Bretherton, Camaioni, & Volterra, 1979; Bates et al., 1991) discovered the following pattern: Word comprehension at about 9 to 10 months correlates with the development of fifth-stage imitation and causality but not with object permanence and spatial cognition. Behaviorally, word comprehension correlates with the emergence of tool use, gestural routines, and deictic gestures (giving, pointing, and showing). These object-oriented gestures are used for both request and reference. Early word production at about 12 to 13 months correlates with sixth-stage deferred imitation. The vocabulary burst at 18 to 20 months correlates with use of gestural combinations in symbolic play. The grammar burst from about 24 to 36 months correlates with use of conventionally ordered scripts in symbolic play (e.g., Bates, 1993; Bates et al., 1979, 1991). See Table 5-1 for a summary.

Bates's colleague Virginia Volterra has done a comparative study of language acquisition in hearing and deaf children using consistent definitions of sign and word. This work has revealed that sign acquisition in deaf children undergoes parallel development with word acquisition

TABLE 5-1 MILESTONES IN COGNITIVE AND LANGUAGE DEVELOPMENT

Age in months	Language milestones	Cognitive milestones
9 to 10	Word comprehension	Deictic gestures Gestural routines Causal understanding
12 to 13	Word production	Deferred imitation Recognitory gestures in pretend play
18 to 20	Vocabulary burst	Gestural combinations in pretend play
24 to 36	Grammar burst	Active sequencing in pretend play

Note. Adapted from Bates et al. (1991).

in hearing children between the ages of 9 and 22 months (e.g., Volterra, 1987). From 9 to 13 months all the children showed equipotentiality between gestural and vocal modalities. The first signs and the first words emerged at about 12 months. Symbols were produced one at a time in both modalities. From 13 to 18 months both hearing and deaf children combined gestures and signs (see definitions above). More words than signs were acquired, however. From 18 to 22 months, speaking children combined words with words, and words with signs, but did not combine signs with signs.

In contrast, signing children did combine signs with signs. By 22 months all gestures but pointing tend to be replaced by words in speaking children. As in the studies cited above, the emergence of words and signs and their combinations correlated with the development of symbolic play. It should be noted in this context that sign languages are full-fledged languages as capable as spoken languages of encoding and transmitting information. This point is often missed owing to the complex and multifarious relationships between gesture and language (Ekman & Friesen, 1969). Table 5-2 presents a comparison of development in the two modalities.

A complementary finding of limited homology revealed a specific connection between Piaget's sensorimotor object concept and causality series and the emergence of specific semantic categories. Gopnik and Meltzoff (1987) showed that the use of disappearance words (*all gone, gone*) correlates with the achievement of sixth-stage object permanence and not with sixth-stage causality. Conversely, the use of success and failure words (*uh huh, there, no*) correlates with the achievement of sixth-stage causality and not with the sixth-stage object concept.

TABLE 5-2 DEVELOPMENT OF VOCAL AND GESTURAL SIGNALS IN HEARING AND DEAF CHILDREN

| Age (months) | Hearing children | | | Deaf children | |
	Gestural communication	Vocal communication	Combination	Gestural communication	Combination
11	Gesture Pointing	Vocalization		Gesture Pointing	
12	Sign	Word	Gesture Vocalization Gesture+sign Point Word	Sign	Point Sign
13	Sign	Word	Point+sign Point+word		
14					Sign Point+sign
18		Word	Word+word Sign+word		Sign+sign

Note. Adapted from Volterra (1987, p. 102). Used with permission.

This study was based on the use of Uzgiris and Hunt's (1975) neo-Piagetian developmental scale. Discovery of these specific relationships depended upon the existence of a developmental gap between the achievement of sixth-stage object permanence and causality. Contrary to Piaget's claim of synchronous development across all the sensori-motor series, this gap, which occurred in both directions, averaged 60 days.

DEVELOPMENT OF SEMANTICS AND GRAMMAR

Language acquisition entails the acquisition of semantics and grammar as well as words or signs. *Semantics* is the study of meanings encoded in words and classes of words. *Grammar* is the study of devices for distinguishing classes of signs or words and for associating them into sentences. R. Brown (1973) summarizes the five major processes of sentence construction: (1) designation of semantic roles such as agent, patient, instrument, locative expressed in linear order, and syntactic relations; (2) designation of such semantic modulators as number, tense, aspect, and mood expressed by inflections or free forms; (3) designation of modalities such as yes-no, interrogation, negation, and imperative; (4) embedding of one sentence as a constituent of another; and (5) coordination of sentences. Both semantics and grammar involve hierarchical classification of elements. According to Piaget, the capacity for hierarchical classification begins to emerge during the preoperations period from 2 to 6 years.

Most models of adult grammar have proved inadequate to describe the transition from individual words to complete sentences. Consequently, many investigators have turned from syntactic to semantic or meaning-based models to describe the development of grammar. In his comprehensive review of early studies of grammatical development, R. Brown (1973) discusses various approaches and suggests that Filmore's case grammar could be useful for studies of language acquisition. Filmore's case model abstracts a universal set of semantic relations that can be distinguished from specific grammatical processes that express them. Table 5-3 presents Brown's adaptation of these categories.

Brown reviews and analyzes his own and others' developmental data on children from four languages. He distinguishes the following stages of semantic and grammatical development, which are defined both in terms of *mean length of utterance* (MLU) and in terms of diagnostic semantic and grammatical achievements:

Stage 0, one-word utterances.

Stage I, two-word utterances expressing "semantic roles and syntactic relations," has an MLU of 1.75 with an upper bound of 5 words and occurs from about 18 to 30 months of age in his sample.

Stage II, two-word utterances expressing "grammatical morphemes and the modulation of meaning," has an MLU of 2.25 and an upper bound of 7 words and occurs from about 20 to 36 months in his sample.

Stage III, two- to three-word utterances expressing "modalities of the simple sentence," has an MLU of 2.75 with an upper bound of 9 words and occurs from about 20 to 38 months in his sample.

Stage IV, three-word utterances, has an MLU of 3.5 and an upper range of 11 words and occurs from 24 to 42 months in his sample.

Stage V, four-word utterances, has an MLU of 4 and an upper range of 13 words and occurs from 26 to 48 months in his sample.

Stage I in Brown's model occurs at the boundary between Piaget's sixth-stage sensorimotor and the onset of the symbolic subperiod of preoperations. Children in this stage are able to express the following

TABLE 5-3   CASE GRAMMAR IN CHILD LANGUAGE

Case name	Definition	Example (italicized noun is in designated case)
Agentive (A)	The typically animate, perceived instigator of action	*John* opened the door. The door was opened by *John*.
Instrumental (I)	The inanimate force or object causally involved in the state or action named by the verb	The *key* opened the door. John opened the door with the *key*.
Dative (D)	The animate being affected by the state or action named by the verb	*Adam* sees Eve. John murdered *Bill*. John gave the book to *Bill*. *Daddy* has a study.
Factitive (F)	The object or being resulting from the state or action named by the verb	God created *woman*. John built a *table*.
Locative (L)	The location or spatial orientation of the state or action named by the verb	The sweater is on the *chair*. *Chicago* is windy. John walked to *school*.
Objective (O)	The semantically most neutral case: anything representable by a noun whose role in the state or action named by the verb depends on the meaning of the verb itself	Adam sees *Eve*. The *sweater* is on the chair. John opened the *door*.

Note. From R. Brown (1973, p. 133). © 1973 by the President and Fellows of Harvard College.

two-term semantic relations: (1) agent and action, (2) action and object, (3) agent and object, (4) action and location, (5) entity and location, (6) possessor and possession, (7) entity and attribute, and (8) demonstrative and entity. They can also express modalities of affirmation, declaration, and, less clearly, interrogation (so-called *wh* questions), negation, and imperative semantically but not syntactically.

Stage II is marked by the gradual appearances of noun and verb inflections, articles, spatial prepositions, and the linking or copular verb *be* and its auxiliary forms. After a detailed study of the development of 14 morphemes, Brown concludes that they are acquired in the order of their relative grammatical and semantic complexity. Table 5-4 lists these morphemes and their order of acquisition. Stage II and the subsequent stages develop well into the preoperations period and are beyond the scope of this review because they have little relevance to comparative studies.

In a more radical use of semantic models to study early language development, Greenfield and Smith (1976) have argued that even one-word utterances (Stage 0) express semantic relations. Greenfield bases her study on an analysis of semantic relations children express to caretakers in the context of shared events through the combined use of gestures, intonations, and words. Her study suggests that children learn language by expressing semantic meanings in shared social contexts. Young children, like older children and adults, encode aspects of an event that are most uncertain from their perspective.

Like Brown, Greenfield and Smith use Filmore's semantic functions as a framework for their analysis. They also add Austin's performative function. These functional utterances are:

1. performatives or words that themselves constitute actions, for example, *bye-bye*

2. volitional performatives designed to obtain a particular response, for example, *mama* + reaching for an object

3. indicative object, referring to the object of an indicative act, for example, pointing + *doggie*

4. volitional object, referring to the object of a demand, for example, reaching + *nana*

5. agent, referring to animate instigator of an action

6. action or state of agent

TABLE 5-4 MEAN ORDER OF ACQUISITION OF 14 MORPHEMES ACROSS THREE CHILDREN

Morpheme	Average rank	Morpheme	Average rank
1. Present progressive	2.33	9. Past regular	9.00
2–3. *in, on*	2.50	10. Third person regular	9.66
4. Plural	3.00	11. Third person irregular	10.83
5. Past irregular	6.00	12. Uncontractible auxiliary	11.66
6. Possessive	6.33	13. Contractible copula	12.66
7. Uncontractible copula	6.50	14. Contractible auxiliary	14.00
8. Articles	7.00		

Note. From R. Brown (1973, p. 274).

TABLE 5-5 ONSET OF SEMANTIC FUNCTIONS: ORDER AND FIRST OCCURRENCE FOR NICKY

Semantic function	Age	Instance
Performative	8 (19)[a]	*Dada*, to accompany every action
Performative object	9 (8)	*Dada*, looking at father
Volition	11 (28)	*Na, na*, crawling to forbidden bookcase
Agent	13 (3)	*Dada*, hearing someone come in
Action or state of agent	14 (21)– 15 (18)	*Do(wn)*, when he sits or steps down
Object	16 (19–25)	*Bar* (fan), demanding fan to be turned on or off
Action or state of object	18 (1–8)	*Down*, shutting cabinet door
Dative	18 (4)[b]	*Mama*, when he gives book to mother
Object associated with another object or location	18 (8–16)	*Poo*, putting his hand on bottom while being changed, usually after a bowel movement
Animate being associated with object or location	18 (19–25)	⌐ *Lara* (Lauren), upon seeing her empty bed[c]
	18 (19–25)	⌐ *Bap* (diaper), to indicate location of feces
Modification of Event	19 (29)[b]	*More*[d] *record*, pointing to record playing

Note. From Greenfield and Smith (1976, p. 70).
[a] Numbers in parentheses represent age of the second of two children studied.
[b] Occurred in formal observation session. All other examples are from Nicky's diary.
[c] Bracketed examples were considered simultaneous.
[d] First example occurred in context of two-word utterance.

7. object, referring to an entity whose state was changed by the action

8. state or action of the object

9. dative, referring to an animate being who experiences an action of another

10. associated object, referring to absent objects involved in some nonspecific relation to an object or event, for example, *kite* in reference to a string

11. associated being, referring, for example, to possession

12. location of an action

13. modification of event referring to an entire event, for example, *again*.

Based on their analysis of data from formal sessions and mother's diaries, Greenfield and Smith classified the one-word utterances of her three subjects, aged 7 to 22 months, into these semantic categories.

They also traced the order of their acquisition. Interobserver reliability of classification into semantic categories was tested and found to range between 66% and 100%, depending on the specific category. Table 5-5 gives the order of acquisition for one of their subjects.

Scripts and Event Representation. The idea that children's grammar represents events or aspects of events has been elaborated by Nelson and her colleagues. Nelson argues that young children linguistically represent persons, objects, and sequences of action that recur in such familiar events as bathing, eating dinner, and going to bed. She characterizes these highly generalized and ritualized representations as *scripts* (Nelson, 1983; Nelson & Gruendel, 1986). Children apparently use these linguistic scripts as frames in which they substitute new items (e.g., new foods into the dining script) and use one item in several scripts or frames. This process helps them differentiate, disembed, and decontextualize linguistic meanings. This process has strong parallels in symbolic play (Bretherton & Bates, 1984).

Combinatory Strategies. Bates has shown correlations between symbolic play and the emergence of words and grammar. Similarly, Greenfield has shown an association between hierarchical strategies of syllable and word combination in early language development and strategies of object combination in play (Greenfield & Savage-Rumbaugh, 1990; Greenfield & Smith, 1976). Greenfield begins by describing three sequentially developing strategies of object combination in children from 11 to 36 months of age.

The first strategy, the pairing strategy, involves putting one cup into or on top of another. The second, the pot strategy, which becomes dominant by 16 months, involves combining three objects by placing two objects one at a time into a third static or immobile object. The third, the subassembly strategy, which first appears at 20 months, involves combining three objects by first placing one object into a second object, then placing the combined set in or on the third object. The first and second strategies correspond to the object combinations seen in the fifth stage of Piaget's sensorimotor spatial series. Figure 5-1 depicts the three strategies.

Greenfield argues that children show a parallel sequence in the development of strategies for combining syllables and words. Strategy one, the pairing strategy, involves combining two identical sounds (e.g., *dada*), combining a consonant and a vowel sound (e.g., *ma*), and, finally,

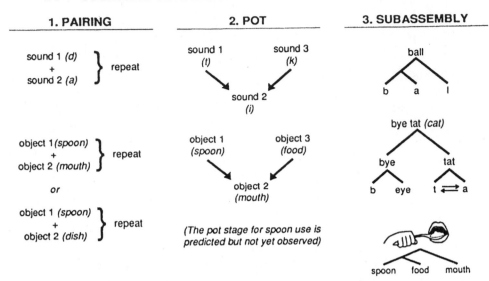

FIGURE 5-1
Depiction of three object manipulation strategies in sequence of development
(from Greenfield, 1992, p. 532). Reprinted with permission of Cambridge
University Press.

combining the same consonant with two different vowels (e.g., *baby*).
Strategy two, the pot strategy, involves combining two sounds in which
the initial consonant differs whereas the vowel sound remains constant
(e.g., *tinny*). Strategy three, the subassembly strategy, involves combining sounds to form a phonologically complex word (e.g., *ball*) or combining two previously constructed sounds to make a two-word sentence.
She also outlines a third parallel construction in the development of the
use of the spoon to get food to the mouth.

Greenfield also argues that development of the three combinatory
strategies in the domains of language, object combination, and tool use
(in this case, use of a spoon) are homologous. She notes that, whereas development in these three domains is parallel during the first 36 months
of life, it diverges and develops independently thereafter. Thus, Greenfield provides another example of local homology as well as an example
of developmental discontinuity.

Discontinuity of Language Comprehension and Production. Another
striking example of discontinuity or asynchrony occurs in the development of language comprehension and production. Large-scale studies
have demonstrated that comprehension precedes language production

developmentally by 2 to 4 months. These gaps occur first in word and then in grammar comprehension and production (e.g., Bates, 1993).

Studies contrasting early and late language learners (Bates et al., 1991) clearly show that language comprehension is a necessary but insufficient condition for the development of production (Bates et al., 1991). This is an important issue in comparative studies of symbol use in great apes, as we will see.

A NEO-PIAGETIAN MODEL FOR LANGUAGE ACQUISITION

In contrast to those who emphasize local homologies and developmental asynchronies in language acquisition, Case (1985) emphasizes parallels in the structures underlying the development of language and other cognitive domains. He proposes the following stages of language acquisition during the relational period (preoperations) using the same model he uses for physical, logical, and social knowledge:

1. Operational consolidation of sensorimotor schemes (1 to 1½ years) during which children first begin to use words for pragmatic purposes. This reflects the ability to construct a relationship between the listener and an event.

2. Operational coordination of relational schemes (1½ to 2 years) during which children begin to use words in the absence of their referents and create two-word utterances. This reflects the ability to construct a relationship between elements within an event (e.g., between an action and an object), as well as a relationship between listener and event.

3. Bifocal coordination (2 to 3½) during which children first begin to construct three-word and longer utterances with simple grammars. This reflects their ability to express two relationships within one utterance, combining, for example, *Daddy hit* and *hit ball* into *Daddy hitted the ball.*

4. Elaborated coordination (3½ to 5 years) during which children begin to use well-formed grammatical utterances that elaborate the utterances of the previous stage, e.g., *Daddy kicked the ball in the garden* (Case, 1985, pp. 169–73).

Consistent with his central conceptual structure model, Case (1996) emphasizes that parallel developmental transitions between relational and dimensional thinking occur at about 6 years of age in language, spatial, and numerical reasoning. At the beginning of the dimensional period, children become capable of constructing narratives based on the coordination of prior knowledge about the structure of events with knowledge about the role these events play in mental states. This ability parallels the ability to use Cartesian coordinates and perspective in the spatial domain and the ability to understand number systems. (See chapter 5.)

It is important to note that Case's formulation is aimed at a much higher level of generality than the studies of local homologies between language and particular cognitive domains. He is describing relatively nonspecific coordinations according to their information-processing capacities. Therefore, we see no contradiction between his approach and the pinpointing of specific local homologies.

CONCLUSIONS ABOUT COGNITION AND LANGUAGE

Language is the primary, universal means humans use for representing events. Although humans also use miming, pictures, maps, and mathematical formulas to represent events, they are uniquely specialized anatomically for language. Language differs from these other forms of event representation in the number and breadth of cognitive capacities it employs and in the complexity of the relationships between these capacities and various facets of language.

As Bates et al. (1979) have indicated (see this chapter's epigraph), language draws on knowledge from virtually every other cognitive domain. Grammar encodes object, temporal, spatial, and causal relationships. The categories of case grammar, for example, reflect agent, action, object, locative, etc., Grammar also draws on logical-mathematical knowledge, including classification, number, and logical possibility and necessity. These relationships are reflected, for example, in number (plural vs. singular nouns), verb mood (perfect vs. imperfect), and aspect (conditional vs. unconditional). Obviously, parts of speech and noun and verb categories reflect classification.

Language also draws on social cognition, representing relationships among actors as well as intentionality. Verb conjugations (I see, you see, he/she sees, we see, they see) represent various social perspectives and alignments as well as genders, and verb modality (imperative, interrogative) represents relationships between the speaker and the listener. Language acquisition depends upon social imitation as well as cognitive construction.

It should be acknowledged, of course, that as well as calling on a multitude of cognitive functions to construct language, speakers use these linguistic constructions to describe real and imaginary, true and false, physical, logical, and social relationships in the world (event representation). Language, the product of cognition, then becomes language, the tool of cognition, which abets the further development of the mind.

Likewise, in humans, language, and other symbolic systems, the expressions of cultural capacity, become generators of cultural innovation and change (Goodenough, 1981). Symbol-trained apes, like children, have "upgraded minds" that are the products of "historical process(es) involving the external organization of symbolic experiences by other beings, and in the chimpanzee's case, another species" (Thompson et al., 1997, p. 42).

The dual nature of language as product and producer reflects its dual nature as an outgrowth of both central and specific cognitive structures. These dual relationships probably account for the debate over the modularity or generality of language abilities. Before turning to this issue, however, we address comparative data on the symbolic capacities of the living great apes.

DEVELOPMENT OF SYMBOLIC COMMUNICATION IN GREAT APES

Studies of the symbolic abilities of great apes became a major focus in primate research in the sixties beginning with the Gardners' (R. A. Gardner & Gardner, 1969) landmark study. Following the Gardners' initial study with Washoe, there were several other studies of chimpanzees (Fouts, 1973; R. A. Gardner, Gardner, & van Cantfort, 1989; Terrace, 1979), a study of two gorillas (Patterson, 1980), and two studies of orangutans (Miles, 1983; Shapiro, 1982). Some chimpanzees have learned sign language; others have learned other kinds of symbol systems (Premack, 1976; Rumbaugh, 1977; Matsuzawa, 1985a). In addition, two bonobos and two chimpanzees have learned a lexigram language (Savage-Rumbaugh et al., 1993; Savage-Rumbaugh, Romski, Hopkins, & Sevcik, 1989).

In accord with the comparative developmental evolutionary theme of this book, we focus primarily on those symbol acquisition studies that have simulated conditions of language acquisition in human infants and children. These naturalistic cross-fostering studies share several characteristics that distinguish them from operant conditioning studies (R. A. Gardner & Gardner, 1989).

First, they have focused on spontaneous communication of meanings between teacher and pupil (and later between pupils) in natural settings rather than on rote learning. Second, they have therefore avoided the use of extrinsic rewards and punishments. Third, they have been longitudinal developmental studies. Fourth, they have explicitly compared symbol acquisition in their subjects to that of human children. Fifth, in most cases they have enlisted linguists as collaborators.

Finally, they have included other developmental studies such as studies of imitation and self-awareness (Miles, 1994; Patterson & Cohn, 1994), tool use (Savage-Rumbaugh & MacDonald, 1988), and sensorimotoric and locomotoric development (B. T. Gardner & Gardner, 1989a). Specifically, these include studies of chimpanzees by the Gardners and associates, studies of gorillas by Patterson and associates, studies of an orangutan by Miles, and the studies of bonobos by Savage-Rumbaugh and associates.

Ideally, comparative studies of symbolic ability focus on acquisition of a natural language, that is, a fully functional language that is used as a full purpose means for symbolic communication among members of a natural community. Given the vocal limitations of great apes, this language would necessarily be ASL or some other equivalent sign language used by deaf signers in other countries.

APE STUDIES USING AMERICAN SIGN LANGUAGE

Because ASL has been used in several studies of ape symbol acquisition, it is important to describe it. ASL is a natural, full purpose language capable of encoding any meaning that can be encoded by any natural spoken language. Although ASL includes some iconic gestures, the majority of ASL signs are arbitrary, conventionalized signs analogous to words in any natural spoken language. Both spoken and signed languages are produced through gestures. One involves gestures in the vocal-auditory modality; the other, gestures in a manual-visual modality. In both modalities, combinations of distinctive features embody semiotic functions (Armstrong, Stokoe, & Wilcox, 1995). The gestural nature of ASL, however, imposes constraints on and opportunities for expression of meanings that differ from those in spoken languages.

Whereas in a spoken utterance, phonological, syntactical, and semantic levels are clearly distinguishable (e.g., in speech sounds, syllables, words, parts of speech, and sentences), in a signed utterance, these levels are less easily distinguished. The same sign can serve as either a noun or a verb depending on its movement. Likewise, the same sign can serve either as a word or a sentence depending on accompanying actions of the arm or hand or additional movements of the face or head (Armstrong et al., 1995).

According to the DASL, cited by B. T. Gardner et al. (1989), a sign has several components. The first component is the place on the body or in space where it is formed. The second is the configuration of the hand (including its orientation, its direction, and the part that contacts a surface). The third is the movement pattern of the hand. The DASL lists 26 terms for place of articulation, 19 terms for configuration of articulation, and 9 terms for direction of movement. The sign for *girl*, for example, is a thumb extended from fist (configuration), tip of the thumb rubbing down (movement) on the cheek (place) of the signer. As in spoken language, individuals show slight differences in articulation that do not affect the intelligibility of their signs. Infants and young children use baby signs that differ in their placement and other features from those of adults.

ASL relies heavily on inflectional devices for modulating grammatical, semantic, and pragmatic meanings. Direction of movement along the line of sight between signers, changes in placement, eye gaze, facial expression, and reiteration are such inflectional devices. When a signer signs *see* or other verbs while moving the sign toward the observer, this means, *I see you*. If he signs *ask question* toward the listener it means *I ask you*; if he signs it toward himself, it means *you ask me* (Rimpau, Gardner, & Gardner, 1989). Thus the same sign, for example, *drink*, can serve as a noun or a verb depending on its inflections. When a signer repeats a sign, reiteration may indicate many items, as when *tree* reiterated means *forest*. Alternatively, especially with a verb, reiteration may indicate emphasis. Use of these inflectional devices explains how a single sign can function as either a word or a sentence.

In their ASL-based comparative study, the Gardners used only the ASL signs in the DASL. In their book *Teaching Sign Language to Chimpanzees*, they have compiled a summary table containing a glossary of all the 430 ASL signs used by the three subjects of their second project, Dar, Tatu, and Moja. Each sign is described in terms of the PCM system of *place, configuration,* and *movement* involved in its articulation. When relevant the Gardners also describe the hand orientation, contactor, and direction of movement. Each sign included in the vocabulary had to meet the criteria of (1) having been reported in three independent observations by three different observers and (2) having been spontaneously used in an appropriate manner for 15 consecutive days. Signs from the earlier study with Washoe were omitted because the records had not included standardized descriptors for PCM of signs. Included signs were

tabulated from qualifying records by the three authors and by a gradu-
ate student who had been a member of a foster family for Tatu. They
achieved roughly 98% agreement.

A study of deviations from the standard or citation ASL forms by
chimpanzees revealed some consistent patterns. First, all of the modifications preserved the basic PCM sufficiently to allow comprehension.
All modifications were used by all the subjects. Second, modifications
of placement of signs occurred when young chimpanzees placed the
signs on the bodies of their addressees or near inanimate referents rather
than on their own bodies. Third, modifications of configuration occurred
when young chimpanzees molded the hand of the addressee rather than
shaping their own hand. Fourth, modifications occurred when they reiterated, enlarged, or invigorated the movement of the sign. Interestingly, many of the same place modifications are seen in young human
children and in adults signing to young children (Rimpau et al., 1989).

DEVELOPMENTAL DATA ON APE SYMBOL USE

Of all the aforementioned cross-fostering studies, only the Gardners' has
provided strictly developmental data on symbol acquisition from birth
onward. For a variety of reasons, most studies have been quasi developmental rather than strictly developmental. Typically they began when
their subjects were 10 months old or older. The single exception to this
seems to be the second study by B. T. Gardner et al. (1989). Although
most of these studies began when their subjects were about a year old,
they do offer comparative developmental data on the sequence and rate
of acquisition of various linguistic abilities, if not data on the first possible ages of acquisition.

Ideally, comparative developmental studies of symbol acquisition
would directly compare the rate and content of language development in
great apes and human infants and children growing up in the same natural language environment from birth onward. These studies would systematically compare developmental rates and contents using the same
kinds of analysis. Necessarily, they would compare the abilities of great
apes to those of infants and children rather than adults. Among the
various comparative developmental studies that approximate the methodological ideal outlined above, we focus on four.

SIGN ACQUISITION IN GORILLAS

The first study is Bonvillian and Patterson's (1993) comparison of the
rate and content of sign acquisition in the gorillas Koko and Michael

with that of deaf children. Their analysis reveals that very young human children acquired signs at roughly twice the rate that the gorillas did. Using Nelson's criteria of age at acquisition of the first 10 signs as compared to the first 50 signs, they found that children achieved these two landmarks at mean ages of 13.3 months and 23 months, respectively. This works out to a rate of acquisition of 7.8 signs per month for human infants, which compares to 4.0 signs per month for the two gorillas. Koko's acquisition ages were 16 months and 25 months, respectively, for the first 10 and 50 signs. Unlike the human children, she began her sign acquisition at 12 months of age.

Their analysis of the first 10 signs revealed that iconic (pantomimic) signs were more common in the gorillas than in the human children. Their analysis of the grammatical categories of the first 50 signs in human children and gorillas, however, shows strong parallels: Both species showed high frequencies of names, actions, and such modifiers as *mine*, etc. The children—but not the gorillas—also showed some examples of personal-social signs and such functions as *what?* B. T. Gardner and Gardner (1980) also showed that the first 50 signs of their four infant subjects could be substituted for the 50-word vocabulary of human children in Nelson's study of language development.

RESPONSES TO *WH* QUESTIONS IN CHIMPANZEES

The second comparative developmental study is the Van Cantfort, Gardner, & Gardner (1989) study of the development of responses to *wh* questions in four infant chimpanzees, Pili, Moja, Dar, and Tatu, ranging from 18 to 74 months. They categorized *wh* questions into the following question categories and target response categories: *Who* demonstrative = proper nouns; *Who* subject, object, trait = proper nouns or pronouns; *Who* possessive = possessives; *What* quality = attributes; *What* demonstrative = common nouns; *What* object of action = common nouns; *What* want = common nouns in verbs; *What* predicate = verbs; *Where* action = locatives; *Where* nominal = locatives; and *How many* = quantitative.

As they developed, these young chimpanzees responded appropriately to *wh* questions with the functional equivalent of proper nouns (including pronouns), verbs, locatives, attributes, and possessives. Table 5-6 summarizes Tatu's responses at four ages.

Like human children, their three surviving infant chimpanzees, Moja, Tatu, and Dar, showed the following trends in the development of responses to *wh* questions: First, the percentage of grammatically appropriate responses increased over development. Second, the development

TABLE 5-6 DISTRIBUTION OF SENTENCE CONSTITUENTS IN REPLIES BY THE CHIMPANZEE TATU

18 Months

Question frame	N,P	n	V	L	Total
Who questions	3*		1	3	7
What demonstrative		6*			6
What want		8*		2	10
Where action/nominal				6*	6

chi-square = 33.519, df = 9, p = .0001

28 Months

Question frame	N,P	n	V	L	Total
Who questions	3*	1			4
Who possessive	5	1			6
What demonstrative			9*		9
What want			9*		9
Where action/nominal	2		1	3*	6

chi-square = 44.058, df = 12, p < .0005

34 Months

Question frame	N,P	PS	n	V	L	Total
Who questions	17*			1		18
Who possessive	3	2*	1	1		7
What demonstrative			9*	1		10
What object of action			3*			3
What want			6*	2*		8
What predicate			2	3*		5
Where action/nominal			2	2	13*	17

chi-square = 136.263, df = 24, p < .0005

49 Months

Question frame	N,P	PS	A	n	V	L	Tr	Total
Who questions	16*					1	1	18
Who possessive	1	7*						8
What quality			8*					8
What demonstrative				1	8*			9
What object of action	1				5*			6
What want					6*	1		7
What predicate				1	6*			7
Where action/nominal					3	13*		16

chi-square = 301.592, df = 42, p < .0005

TABLE 5-6 *(Continued)*

Question frame	N,P	PS	A	n	V	L	Tr	Total
			61 Months					
Who questions	24*				2			26
Who possessive	4	5*						9
What quality	1		7*					8
What demonstrative			1	9*				10
What object of action				9*				9
What want				2*	1	5		8
What predicate				1	7*		1	9
Where action/nominal				1		12*		13

chi-square = 321.136, df = 42, p < .0005

Notes. From Gardner, Gardner, and Van Cantfort (1989). Reprinted by permission of the State University of New York Press. © SUNY Press. All rights reserved. Cells for appropriate constituents are marked with an asterisk. N = proper noun, P = pronoun, PS = possessive, A = attribute, n = common noun, V = verb, L = locative, Tr = trait.

of responses to specific categories of *wh* questions occurred in the same order as in human children:

> Both children and chimpanzees, initially, provide nominals for What questions and locatives for Where questions, followed by verbs for What do (predicate) questions and proper nouns and pronouns for Who questions, and followed still later by appropriate replies to How questions. Appropriate answers to the other types of Wh-questions, such as When questions, appear much later in the developmental pattern of children and have not yet been studied for chimpanzees. (Van Cantfort et al., 1989, p. 236)

Similar results were obtained from analysis of spontaneous responses of Dar to *wh* questions in nontest sessions (Rimpau et al., 1989).

GRAMMAR IN A BONOBO

The third comparative developmental study is Greenfield and Savage-Rumbaugh's (1990) comparative study of grammar in the bonobo Kanzi and human children. They begin their study by stipulating five criteria for a grammatical rule:

1. independent symbolic status of each component of a combination

2. reliable and meaningful relationships among the symbols

3. specification of relations among categories of symbols rather than individual symbols

4. some formal device relating symbol categories across combinations

5. spontaneous productive use of this device

Grammars can be based on two different kinds of devices, morphological inflections (see above) and rule-based ordering of elements.

Greenfield and Savage-Rumbaugh used semantic roles used in child language studies to characterize Kanzi's two-element combinations. Their data base consisted of all two-element combinations Kanzi produced over a 5-month period in 1986 when he was 5½ years old. This comprised 1,422 combinations of lexigrams or lexigrams and gestures.

Their analysis revealed that Kanzi's two-element "utterances" fell into ten major categories of semantic relations: conjoined actions, agent-action, action-object, object-agent, agent-object, entity-demonstrative, goal-action, action-goal, conjoined entities, conjoined locations, location-entity, and entity-attribute. (Kanzi also produced some nonredundant three-element combinations that were not analyzed in this article.) Seven of these are the same as those accounting for most of human children's utterances in the two-word stage. Kanzi failed to produce one relation—possession—frequently found in children. On the other hand, he produced a category of relations—conjoined actions—that children fail to produce. He also produced a much smaller proportion of statements to requests than human children do.

The authors show that these ten semantic relations reflect protogrammatical rules. The rule ordering of action-object relations, for example, reflected the order used by his models. The rule ordering of action-action relations, for example, *chase hide, chase bite,* which was Kanzi's own rule, reflected the behavioral order. Use of these ten relations fulfill the five criteria Greenfield and Savage-Rumbaugh set out for grammatical rules. For example, Kanzi used the same symbol in a variety of semantic relations (criteria 3 and 5); he also showed generality and consistency in his use of his grammatical rules (criterion 4).

In conclusion, these authors note that whereas Kanzi produced some of the same grammatical relations as human children at the two-word stage, he produced a much smaller number of combinations than human children do. He also showed a much slower rate of development within that stage than human children do.

Subsequently, Greenfield and Savage-Rumbaugh (1993) analyzed the use of repetition in conversations between bonobos and their caretakers. They found that, like children in stage 1 of language development, these great apes displayed communicative competence by using repetition for a variety of pragmatic functions, including confirmation/agreement, choosing among alternatives, and requesting things. They note that their results call into question the interpretation that all repetition in language-trained apes is mindless mimicry as suggested by Terrace et al. (1979).

SENTENCE COMPREHENSION IN A BONOBO

The fourth study is Savage-Rumbaugh et al.'s (1993) comparative study of sentence comprehension. This study compared comprehension of spoken English sentences by the 8-year-old bonobo Kanzi and a 1½-year-old human child, Alia, both of whom had been exposed to lexigrams from early infancy. Savage-Rumbaugh and her associate tested the two youngsters in parallel settings for appropriate action responses to the following classes of spoken sentences (spoken by an unseen tester):

Type 1A—Put object X in/on transportable object Y.

Type 1B—Put object X in nontransportable object Y.

Type 2 A—Give (or show) object X to animate A.

Type 2B—Give object X and object Y to animate A.

Type 2C—Do action A on animate A.

Type 2D—Do action A on animate A with object X.

Type 3—Do action A on object X with object Y.

Type 4—Respond to information announcement.

Type 5A—Take object X to location Y.

Type 5B—Go to location Y and get object X.

Type 5C—Go get object X that's in location Y.

Type 6—Make pretend animate A do action A on object Y.

Type 7—All other sentence types.

The investigators used the following criteria for scoring correctness of responses: Correct (C, C1–C5 according to immediacy of response);

Partially Correct (PC, OE, I for partially correct, retrieves more objects than requested, uses inverse order of acts); Incorrect (W, NR, M, for incorrect, no response, mistrial owing to absence of appropriate object).

The results were that both youngsters performed quite well, but overall Kanzi did better than Alia, 72% versus 66%, respectively. The two youngsters of the two species showed interesting similarities and differences in their error patterns. Perhaps because of its stringent memory requirements, sentence Type 2B (give or show object X and object Y to animate A) was the most difficult for Kanzi and the second most difficult for Alia. Most Type 2C (do action A on animate A) sentences were easy for both Kanzi and Alia, but the verb *hide* was difficult for both when applied to objects (but not to themselves). Type 6 sentences (make pretend animate A do action A on object Y) were difficult for both youngsters. Neither could seem to pretend that a toy bug was an animate creature, . though they had no trouble pretending that a toy dog and a toy snake were animate.

Savage-Rumbaugh and colleagues note that Kanzi seems to understand several syntactically expressed semantic relations as well as the phrasal modifier *that's,* which she interprets as understanding recursiveness. One of her main conclusions is that comprehension of language and grammar significantly outstrips production in the bonobo. She and Bates (1993) both note that this is also true for children early in language acquisition.

GREAT APE SPECIES DIFFERENCES IN SYMBOLIC ABILITY

Ideally, comparative developmental studies of cognition would focus on differences among species as well as those among genera and families. In practice this is more easily said than done. Identification of differences in the symbolic abilities of great apes has been compromised by differences in rearing practices and pedagogy in the few ongoing studies of symbol use in slow-developing ape species. An early effort to compare symbolic abilities in two bonobos and two chimpanzees suggested that bonobos displayed greater ease of acquisition, more rapid grasping of referential function, and greater variety and flexibility of usage (Savage-Rumbaugh et al., 1989).

Beginning in 1985, in an effort to create the conditions for a fair comparison, Savage-Rumbaugh and her colleagues devised an experiment in corearing of infants of two *Pan* species, Panbanisha, a bonobo, and Panpanzee, a chimpanzee. From the time that Panpanzee was less than 2 months old and Panbanisha was less than 1 month old, they were reared together in a rich communicative environment designed to approximate that of human infants. Throughout their daily routines, they were addressed vocally and with lexigrams in an effort to symbolically label objects and events contextually. Joint attentional activities included looking at and labeling objects in the environment and in picture books (Brakke & Savage-Rumbaugh, 1995).

Throughout the 4-year study, virtually all instances of symbol production and speech comprehension were recorded. Periodic tests of comprehension were administered throughout. By 6 or 7 months of age, both infants began to respond to regulatory utterances (*no* and *come here*). Between 6 and 8 months of age, they frequently stared at and touched a lexigram symbol immediately after the caregiver touched it. Shortly after 1 year of age, both began to respond appropriately to a variety of utterances. Nevertheless, Panbanisha, the bonobo, consistently showed greater understanding of both lexigrams and utterances, and, concomitantly, the caretakers addressed more utterances to her and expected more from her.

During the 3-year period in which records of utterances were kept, Panbanisha responded appropriately to 92% (2616/2852) of the utterances of caretakers, whereas Panpanzee responded to only 81% (1596/1825). A sample of these utterances were analyzed into *action grammar* categories (action, agent, object, recipient, and location). When her responses were analyzed, it became apparent that Panpanzee did better with most two-term utterances than most three-term utterances. For example, she responded correctly most of the time to action-recipient and action-location utterances but less often to action-object-location and action-recipient-location utterances. She also had trouble with action-object utterances (Brakke & Savage-Rumbaugh, 1995).

Brakke and Savage-Rumbaugh suggest that the chimpanzee infant's difficulties arose from the demands on memory. They point out that three-term utterances require remembering and integrating three terms. They report that Panpanzee often acted on the appropriate object but performed the wrong but familiar action on it and/or took it to the wrong location. This did not happen if she already had the object in hand at the time of the utterance. They suggest that her difficulty with action-object utterances reflected her limited vocabulary. Their tests with single-word comprehension revealed that, at 3 years of age, Panpanzee's receptive vocabulary was only about a third the size of Panbanisha's at the same age (Brakke & Savage-Rumbaugh, 1995).

Finally, they note that the same processes of attention, memory, and information processing and integration that apparently distinguish bonobos from chimpanzees distinguish humans from both *Pan* species, but in greater degree.

SYMBOLIC COMMUNICATION AMONG CAPTIVE APES

More compelling than demonstrations of symbol production and comprehension in interactions with humans is evidence that language-immersed great apes have appropriated symbols for themselves. First, they use their linguistic capacities when they are alone to communicate with themselves (B. T. Gardner et al., 1989). Second, they use these capacities to communicate with one another (Savage-Rumbaugh et al., 1977; de Waal, 1983; B. T. Gardner et al., 1989; Fouts, Fouts, & Schoenfeld, 1984; Tanner & Byrne, 1996). Third, they spontaneously teach signs to their own offspring.

Savage-Rumbaugh et al. (1977) reported that captive bonobos used iconic signs to request that their sexual partners assume particular positions during copulations. Tanner and Byrne (1996, in press) reported a similar use of iconic signs by a captive gorilla. De Waal (1983) reported that a captive female chimpanzee, Gorilla, used a touching gesture as a good-bye signal every day before retiring to bottle feed her infant, Roosje. Unlike deaf children with no exposure to sign language who invent iconic signs (Goldin-Meadow, 1993), untutored great apes fail to combine their iconic signs into strings.

Washoe apparently taught signs to her adopted son Loulis (Fouts et al., 1989). In an analysis of "conversations" among signing apes, Fouts et al. (1984) discovered that more than 88% of the signing fell under the categories of social interaction, play, and reassurance. The remaining 12% fell under the categories of feeding, grooming, cleaning, discipline, and signing to the self.

SYMBOLIC COMMUNICATION AMONG WILD APES?

Most primatologists have assumed that the capacity for symbolizing is unexpressed in great apes in the wild. (Indeed, some primatologists believe that it is only induced by human training [Tomasello et al., 1993].) Several recent reports, however, suggest that chimpanzees and bonobos may use symbolic means to communicate information in the wild.

First, Boesch (1991a) has reported incidents in which a male in his Tai Forest group used banging on tree trunks to convey information. He loudly banged on tree trunks before group resting, the number of hits apparently corresponding to the length of the rest stop. This same male used banging on two adjacent trees to denote the direction of travel after resting.

In the first case, one hit symbolized about an hour's rest, whereas two hits symbolized about a two-hour rest. In the second case, two hits on adjacent trees denoted the intended line of future travel for the group. (Each meaning depended on the context.)

Second, two observers of wild bonobos have reported iconiclike signaling in this species. Ingmanson (1996) has described repeated branch dragging as an apparent signal to initiate group movement in a new direction. Savage-Rumbaugh (1995) has reported circumstantial evidence that bonobos use leafy branches before and after a Y in trails through the forest to indicate the direction of travel to subsequent groups. The spatial pattern of leafy markers was highly consistent and always indicative of the direction of group travel. Other observers at this site have been unable to confirm this (Ingmanson, 1997).

Two earlier reports of peculiar behaviors suggest spontaneous use of symbolic communication in wild chimpanzees. The first behavior is leaf grooming in which a chimpanzee picks a leaf, examines it closely, lip-smacks, and picks at it. McGrew (1992) cites an unpublished manuscript by Wrangham showing that leaf grooming was used by chimpanzees as a signal to initiate or resume grooming sessions. Nishida (1986) describes a leaf-clipping display in which chimpanzees grasp a thick leaf by the stem between his index finger and thumb and pulls it from side to side in his mouth removing the blade with his incisors. This display is often directed toward females when males are frustrated in their courtship overtures.

These reports and those cited in captive great apes suggest that these creatures may be inventing and using simple iconic gestures to communicate their requests and desires to conspecifics. It is notable that all of these reports occur in contexts of attempted persuasion of others to move in a new direction, to copulate or assume a copulatory position, or to begin grooming.

SUMMARY AND CONCLUSIONS REGARDING APE LANGUAGE STUDIES

The main conclusion we draw from these quasi-developmental comparative studies of the symbolic abilities is that all the great apes have traversed the same stages of language acquisition as human children through Brown's stage I. That is, they traverse the one-word and two-word stages characteristic of human children aged 12 to 18 months and

TABLE 5-7 DEVELOPMENTAL STAGES IN "LANGUAGE" ACQUISITION IN APES AND HUMANS IN MONTHS

Species	First words or signs	Word + sign or sign + sign with semantic roles	Two-plus words with grammatical morphemes
		Piagetian corollary	
	Sensorimotor: fifth stage	*Sensorimotor: sixth stage ?*	*Symbolic subperiod pretend play*
Humans	12 to 13	18 to 22	20 to 36
Chimpanzees[a]	10	10 to 12	?
Bonobos	?	?	?
Gorillas	?	?	?
Orangutans	?	?	?

Note. Studies began after subjects were a year or more of age. ? = No data available.
[a]R. A. Gardner et al. (1989).

18 to 30 months, respectively. The one-word stage involves one-word utterances expressing basic semantic categories (Greenfield & Smith, 1976). The two-word stage is Brown's stage 1: two-word utterances expressing semantic roles and syntactic relations with an MLU of 1.75 and an upper bound of 5 words (R. Brown, 1973). See Table 5-7 for a summary.

During stage 0, the signing apes make signs that are specific to particular exemplars, but gradually generalize to whole classes of exemplars. Their early signs are babyish in form but become more specific and elaborated as they grow older (B. T. Gardner et al., 1989). Their early signs fall into the semantic categories of performative, agent, action, object, location, etc.

During the two-word stage (stage I in Brown's sequence) they combine signs in an elementary functional grammar expressing the following two-term semantic relationships (agent and action, action and object, agent and object, action and location, entity and location, possessor and possession, entity and attribute, and demonstrative and entity) (e.g., Patterson, 1980; B. T. Gardner et al., 1989; Greenfield & Savage-Rumbaugh, 1990; Miles, 1990). They also express affirmation, declaration, and interrogation (wh?) (Van Cantfort et al., 1989).

As adults, all the language-immersed great apes display vocabularies in the range of 150 to 300 signs (as many as 1,000 signs in some cases, Patterson, personal communication), the range in which human children begin multiword speech. They generally show MLU approximating

1.5 to 1.75. Recent research suggests a powerful link between grammatical and vocabulary development in human children between the ages of 17 and 30 months when they have mastered 300 to 400 words. The two functions are separated by a short lag between the two (Bates & Goodman, 1996).

Unlike human children, great apes fail to develop more complex grammatical categories and longer utterances. Rather, they peak at levels similar to those of 2-year-old children. As R. Brown (1973), Bates et al. (1979), and others have noted, the one-word stage coincides roughly to the end of Piaget's sensorimotor period, whereas the two-word stage coincides roughly with the onset of his symbolic subperiod of preoperations and particularly the onset of pretend play. This suggests that great apes peak at a lower developmental level in language than in other social domains. They also peak at a lower level in language than in logical-mathematical knowledge.

Much remains to be learned about species differences in symbol acquisition in the four species of great apes. Systematic comparisons are hampered by the lack of common analytic methodologies, but in their comparative study of repetition, Greenfield and Savage-Rumbaugh (1993) found that bonobos and chimpanzees failed to show the following three patterns common to human children at a comparable stage: (1) use of language to elicit a linguistic response, (2) use of language to describe the self in relation to previous speaker's utterances, and (3) increasing mean length of utterance.

COMMENTS ON THE CONTROVERSY OVER GREAT APE LANGUAGE STUDIES

The data reviewed here support the claim that symbol use is a robust phenomenon in great apes that depends upon considerable cognitive sophistication. Given the controversy that has attended research into the linguistic capacities of great apes, it is important to address the criticisms and counterclaims of those who discount the linguistic significance of utterances of language-immersed great apes.

First, early claims regarding unavailability of a full corpus of utterances are no longer valid after the publication of *Teaching Sign Language to Chimpanzees* (B. T. Gardner et al., 1989). Second, claims that investigators have been duped by their own naive belief in their subjects' abilities are rebutted by controlled studies of comprehension (R. A. Gardner et al., 1989; Savage-Rumbaugh et al., 1993). Third, claims that signing is simply a conditioned response or mere mimicry without understanding of generalized semantic categories are rebutted by studies of great apes who acquired signs spontaneously through interactions with signing models. Fourth, claims that strings of signs carry no grammatical meaning are undermined by Greenfield and Savage-Rumbaugh's (1990) analysis of Kanzi's action grammar.

Learning theorists imply that conventional operant conditioning training techniques provide greater access to great ape abilities than spontaneous cross-fostering techniques (Terrace, 1979). This view must be re-examined in light of recent assessment of outcomes of the two methodologies. A recent study of Nim signing under two contrasting conditions, operant conditions and spontaneous conditions, reveals that he behaved differently under the two conditions. (Nim is the famous chimpanzee pupil that Herbert Terrace used to demonstrate that chimpanzee signing was a rote or at best imitative performance devoid of understanding [Terrace et al., 1979].) When prompted, Nim refused to sign and became aggressive. When spontaneously interacting, he used signs productively and appropriately (Miles, 1983; O'Sullivan & Yeager, 1989).

B. T. Gardner and Gardner (1989b) note that great apes spontaneously learn, generalize, and productively use signs without extrinsic rewards when they are raised in rich communicative environments. More surprising, when they are subjected to operant conditioning, they fail to use signs spontaneously and productively to communicate with humans and each other as they do when they are immersed in naturalistic communicative environments. The Gardners note that "negative effects of rewards are commonly seen in teaching and training situations of all kinds" (R. A. Gardner & Gardner, 1989, p. 21). Perhaps language acquisition is a problem space in which anthropomorphism is a better guide to teaching than behaviorist objectivity: "It is an irony of ape language research that . . . one must assume and attribute language abilities to apes in order to produce those abilities" (Miles, 1997).

CONCLUSIONS

Much work remains to compare the development of linguistic abilities in the great apes, especially with regard to developmental timing. Even so, the common abilities revealed by these four cross-fostering studies of symbol acquisition in three great ape species are striking. Moreover, these careful studies demonstrate the validity of more informal interpretations of all the investigators who have used naturalistic cross-fostering methodologies in their studies of symbol acquisition in great apes. The unanimity of the descriptions and interpretations of these investigators in all four species is remarkable. In addition to revealing the symbolic abilities of great apes, these naturalistic cross-fostering studies have provided independent confirmation of abilities in other cognitive domains.

Consider, for example, the following cognitive entailments of ape language use. First, the use of such simple semantic relations as agent and object, agent and action, and action and object implies relational understandings of causality. Second, the use of such attributes as colors, sizes, and other categories implies a capacity for classification. Third, the use of locatives designating such relations as in, on, and under implies an understanding of spatial relations. Fourth, as indicated earlier, the use of nouns to label absent objects implies object permanence. Fifth, use of pronouns implies understanding of self and others. Sixth, use of possessives implies a rudimentary understanding of ownership. Seventh, use of emotion terms implies an understanding of desire psychology. Eighth, use of judgment terms such as *good, bad,* and *dirty* implies some rudimentary understanding of social standards. Ninth, conversational turn-taking reveals some rudimentary understanding of role of complementarity and intentionality. Finally, and most obviously, sign acquisition implies the capacity for gestural imitation. Reciprocally, the expression of symbolic abilities in the domains of number, drawing, map reading, and pretend play lends credence to the claims for symbolic communication in great apes. The overall picture is highly consistent.

The obvious implication of these entailments is that language depends upon many cognitive domains. This fact confirms Bates et al.'s (1979) contention that language is a collage that could only evolve after its component parts had evolved. It also casts doubt on the notion that there is a "language module" (Pinker, 1994) that is essentially isolated from other cognitive modules (Elman et al., 1996; Deacon, 1997).

People sometimes ask why chimpanzees have evolved such complex intellectual powers, when their lives in the wild are so simple. The answer is, of course, that their lives in the wild are not so simple! They use—and need—all of their mental skills during normal day-to-day life in their complex society.

GOODALL, 1990, P. 23

CHAPTER 6
COMPARING PRIMATE COGNITION ACROSS DOMAINS: INTEGRATION OR ISOLATION?

The preceding chapters have surveyed current evidence on cognitive development of monkeys, great apes, and human infants and children in physical, logical-mathematical, social, and symbolic domains. This chapter summarizes this evidence and introduces some parallel data on the cognitive abilities of early hominids. Finally, it reviews some reports of intelligent behaviors in the daily lives of wild chimpanzees to place their abilities in context.

COMPARATIVE PATTERNS OF PRIMATE COGNITIVE DEVELOPMENT BY DOMAIN

COGNITIVE DEVELOPMENT IN HUMAN CHILDREN

Human children achieve the highest sensorimotor stage, the sixth stage, at about 2 years of age in all six series. They achieve mental representation of simple spatial, temporal, and causal relations among objects in the physical domain, mental representation of simple logical and numerical relations in the logical-mathematical domain, and mental representation in the form of deferred imitation in the social domain (Piaget & Inhelder, 1969). In other words, sixth-stage achievements all occur roughly synchronously across domains.

The first subperiod of preoperations, the symbolic subperiod, begins at the end of the sensorimotor period with the onset of pretend play, simple drawing, and understanding of simple topological properties of space. Words are first combined during this subperiod. The second sub-period of preoperations, the intuitive subperiod, begins at about 4 years of age with the onset of trial-and-error classification and seriation and the beginning understanding of Euclidean spatial concepts of angles (Piaget & Inhelder, 1969). Simple grammar is well developed by this time.

The preoperations period is succeeded by the concrete operations period, characterized by understanding of the logical necessity of conservation of quantities under transformations owing to reversibility and compensation. It is also characterized by logical understanding of classification, seriation, and arithmetic operations, as well as understanding of the role of forces mediating causation (Piaget & Inhelder, 1969).[1]

COGNITIVE DEVELOPMENT IN GREAT APES

PHYSICAL SUBDOMAINS

Great apes complete the highest sensorimotor stages in the physical subdomains of space and causality at about 3 and 4 years of age, respectively. This compares to 2 years for human infants. The highest level regularly achieved in these subdomains is the symbolic level, equivalent to that of human children 3 or 4 years of age. Unfortunately, there are no strictly developmental data for these postsensorimotor periods. Anecdotal evidence suggest that great apes may achieve their highest levels in these domains as late as 7 or 8 years of age or even later (see chapter 2).

LOGICAL-MATHEMATICAL SUBDOMAINS

In the logical-mathematical subdomains of spontaneous classification and logic, nonsymbol-trained chimpanzees achieve sixth-stage sensorimotor–early symbolic levels of understanding. They do so relatively late, by about 4 years of age, as compared with 2 years in human infants. In the subdomains of number, multiplicative classification, seriation, and conservation, symbol-trained chimpanzees apparently achieve early intuitive substage levels. In the subdomain of number, they achieve these levels by 3½ to 5 years of age. No data exist on the development of achievements in conservation and multiplicative classification. (Symbol

1. For consistency, we retain Piaget's terminology for the two subperiods of pre-operations though we think the evidence supports the neo-Piagetian reformulations of Case (1985) and Perner (1991). Specifically, in relation to representation and symbolic capacities, we favor Perner's distinction between primary, secondary, and metarepresentations, according to which symbolic play and other symbolic-subperiod abilities precede the capacity for symbolic, that is, metarepresentational thought.

training refers to training in lexical, numerical, or other iconic or arbitrary symbols—see chapters 3 and 5.)

Great apes display the following patterns in the social subdomains. First, all four species achieve the sixth sensorimotor stage in the imitation series (deferred imitation of novel actions) in the manual/gestural and facial (but not the vocal) modalities. They seem to do so by about 2 or 3 years of age, which is not much later than human children.

Second, some individuals of all the great ape species achieve early symbolic-subperiod preoperational abilities to pretend play as measured by pretend actions on a passive partner by 3 years of age. This corresponds roughly to the level of 2½-year-old human children. A few great apes have been reported to use substitute objects in pretend play by 8 years (in nondevelopmental studies), thereby fulfilling Inhelder's more rigorous criterion for pretend play. This is equivalent to the level of 2-year-old human children.

Third, at least some individuals of each great ape species display MSR. They typically achieve stable MSR by about 4 years of age as compared to about 2½ years in human children. Stable MSR probably represents sixth-stage sensorimotor abilities. Some individuals of great apes of each species trained in lexical symbols also manifest symbolic-level self-awareness as indicated by use of self-evaluative terms. They seem to do so by about 3 or 4 years of age.

Fourth, some individuals of all great ape species show evidence of symbolic-subperiod level of theory of mind as measured by such intentional acts of deception as hiding objects. Some chimpanzees also manifest this level of theory of mind as manifested by misleading uncooperative interactants. These, of course, are scaffolded performance levels.

SYMBOLIC COMMUNICATION DOMAIN

In the domain of symbolic communication, great apes peak at levels similar to those of 2-year-old children, at the two-word stage in production. As R. Brown (1973), Bates et al. (1979), and others have noted, the one-word stage coincides roughly to the end of Piaget's sensorimotor period, whereas the two-word stage coincides roughly with the onset of his symbolic subperiod of preoperations and particularly the onset of pretend play. Great apes achieve these levels at about 3 years of age. Overall, these data suggest that great apes peak at a lower developmental level in language than in other social domains. They also peak at a lower level in language than in logical-mathematical knowledge (see chapter 5).

CONCLUSIONS ON GREAT APE COGNITIVE DEVELOPMENT BY DOMAINS

First, the highest cognitive level great apes achieve spontaneously in any domain is the symbolic level of preoperations. The highest they achieve under tutelage is the intuitive level. Overall, the highest level is achieved in the domain of logical-mathematical reasoning, the next highest in the domains of physical and social cognition, and the lowest in symbolic communication.

Second, the rates at which great apes achieve their highest spontaneous level of development vary from domain to domain. They may achieve higher levels with training, especially in the domain of logical-mathematical cognition. Specifically, great apes show the following mosaic pattern of completion of the sensorimotor period: (1) In the social domain of imitation, they achieve the end of the sensorimotor period and the beginning of the symbolic period at the earliest at about 3 years of age. (2) In various physical subdomains, they show a mixed pattern: the object concept develops by about 1 year, earlier than in human infants, and the causality and spatial subdomains develop as late as 3 and 4 years (using the rigorous measure of tool use and block stacking). (3) In the neo-Piagetian logical-mathematical subdomains of classification and logic, they achieve this transition by about 4 years. These ages compare to development at about 2 years across the board in human children.

Third, the timing of development in the symbolic subperiod has not been studied in great apes in any domain. Anecdotal evidence suggests that development peaks earlier in social domains than in physical domains of causality and space in advanced tool use. Development in these domains may extend to 7 or 8 years. Only a few individuals of two species have been tested for their advanced tool-using abilities, and these show considerable individual variation in their performances.

Fourth, in the logical subdomains of number, seriation, and conservation, symbol-trained great apes may achieve the level of the intuitive subperiod in scaffolded settings. This level has been seen in some adult chimpanzees and orangutans. The number of individuals studied is small, and individual variation is great, suggesting that the higher-level performances are at the edge of species' abilities.

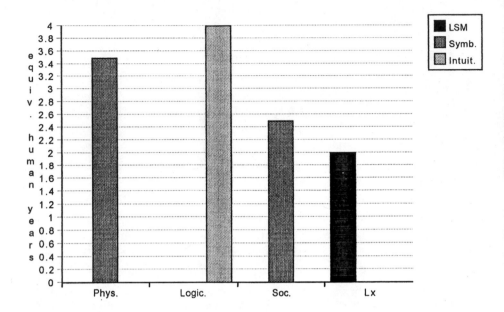

FIGURE 6-1
Levels achieved by chimpanzees in various cognitive domains by subperiod and approximate equivalent age in humans.

Fifth, symbolic-level development in great apes follows a temporal course roughly similar to that in human infants, until it truncates at the beginning of the symbolic subperiod at the equivalent of 2½ years in humans. It is interesting to note that the developmental level great apes achieve in language acquisition in heavily scaffolded settings is slightly lower than their spontaneous achievements in physical, social, and logical domains. These patterns can be verified only with strictly developmental data on juvenile great apes aged 3 to 10 years.

Sixth, these comparisons, in turn, suggest that the rates of development in the social and nonsocial domains are asynchronous, social cognition developing earlier than nonsocial cognition. See Figure 6-1 for a depiction of the relative levels of knowledge achieved in various cognitive domains by the chimpanzees, the best studied of the great apes, by humans, and by macaques.

SPECIES DIFFERENCES IN COGNITION AMONG GREAT APES

Although all four species of great apes seem to display similar levels of abilities and developmental trajectories in the domains in which they have been tested, sampling is highly uneven. More chimpanzees have been studied in more domains than have other great ape species. Fewer gorillas have been studied in fewer domains than any other great ape species. Next to chimpanzees, orangutans have been studied in more domains than any other great ape species. Moreover, few studies have addressed species differences explicitly. If we look domain by domain, we see some clues to species differences.

In the physical domain, there are few studies or reports of gorilla cognition above the sensorimotor level. Captive gorillas use tools in a variety of situations (Parker, Markowitz, Gould, & Kerr, in press), but we know of no reports of such advanced practices as use of anvils and wedges or tool kits. On the other hand, captive gorillas engage in drawing, which is typical of other great apes at the onset of symbolic-level understanding of space.

In the logical-mathematical domain we know of no studies of classification, seriation, number, or conservation in gorillas or bonobos. In the social domain, studies of self-awareness in gorillas have yielded only occasional evidence of MSR (Gallup, 1977; Parker et al., 1994; Patterson & Cohn, 1994; Swartz & Evans, 1994). On the other hand, there are several reports of symbolic-level deception in gorillas (Mitchell, 1991; Parker et al., 1994; Tanner & Byrne, 1993). There are also reports of symbolic play (Patterson & Cohn, 1994).

The one study that focuses on species differences in symbolic ability has shown greater symbolic abilities in bonobos as compared to chimpanzees (Brakke & Savage-Rumbaugh, 1995). Specifically, bonobos achieve the level of communicative competence similar to 2½-year-old children as compared to that of 2-year-old children achieved by chimpanzees.

Clearly, more research on gorilla and bonobo cognition is necessary to fill out the picture of species differences in great apes within and across domains. More research on cognitive development is necessary in all great ape species. This is particularly true of post–sensorimotor-period development.

COGNITIVE DEVELOPMENT IN MONKEYS

The following mosaic pattern of cognitive development emerges from the data on macaque monkeys. First, macaques fail to complete the sixth stage of the sensorimotor period in any domain. Second, they complete different levels in various sensorimotor series: They complete their highest level in the object concept series (fifth stage), their lowest level in the imitation series (third stage), and intermediate stages in the other sensorimotor series and domains (fourth or fifth stage in space and causality and fourth stage in logic). Third, they complete their terminal levels in the various series at different ages: They complete the fifth stage of the object concept and the fourth or fifth stage of the spatial series by about ½ year and fourth stage of logic by about 2 to 3 years. They display neither secondary nor tertiary circular reactions.

Cebus monkeys, though more distantly related to apes than are macaques, achieve higher levels of sensorimotor-period development in causal and logical domains. They also develop at a slower rate. Cebus achieve the fifth stage in the object concept, the causal and spatial domains, but not in imitation. They complete object concept development at 8 to 12 months and causal development at approximately 1½ years. They display secondary but not tertiary circular reactions. They achieve their fifth stage in logic between 3 and 4 years.

SUMMARY OF COGNITIVE DEVELOPMENT IN HUMAN AND NONHUMAN PRIMATES

The following overall pattern emerges from the comparative data on intellectual development in monkeys, great apes, and humans. First, humans undergo three major transitions: from sensorimotor to symbolic at 2 years, from symbolic to conceptual at 6 years, and from conceptual to abstract at 12 years. (As we see below, each of these major transitions breaks down into two minor transitions between subperiods.) Second, chimpanzees and the other great apes undergo one major transition from sensorimotor to symbolic thinking at about 4 years of age. Third, macaques and cebus monkeys undergo none of these transitions, stopping short of symbolic thought.

The following pattern of development across domains emerges. First, *humans show more or less synchronous development* across physical, logical, and social domains of cognition, particularly during the sensorimotor period. Second, *great apes show asynchronous developmental*

TABLE 6-1 SUMMARY OF COGNITIVE DEVELOPMENT OF MONKEYS, APES, AND HUMANS

	Old World monkeys	Great apes	Humans
Highest levels achieved	Sensorimotor up to fourth or fifth stage	Preoperational up through symbolic	Formal operations
Sequences traversed in stages and periods	Same as great apes and humans up to terminal level except for absence of secondary and tertiary circular reactions	Same as humans up to terminal level except for absence of vocal imitation and low frequency of gestural imitation and tertiary circular reactions	Human standard
Displacements or asynchronies between highest levels of series and subdomains	Yes, object concept higher than imitation, and causality series	Yes, physical and logical-mathematical domains higher than social domains	Very minor
Displacements or asynchronies between rates of development of series and subdomains	Yes, logic later than object concept	Yes, nonsocial later than social domains; causality and space later than object concept	

levels and rates across domains and within major domains. Third, *monkeys show asynchronous developmental rates and terminal levels* between the object concept and other sensorimotor series. Fourth, *all nonhuman primates show more rapid development of the object concept than of any other sensorimotor series.* Human infants also traverse this series at a slightly faster rate than the other sensorimotor series. Monkeys achieve a higher terminal level in the series than they do in any other. Table 6-1 summarizes cognitive development in these taxa in various domains.

At a slightly finer level, these comparative data can be represented in terms of the number of subperiods of cognitive development completed in macaques, great apes, and humans (Table 6-2). These patterns, described below, are useful for reconstructing the evolution of developmental patterns, which we explore in the following chapter.

Human children traverse four periods and seven or eight subperiods of cognitive development: in several domains: (1) early (stages 1–4) and late (stages 5 and 6) sensorimotor-period development, (2) early (sym-

TABLE 6-2 ONTOGENETIC CHARACTERS FOR COGNITIVE
DEVELOPMENT IN ANTHROPOID PRIMATES

Taxon	Ontogeny
Humans	ESM - LSM - EPO - LPO - ECO - LCO - FO
Homo erectines	ESM - LSM - EPO - LPO - ECO
Great Apes	ESM - LSM - EPO
Macaques	ESM - LSM

Note. ESM = Early sensorimotor, LSM = Late sensorimotor, EPO = Early
preoperations, LPO = Late preoperations, ECO = Early concrete operations,
LCO = Late concrete operations, FO = Formal operations.

bolic) and late (intuitive) preoperational-period development, (3) early
and late operational-period development, and (4) early and perhaps late
formal operations–period development.

Great apes traverse three or four of these subperiods of cognitive development: (1) early sensorimotor (ESM) period development, from birth
to almost 12 months; (2) late sensorimotor (LSM) period development,
from 1 to 3½ years (12 through about 40 months) depending on the
domain; (3) early preoperational-period (EPO) symbolic development,
from 3½ years to 8 years (40 to about 96 months), and, in some cases
under tutelage, (4) late preoperational-period (intuitive) development.

Macaque monkeys traverse the first subperiod, the early sensorimotor
period, from birth to about 5 months and partially traverse the late
sensorimotor period to the fifth stage in one domain. They develop at
different rates in different domains. They complete the fifth-stage object
concept and the fourth or fifth stage in the spatial series at about ½ year
and the fourth stage in the logical domains at about 2 to 3 years. Each
pattern is characteristic of a particular taxonomic group and originated
in its ancestor.

Perusal of these data reveal the following patterns. First, *all these anthropoid species traverse the same sequence of stages and subperiods
within each domain* up to the terminal level for their group (except for
the omission of secondary and tertiary circular reactions in macaques).
Second, *different taxa achieve different terminal levels of achievement;*
monkeys achieve the lowest level, apes the next lowest, and humans
the highest. Third, *development in monkeys and apes is asynchronous
across domains* compared to that of humans. Fourth, *different taxa develop at different rates,* monkeys the most rapidly and great apes the
most slowly.

Specifically, macaques reach their highest level in the third stage of the imitation series, fourth or fifth stage of the causality and spatial series (depending on the species), and fifth stage in the object concept series of the sensorimotor period. In contrast, chimpanzees complete all six sensorimotor-period stages and reach their highest level in the symbolic subperiod of preoperations. Humans reach their highest level in formal operations.

Macaques complete the sensorimotor period by 2 or 3 years of age, earlier than great apes, who complete this period by 4 years, which is 2 years later than human infants. Consequently, they also begin and end the symbolic subperiod of preoperations later than human children.

Finally, given that chimpanzees and bonobos are the closest living relatives of humans, these ontogenetic patterns reveal that since the divergence of chimpanzees and humans, *human ancestors have added four new subperiods of cognitive development:* late preoperations (LPO), early and late concrete operations (ECO and LCO), and early formal operations (FO). As we discuss below, analysis of stone tools associated with late Homo erectines suggests that they may have displayed early concrete operational cognition, at least in the domain of spatial reasoning (Wynn, 1989).

SUMMARY OF DIFFERENCES IN SENSORIMOTOR AND SYMBOLIC SCHEMES OF APES AND HUMAN CHILDREN

The preceding section of this chapter summarizes similarities in the sequence of cognitive development in great apes and human children. This section summarizes differences between the two groups that occur within the common sequential stages. A clear exposition of differences in the quality and quantity of schemes as well as in their developmental rate is crucial for understanding the evolution of human cognitive development.

Compared to human infants and young children, great apes show the following limitations across all the sensorimotor series and stages:

1. a more limited repertoire of schemes

2. limited and impoverished tertiary circular reactions

3. less differentiation and recombination of schemes

4. fewer applications of schemes to objects

5. lower frequency of scheme production

6. lower frequency of stable object constructions by manual schemes

7. little joint attention between mother and infant

8. lack of the face-to-face reciprocal vocal and gestural schemes of social interactions with caretakers

Similarly, great apes show even greater limitations, particularly in social and communicative domains, compared to young human children during the preoperations period, their terminal level:

1. lower frequency of imitation of caretakers

2. restricted attention to the gaze of caretakers and others

3. restricted use of gestural schemes of request and reference (i.e., referential pointing, object showing, and "gimme" gestures typical of young children)

4. restricted use of requests for caretaker to act as surrogate agent

5. fewer, and relatively impoverished, interactional routines (scripts) with caretakers

6. limited and impoverished pretend play

7. limited mean length of utterance and hence limited grammar

These differences correlate with and reflect striking differences in the behaviors of adult caretakers in the two taxa. In contrast to human mothers, great ape mothers show little interest in reciprocal face-to-face interactions with their infants. They fail to imitate their infants' facial and vocal schemes, fail to follow the visual attention of their infants, and show little interest in responding to their infants' requests for objects or for help (Parker, 1993).

These differences are just those we might expect to exist between a species that uses language and those that do not. They strongly suggest that the evolution of hominid cognitive development involved a subsequent elaboration of these language-related elements in early development as well as acceleration of early stages.

COGNITIVE CAPACITIES OF EARLIER *HOMO* SPECIES

When we compare the mental abilities of great apes and humans, the most striking feature of this picture is the tremendous gap between the highest level of cognitive development in great apes and modern humans. Great apes stop at the level of the early preoperations period characteristic of human toddlers, whereas humans stop at the level of formal operations characteristic of human adolescence. This gap implies that evolution of the intervening periods, late preoperations, concrete operations, and formal operations, occurred after the divergence between African apes and hominids. But how can we diagnose the cognitive abilities of ancestral hominids?

Diagnosing the cognitive abilities of fossil hominids is tricky. Fortunately, Piagetian stages in the domain of spatial knowledge provide a framework for assessing the cognitive abilities of fossil hominids from tools and other artifacts. Wynn (1989) used Piaget's stages of development of spatial cognition (see Piaget & Inhelder, 1967) to analyze Oldowan and Acheulian tools from three sites in East African spanning a period of 1.5 million years of hominid evolution.

Wynn found support for the following conclusions:

1. Oldowan stone knappers (beginning about 2 million years ago) used topological notions of space that are typical of human children in the symbolic substage of the preoperations period. These notions include proximity, separation, and order reflected in a series of blows made in proximity to one another and in an ordered series.

2. Early Acheulian stone knappers (beginning about 1.2 million years ago) added projective notions of interval and symmetry based on diameter and radius typical of modern children in the intuitive stage of the preoperations period. These suggest an overall sense of design, though this notion remained internal to the object at hand.

3. Late Acheulian stone knappers (beginning about 300,000 years ago) added the projective notion of parallels and Euclidian notions of cross section. Cross sections are based on perspective taking, that is, an external frame of reference typical of modern children in the concrete operations period. A drawing of an Acheulian hand ax is shown in Figure 6-2 (Oakley, 1959).

The earliest hominids, the Australopithecines (*Australopithecus, Paranthropus*, etc.) who lived from about 4.3 to about 1.5 million years ago, certainly used tools at least as complex as those of chimpanzees. Whether they made flaked stone tools is a matter of dispute (Susman, 1988). Primitive choppers and flakes, the oldest flaked tools, date from about 2.4 million years ago (Semaw et al., 1997). Most anthropologists believe that these early flaked tools were made by early *Homo* species

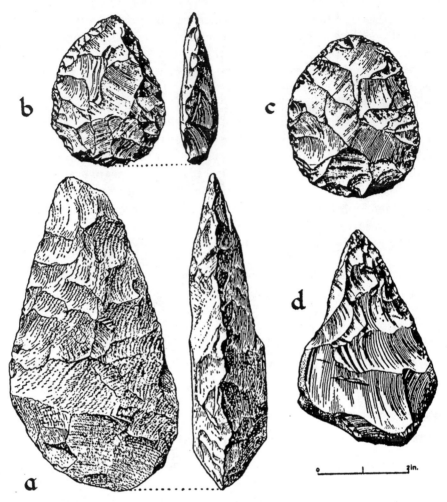

FIGURE 6-2
Acheulian tools (a) Lava hand ax, 01 Orgesailie, Kenya; (b) Twisted ovate, argile rouge on 30-m terrace, St. Acheul, near Amiens (Somme); (c) Ovate hand ax, south of Wady Sidr, Palestine; (d) Hand ax, of Micoquian type, brick earth, H oxne, Suffolk. (from Oakley, 1959).

(*Homo habilis, Homo rudolfensis*), which descended from one of the Australopithecines. One of these early *Homo* species gave rise to Homo erectines (*Homo erectus, Homo ergaster,* and others). (The taxonomy and phylogeny of fossil hominids continues to generate controversy [Straight, Grine, & Moniz, 1997]. The general trend, however, has been to recognize two or more species living at any given time period up until about 27,000 years ago [Tattersall, 1986; Wood, 1992].)

Homo erectines lived from about 2.4 million years ago in East Africa (F. H. Brown, Harris, Leakey, & Walker, 1985). According to Wynn and McGrew (1989), Oldowan tools displayed primitive spatial notions of proximity, boundary, and order typical of human children in late pre-operations. They argue that these are the same spatial notions understood by chimpanzees. Hence, they suggest that these creatures had approximately the same spatial knowledge as chimpanzees (see Parker & Gibson, 1979, for a similar conclusion). The apparent reason that early *Homo* species made stone tools whereas chimpanzees do not is to be found in their subsistence strategies. Unlike chimpanzees, early *Homo* species scavenged for meat and broke open bones for marrow. They had substantially larger brains than the Australopithecines, and they made stone tools and possibly rude shelters. For these reasons, we suspect they may have had intuitive-subperiod preoperational abilities. Certainly, early Homo erectines must have achieved this level.

Acheulian tools (including bifaced hand axes) were made by Homo erectines, the ancestors of *Homo sapiens*. These creatures lived from about 1.8 million years ago until about 200,000 years ago in Africa, Asia, and Europe. Wynn's reconstruction pegs the highest cognitive attainments of early *Homo* species at the level of late preoperations, and those of late Homo erectines at the level of concrete operations, at least in the realm of spatial cognition. This makes sense in terms of the relatively large adult brain size of Homo erectines, approximately 909 cm^3 (Begun & Walker, 1993). The brain size of Australopithecines was only slightly larger than those of living great apes, but early *Homo* species have a brain size of approximately 600 to 700 cm^3, which must have reflected greater cognitive abilities in at least some domains. See chapter 12 for brain sizes of hominids.

From a psychological perspective, cognitive domains are useful categories for studying the development of problem-solving abilities in human children or comparing the problem-solving abilities in animal species. Historically, these domains have arisen adventitiously from the research goals and strategies of developmental and comparative psychologists, much of which was framed in terms of philosophical categories of object, space, time, and causality going back to Kant (Piaget, 1970).

In contrast, from our evolutionary perspective, cognitive abilities are viewed as powerful, flexible strategies that have evolved for solving complex, unpredictable problems arising in social and ecological environments.[2] Whereas an evolutionary perspective does not deny the validity of the psychological perspective, it suggests the importance of relating cognitive domains to naturalistic problem solving.

PRIMATE COGNITION IN NATURAL CONTEXTS

Having compared the cognitive abilities of monkeys and great apes and charted their distribution among related species, we turn to the question of how these abilities are manifested in the wild. Placing behaviors in their natural contexts is a first step toward discovering their adaptive significance or evolutionary functions (Williams, 1992). Attempts to do so must be speculative, however, because most studies of cognition have been done on laboratory-reared or home-reared individuals. Nevertheless, it is important to relate these cognitive abilities to descriptions of life in the wild and/or in naturalistic social groups. We begin by considering manifestations within each cognitive domain and finish by considering their manifestations across domains in the daily routine of chimpanzees.

SITUATED EXPRESSIONS OF PHYSICAL KNOWLEDGE IN APES

Although psychologists treat the various subdomains of physical knowledge separately, field studies suggest that wild monkeys and apes combine their knowledge of objects, space, and causality in a variety of subsistence activities. Among apes, these activities include foraging and feeding, hunting, nest-building, and self-grooming. All great apes build nests, but only chimpanzees and orangutans use tools for foraging and feeding, and only chimpanzees engage in hunting. Apes also use physical knowledge in defensive and social contexts, as in throwing objects at predators and in dominance displays to conspecifics.

Wild chimpanzees use tools in a variety of contexts (McGrew, 1992).

2. Arguments over the question of whether human intelligence (and/or the brain) should be viewed as adaptive go back to the debate between Darwin and Wallace (1864) (Kottler, 1985). Wallace believed that intelligence allows humans to transcend nature and puts them above natural laws, whereas Darwin believed that this was untrue and contradicted the theory of natural selection. In a modern version of this debate, Gould and Lewontin (1979) argue that intelligence is a side product of other features rather than a specific adaptation. We believe the high cost of the human brain alone militates against such an interpretation (Parker, 1990b). See chapters 9 and 10 for discussion of the adaptive significance of cognition and chapter 11 for a discussion of various interpretations of adaptation. See chapter 12 for a discussion of brain evolution.

Tool-mediated foraging on such embedded and encased food sources as termites and ants, nuts, hard-shelled fruits, and honey is the most common form of tool use in wild chimpanzees. All of these activities require considerable dexterity and understanding of causality. Termite fishing, for example, requires identification of the probable location of a tunnel in a termite mound before the nuptial flight of termite nymphs.

> The chimpanzee first opens a hole on the bare surface of the termite earthen mound. The ape then inserts into the mound a long, thin probe made of plant material such as a blade of grass, strip of bark or segment of vine. Most of these simple probes have been modified by the chimpanzee by shortening, splitting, stripping, etc. . . . The chimpanzee then carefully withdraws the tool and uses her lips to nibble the insects from it, often one at a time. (McGrew, 1992, p. 90)

Ant dipping is more difficult than termiting owing to the biting defense of these insects.

> The chimpanzee finds an underground nest of driver ants and digs into it by hand. The ape then makes a long smooth wand of woody vegetation by modifying a branch. When the tool is inserted into the nest, the ants stream up it and attack. The chimpanzee quickly withdraws the tool and, while holding it in one hand, sweeps the length of the wand with the other in a loose grip. The ants are momentarily collected in a jumbled mass on the sweeping hand and are directly popped into the mouth. (McGrew, 1992, p. 93)

Probing with sticks for honey and honeycomb is another use of tools in wild chimpanzees. In some cases, they use a series of tools or a *tool set* in the process.

> A female used a tool-set of four components (stout chisel, fine chisel, bodkin, dip stick) in sequence to extract honey from a stingless bees' . . . nest in a hollow tree. She first tried to reach the honey by dipping directly from a flight entrance, but when this failed she attacked the involucrum of the nest with a stout chisel. Having started an indentation, she continued to work on it with a finer-pointed chisel. This eventually allowed her to puncture the nest with a sharp-pointed bodkin, and then a longer, flexible dip-stick was used to dip out the dripping honey. (McGrew, 1992, pp. 164–65)

Extraction of nuts depends first on location and transport of these foods and of the tools for opening them (Boesch & Boesch, 1984). Second, it depends upon appropriate choice and preparation of implements according to their shape, size, strength, and flexibility. Third, it depends upon accurate arrangement of food source relative to substrate, and accurately aimed and calibrated force in hitting them or pulling them out.

> Typically, the chimpanzees collect nuts on the ground or in the trees, carry them to a root which is used as an anvil, sit and place their provisions on the ground, and start pounding. For this, a nut is placed in a depression of the root . . . and opened by hitting it repeatedly with a hammer tool. Each nut is eaten at once and, having consumed what they have collected, the chimpanzees may search for more nuts and resume cracking at the same anvil. . . . Both the nuts and the hammers are transported. (Boesch, 1993, p. 173 [reprinted with permission of Cambridge University Press])

In some cases, chimpanzees use a stone rather than a tree root as an anvil for holding the nut. They have also been observed using a wedge to stabilize a stone anvil. "One of the most interesting findings is how the chimpanzees make a metatool, that is a tool that serves as a tool for another tool. . . . For example, a *metatool* is a third stone placed beneath the anvil stone as a wedge to keep the surface of the anvil stone flat and stable. So far, I have observed three instances in which three separate individuals have made such complex anvils" (Matsuzawa, 1994, p. 361). Table 6-3 lists various kinds of tool use observed in wild chimpanzees.

The cognitive complexity of these extractive technologies is suggested by the long apprenticeship young chimpanzees require to become proficient tool users. Juvenile chimpanzees master termite fishing by about 5 years of age and ant dipping and nut cracking at about 8 years (McGrew, 1992). As we shall see, apprenticeship in these technologies involves social as well as physical cognition (Boesch & Boesch, 1989).

Recent field studies have revealed that wild Sumatran orangutans also use tools in *extractive foraging* on insects (termites, ants, and bees), honey, and hard-shelled fruits (Fox & van Schaik, in press; van Schaik, Fox, & Sitompul, 1996). Like chimpanzees, they used a series of different tools in a single foraging episode, for example, hammering to break open termite and bee nests, poking to rupture the nests, and probing or scraping to obtain ants or termites. Like chimpanzees, they modified sticks for use as tools.

Cracking the superhard *Panda* nuts entails foraging for and transporting stone tools considerable distances. In their study, Boesch and

TABLE 6-3 TOOL-USE PATTERNS IN VARIOUS POPULATIONS OF WILD CHIMPANZEES

							Site		
Pattern	Gombe	Bossou	Kasoje	Tai	Kanka Sili	Assirik	Kanton, Sapo, Tiwai	Campo, Okorobikó	Kibale
Termite-fish	X		X			X			
Ant-dip	X	X		X		X			
Honey-dip	X			X					
Leaf-sponge	X	X							
Leaf-napkin	X								X
Stick-flail	X	X	X		X				
Stick-club	X	?X	X		X				
Missile-throw	X	X	X		X				
Self-tickle	X								
Play-start	X		X						
Leaf-groom	X		X						
Ant-fish			X						
Leaf-clip		X	X						
Gum-gouge		X							
Nut-hammer		X		X			XXX		
Marrow-pick				X					
Bee-probe				X					
Branch-haul		X							
Termite-dig								XX	
Total	11	8	8	5	3	2	(3×)1	(2×)1	1

Notes. From McGrew (1992). Reprinted with the permission of Cambridge University Press. X indicates presence of pattern.

Boesch (1984) found that Tai chimpanzees minimize hammer transport distance. "For *Panda*, 40% of the transports are made between trees out of sight of each other, that is, more than 20 m apart. Moreover, the animals crack *Panda* nuts mainly alone or in pairs . . . the group does not provide a fixed direction, so that the crackers choose the transport direction themselves" (Boesch & Boesch, 1984, p. 163).

Regardless of the sequence in which the chimpanzee brings stone and tree together, the fact remains that he or she nearly always brings the stone nearest to a goal tree, even though he or she can only see either the stone or the *Panda* tree at the same time. This requires the following mental operations: (1) *Measurement and conservation of distance* . . . (2) *Comparisons of several distance* [sic]: . . . they begin to seriate them according to their length, in order to know the two shortest ones of which the weight will be compared; (3) *Permutations of objects in this map:* Newly transported stones are placed correctly in their mental map with reference, at least, to all the *Panda* trees to which transports will be aimed; and (4) *Permutation of the point of reference:* The distances associated to a stone are not invariant, but the set of distances between the stones and one tree can be mentally exchanged by a set relating the same stones to another tree. (Boesch & Boesch, 1984, p. 168 [reprinted by permission of the publisher])

Wild apes also use tools for nonsubsistence activities. Wild chimpanzees use objects as aids for grooming themselves, using leaves to clean themselves (Goodall, 1986), for example, for self-care as in the famous case of a chimpanzee using a stick to extract another chimpanzee's rotten tooth (McGrew, 1992). Recently, chimpanzees have been seen using stepping-sticks and seats to protect their feet and rumps from sharp thorns while they feed in kapok trees (Alp, 1997). Both orangutans and bonobos use leaves as umbrellas to protect themselves from the rain (Ingmanson, 1996).

Like many monkeys, great apes throw objects at predators and sometimes at prey. Since the accuracy of their aim is poor, it seems likely that this activity functions primarily as a mobbing display. Throwing objects also occurs as a part of highly ritualized agonistic display designed to intimidate rather than physically damage conspecifics (Goodall, 1986; Schaller, 1963). (We know of no cognitive study of aimed throwing.)

Wild orangutans apparently use considerable knowledge of physical causality and spatial relations in swaying large trees in order to bridge gaps in the forest. Some investigators have suggested that sensorimotor intelligence arose as an adaptation for these activities (Bard, 1995; Chevalier-Skolnikoff, Gladikas, & Skolnikoff, 1982).

Finally, of course, wild great apes use objects in nest construction (Fruth & Hohmann, 1996). The long developmental trajectory for *nest building*—spanning the period from about 8 months to 4 years—suggests that it may involve at least fifth-stage sensorimotor understanding of causality and space. (Fruth and Hohmann, 1996, suggest that tool use in apes had its origins in nest building, though they do not elaborate their argument.)

Hunting by chimpanzees is also notable in its use of physical knowledge. Chimpanzees at the Gombe hunt 13 species of primates, 4 species of ungulates, 3 species of rodents, and 1 species each of insectivores, carnivores, and hyrax. Their hunting strategies vary across prey species. Hunting requires knowledge of the habitats and habits of the prey, including anticipation of the pathways (detours) they are likely to follow when pursued and the places they can be trapped without possibility of egress. Goodall reports that the most sophisticated tactics of cooperative hunting at the Gombe have been seen in attacks on baboons.

> In September 1979, all but one of the Kasakela adult males were traveling in Lower Mkenke Valley when they came upon a female baboon with a tiny black infant. She was feeding in an oil-nut palm, and appeared to be quite on her own. Goblin grinned, squeaked softly and reached to touch Satan. All six males had their hair erect. When the baboon noticed the chimpanzees, she stopped feeding and gazed toward them. After approximately half a minute she began to show signs of unease; she backed away from the predators, gecking softly. Jomeo, moving very slowly, left the other males and climbed a tree close to the palm. At this point the female baboon began to scream but she did not run off. When he had climbed a branch level with her and about 5 meters away, Jomeo stopped. He stared at her, then began to shake a branch—possibly to try to make her run. She screamed even louder, but apparently no other baboons were within earshot. Two minutes later Figan and Sherry, also moving slowly, climbed two other trees. A male chimpanzee was now in each of the trees to which the baboon could have leaped from her palm; the other three, still looking up, waited on the ground. A this point Jomeo leaped over onto the palm. The baboon made a large jump into Figan's tree, where she was easily grabbed and her infant seized. The mother ran off 6 meters or so, where she remained, screaming, then uttering waa-hoo calls for the next fifteen minutes while the chimpanzees consumed her infant. (Goodall, 1986, p. 286 [reprinted by permission of the publisher; © 1986 by the President and Fellows of Harvard College])

SITUATED EXPRESSIONS OF LOGICAL-MATHEMATICAL ABILITIES

Physical knowledge and its expression in the tool use of wild chimpanzees is well known. Logical-mathematical knowledge in chimpanzees is less well known, and its expression in the wild barely addressed. We suggest that these abilities are manifested in both social and subsistence contexts.

First, descriptions of advanced tool use in wild chimpanzee subsistence activities imply certain logical abilities. Seriation, for example, seems to be involved in the selection of a series of tools of different sizes and diameters for excavating honey (McGrew, 1992). As indicated above, it may also be involved in calculating optimal distances in hammer transports (Boesch & Boesch, 1984).

Multiple classification seems to be involved in seeing that the same object can simultaneously serve two or more functions, as, for example, in using a rock as an anvil or a hammer or a wedge, or using a stick as a probe or a weapon (Matsuzawa, 1994), or for that matter, seeing a branch on a tree as a potential probe.

Second, descriptions of *social reciprocity* in chimpanzees suggest the tantalizing possibility that apes keep accounts of favors given and favors received. De Waal's long-term study of reciprocity suggests that chimpanzees (de Waal, 1996b) not only return favors to those who have favored . them, but that they punish those who default on their obligation to return favors received. Not only do they track favors given and favors received within a given domain such as in food sharing, grooming, or alliances, but they do so across domains as in sex for food, or support for grooming or sex. In addition, they do so over long periods of time. Their ability to punish cheaters contrasts with that of macaques, who seem incapable of this degree of social accounting (de Waal, 1996b). The following incident from the Arnheim chimpanzee colony, which occurred after the publication of *Chimpanzee Politics*, reveals the high stakes involved in such social accounting.

In chimpanzees, male access to estrous females is affected by social dominance . . . but also by alliances in the context of sexual competition. As illustrated by the Nikkie-Yeroen alliance at the Arnhem Zoo, such sex-related alliances are not necessarily identical to those formed during dominance struggles.

Nikkie attained alpha rank with the help of Yeroen, and remained dependent on Yeroen in order to dominate their common rival, Luit. Initially, Yeroen gained considerably from supporting Nikkie. His contribution to the mating activity in the group increased from 9% during Luit's alpha period to 39% under Nikkie. This changed during the second part of Nikkie's rule, however. Yeroen's mating success dropped to 19% while Nikkie's rose to 46%. The reason Yeroen's success decreased was that Nikkie had begun to cooperate with the other male in a sexual context. . . . Nikkie's tolerance of Luit increased steadily until one summer day, Luit started to climb towards a sexually receptive female who sat high up in a tree. Yeroen vocalized and looked alternately at Nikkie and Luit. In response, Luit returned to the ground, approached the other two males and joined in their hooting chorus. After a couple of minutes, however, Luit returned to the female. Now Yeroen burst into loud screaming against Luit, while holding out his hand in a request for support from Nikkie. Nikkie, however, walked away from the scene. This led to a highly unusual surprise attack by Yeroen on Nikkie, jumping on him from behind and biting him in the back. (de Waal, 1996a, p. 168 [reprinted with permission of Cambridge University Press])

Two days after this, Yeroen attacked Nikkie again in the night quarters injuring him badly. The breakdown of their coalition led to Luit's ascendance as alpha male. The males were separated in the night quarters for about 7 weeks. At the end of that period, they were rejoined. Shortly after this, Luit was attacked and castrated by the other two males and died of his injuries (de Waal, 1989).

The ability to keep track of favors across domains suggests true counting (of unlike items across events) and multiplicative classification (according to more than one feature). The fact that logical-mathematical abilities operate in social interactions suggests that the cognitive domains of developmental psychologists do not map neatly onto separate behavioral domains.

SITUATED EXPRESSIONS OF SOCIAL COGNITION IN APES

Great apes, particularly chimpanzees, show more complex social cognitive abilities than monkeys do. In particular, they display greater imitative abilities and greater awareness of self and others. The existence of these greater abilities raises the question of their uses in nature. Two arenas in which they seem to use these abilities are in political competition and in teaching offspring.

The clearest example of use of symbolic abilities in *teaching* comes from Boesch's (1991b) description of a Tai chimpanzee mother's demonstration teaching of nut-cracking skills to her youngster and the youngster's imitation of the mother's behavior.

Ricci's daughter, Nina, tried to open nuts with the only available hammer, which was of an irregular shape. As she struggled unsuccessfully with this tool, alternating her posture (14 times), hammer grip (about 40 times), the position of the nut or the nut itself, Ricci was resting. Eventually, after eight minutes of this struggle, Ricci joined her and Nina immediately gave her the hammer. Then, with Nina sitting in front of her, Ricci, in a very deliberate manner, slowly rotated the hammer into its best position for efficiently pounding the nut. As if to emphasize the meaning of this movement, it took her a full minute to

perform this simple rotation. With Nina watching her, she then pro-
ceeded to use the hammer to crack ten nuts. . . .Then Ricci left and Nina
resumed cracking. Thereafter she succeeded in opening four nuts in
15 minutes, and as she still had difficulties, she regularly changed her
own position (18 times) and the position of the nut, but always held the
hammer in the same position as had her mother and never changed her
grip nor the position of the hammer. (Boesch, 1993, p. 177 [reprinted
with permission of Cambridge University Press])

This behavior depends upon the mother's capacity for imitation, pretend play, and awareness of self and other. This interaction suggests that
imitation and teaching play an important role in the social transmission of subsistence skills in chimpanzees. It also suggests that social
and physical cognition work in concert in activities in natural settings.
Other examples of symbolic play with "log dolls" in wild chimpanzees
have been reported by Wrangham (1995) and by Matsuzawa (1995) (see
chapter 4).

Perhaps the clearest explication of the political use of these Machiavellian abilities comes from de Waal's (1983, 1996b) long-term study of
the social behavior of chimpanzees at the Arnheim Zoo in Amsterdam.
Machiavellian maneuvers were especially apparent among the three
adult males, who through a series of changing coalitions, were jockey-
ing for the position of alpha male.

In one prolonged episode, the junior male, Luit, began to challenge
the dominance of Yeroen in three ways. He cultivated the third male,
Nikkie; he punished the females when they associated with Yeroen
("separating interventions"); and he engaged in tantrums and aggressive
displays. Once Luit had succeeded in dominating the older alpha male,
Yeroen, through intimidation displays, he adopted a new "policy" of
supporting the losers in conflicts (he became a "loser-supporter" rather
than "winner-supporter"). "His new policy seemed to be aimed at a completely different objective, namely to stabilize his newly acquired position. He changed his attitude towards the adult females, towards Yeroen
and towards Nikkie" (de Waal, 1983, p. 124).

Subsequent to this, Nikkie began to challenge Luit with dramatic displays involving hooting and hurling of stones. When Nikkie had been
subdued and driven into a tree by the females, he began hooting again.
Luit responded to the resumption of Nikkie's challenge with an invol-
untary fear grimace, but he put his hand to his mouth and pressed his

TABLE 6-4 HYPOTHESIZED USE OF VARIOUS COGNITIVE DOMAINS IN DAILY ACTIVITIES OF CHIMPANZEES

Activity	Logical-mathematical	Physical	Social	Symbolic
Preparing implements and foods for use	Classification of objects; reckoning quantities	Physical causality and spatial relations in tool making and tool use		
Hunting game		Physical causality and spatial relations in predicting escape routes	Awareness of positions and activities of self relative to others	
Political activities	Keeping track of quantities and catego-ries of goods and services rendered	Use of objects in intimidation displays and in giving gifts to others	Awareness of activities and appearance of self to others	Possible use of symbolic gestures in greeting
Coordinating progressions		Spatial maps of key resources and of group members; Ability to take detours	Awareness of effect of activities of self on others	Possible marking of directions of move-ment
Nest building		Use of causal and spatial relations to construct adequate sleeping platform		
Teaching offspring		Use of causal and spatial relations in tool use	Awareness of effect of own actions on others	
Courtship and mating	Keeping account of favors owed		Awareness of appearance in directed sexual displays	Possible iconic gesturing of preferred positions

own lips together to hide his response. These maneuvers involved a variety of cognitive abilities, including understanding of physical and social causality, deception, and self-awareness.

These incidents illustrate the point that various forms of cognition, social, physical, and logical, act in concert when apes behave in natural settings.

AN ORCHESTRAL MODEL FOR SITUATED KNOWLEDGE

The daily lives of wild monkeys and apes are composed of a variety of routine activities and events involved in subsistence, defense, and reproduction. For male chimpanzees, for example, these may include progressing from one location to another, foraging for and sometimes transporting foods and implements, hunting, preparing implements and foods, eating, resting, socializing and engaging in political activities (food sharing, grooming, threatening, submitting, and, occasionally, fighting), courting and mating, patrolling territorial boundaries, nest building, and sleeping. For female chimpanzees these may include many of the same activities (with the exception of hunting and boundary patrolling), in addition to carrying, nursing, or food sharing with offspring and playing with and teaching offspring subsistence skills.

These activities call upon skills from a variety of cognitive domains. Political activities, for example, call upon knowledge of physical and social causality, spatial relations, and logical-mathematical knowledge, as well as self-awareness. Understanding of physical causality is involved, for example, in hurling rocks at opponents during threat displays. Understanding of social causality is involved in separating interventions. Self-awareness and spatial understanding are involved in hiding of fear grimaces as described by de Waal (1983). Nut cracking and foraging for nuts and tools involve understanding not only of space and causality, but also of classification and seriation. Teaching nut cracking to a juvenile offspring calls upon knowledge of physical and social causality and spatial relations, as well as imitation, pretend play, and self-awareness. Table 6-4 lists possible mappings of various cognitive domains onto daily activities of chimpanzees.

An analysis of naturalistic behaviors suggests an *orchestral model* of ape cognition in which particular abilities are called into play in a variety of tasks occurring across a variety of activities during a typical day. Each activity is like a melody that involves its own peculiar combination and sequence of musical notes, sensory, motoric, and cognitive. In other words, daily activities—like language—are symphonies composed of ever-changing combinations of cognitive abilities that flexibly participate in a variety of functional patterns. This model contrasts with the model of inflexible encapsulated cognitive *modules*, each of which has evolved for a specific purpose.

Why study life histories? Life histories lie at the heart of biology; no other field brings you closer to the underlying simplicities that unite and explain the diversity of living things and the complexities of their life cycles.

STEARNS, 1992, P. 9

CHAPTER 7
COGNITIVE DEVELOPMENT IN THE CONTEXT OF LIFE HISTORY

Taxonomists group species into larger categories of genera, subfamilies, families, etc. These so-called higher taxonomic categories are based on closeness of phylogenetic relationship. Techniques for discovering phylogenetic relationships are discussed in chapter 9, which also includes a family tree of primates. Figure 7-1 presents a diagrammatic presentation of anthropoid genera and higher taxonomic categories.

The number of living species of great apes has recently been revised upward from four to six: one or possibly two species of orangutans (*Pongo pygmaeus*), two species of gorillas (*Gorilla gorilla* and *Gorilla beringei*) (Morell, 1994; Ruvolo, 1994), and three species of chimpanzees, *Pan troglodytes*, *Pan verus* or common chimpanzees, and *Pan paniscus*, the bonobos or pygmy chimpanzees (Morin et al., 1994). Orangutans today are confined to the islands of Borneo and Sumatra. All the other great apes live in Africa. Chimpanzees have the widest distribution of the three, living in a variety of habitats in both eastern and western Africa, whereas the bonobos have the more restricted distribution, living only in the Congo Basin forests. Gorillas are spottily distributed across the center of Africa from the mountains of Rwanda to the lowlands of eastern and western Africa, with different subspecies living in different forest habitats. Great apes are capable brachiators when they are young, but spend considerable time walking on their knuckles on the ground. They sleep in nests they construct in trees.

Although the precise figures given in different sources vary, it is clear

Suborder	Infraorder	Superfamily	Family	Subfamily	Genus	Common name
Haplorhini	Platyrrhini	Ceboidea	Cebidae	Alouattinae	Alouatta	Howler monkey
				Aotinae	Aotus	Owl or night monkey
				Atelinae	Ateles	Spider monkey
					Brachyteles	Woolly spider monkey
					Lagothrix	Woolly monkey
				Callicebinae	Callicebus	Titi monkey
				Cebinae	Cebus	Capuchin monkey
				Pitheciinae	Cacajao	Uakari
					Chiropotes	Bearded saki
					Pithecia	Saki
				Saimiriinae	Saimiri	Squirrel monkey
	Catarrhini	Cercopithecoidea	Cercopithecidae	Cercopithecinae	Allenopithecus	Swamp monkey
					Cercocebus	Mangabey
					Cercopithecus	Guenon
					Erythrocebus	Patas
					Macaca	Macaque
					Miopithecus	Talapoin
					Papio	Savanna baboon
					Theropithecus	Gelada baboon
				Colobinae	Colobus	Colobus monkey
					Nasalis	Proboscis monkey
					Presbytis	Langur
					Pygathrix	Douc langur
					Rhinopithecus	Golden monkey
		Hominoidea*	Hominidae	Gorillinae	Gorilla	Gorilla
					Pan	Chimpanzee
				Homininae	Homo	Human
			Hylobatidae		Hylobates	Gibbon
			Pongidae		Pongo	Orang-utan

Source: Adapted from Hershkovitz 1977. Classification of the Lemuroidea follows Tattersall 1982; the Lorisoidea, Charles-Dominque 1977; the Cercopithecidae, Thorington and Groves 1970; and the Hominoidea, Andrews and Cronin 1982.
*This classification is based on recent molecular evidence. Traditional classifications put Gorilla and Pan in a separate family from Homo.

FIGURE 7-1
Graphic depiction of taxonomy of anthropoid primates (Richard, 1985).
© 1985 by W. H. Freeman and Company. Used with permission.

that great ape species differ substantially in their adult body sizes: gorillas are the largest (females 67 to 98 kg; males 130 to 218 kg); orangutans the next largest (females 33 to 45 kg; males 34 to 91 kg); common chimpanzees the next (females 26 to 50 kg; males 32 to 70 kg); and bonobos the smallest (females 27 to 38 kg; males 37 to 61 kg) (from Tuttle, 1986, who gives the largest range of variation for each species). The larger great apes are highly sexually dimorphic. The variation in body size within genera may reflect variation across subspecies or even species that were previously unrecognized or variation in definitions of adult status by different investigators. Such variation in body size within an adaptive array, that is, a group of sister species who have diverged from a common ancestor, is often seen in birds and mammals (Fleagle, 1988).

The *lesser apes*, the gibbons (*Hylobates*), are the closest relatives of the great apes. The lesser apes, members of the family Hylobatidae, include at least eight species of lesser apes: the gibbons (genus *Hylobates*) and the siamangs (classified as genus *Symphalanges* by some and *Hylobates* by others). Gibbon species live throughout Southeast Asia and range from China to Assam and across Indonesia and the Mentawai Islands. Siamangs live in Malaysia and Sumatra (Preuschoft, Chivers, Brockelman, & Creel, 1984). The family of lesser apes branched off their common ancestor with great apes approximately 22 million years ago and differentiated into gibbons and siamangs approximately 6 million years ago (Cronin, Sarich, & Ryder, 1984).

Lesser apes are small, highly specialized brachiators who are strictly arboreal. Although they have lost their tails as part of their brachiation complex, lesser apes have the sitting pads or ischeal callosities that are also typical of Old World monkeys. Unlike great apes, they do not build nests but sleep sitting up on these calluses.

Like the great apes, lesser apes comprise an *adaptive array*, falling into three groups by body weight: about 5 kg, 7 kg, and 10 kg (Raemaekers, 1984). Unlike the great apes, lesser apes show little sex difference in body weight or canine size: Gibbon females average between 5 and 7 kg, whereas gibbon males average between 6 and 7 kg. Siamangs are slightly larger, both males and females averaging 11 kg. Their monomorphism in body size correlates with their monogamous mating systems and joint defense of feeding territories. Although most species are lacking in sexual dimorphism, some species are dichromatic in coat color, whereas other species show two different color phases without regard to sex, and other species show no variation in color.

The next closest relatives are the Old World (*Cattarhine*) monkeys. The Old World monkeys live in Africa, Asia, and the islands off Southeast Asia. They branched off a common ancestor with the apes about 22 million years ago. They include two families, the leaf-eating monkeys (Colobidae) and the cheek-pouch monkeys (Cercopithecidae). The Cercopithecidae include macaques, baboons (*Papio*), guenons (*Cercopithecus*), and mangabeys (*Cercocebus*). These two adaptive radiations branched off a common ancestor approximately 12 million years ago, several million years after that common ancestral species branched off its common ancestor with the apes.

The macaques are a large genus containing as few as 9 or as many as 19 species, depending on the taxonomist (Lindburg, 1980). Their geographic range extends from Morroco in North Africa across India and the Himalayas into Southeast Asia and Japan and China. This enormous range reflects the variety of habitats exploited by members of this genus, from seashore to high mountains, from forests to mangrove swamps to scrublands to cities. In tropical rainforests they are often sympatric with gibbons, with other Old World monkeys, and with gibbons (Lindburg, 1980).

Like great apes, macaques form an adaptive array of body sizes, ranging from the small rhesus macaques of India (females about 3 kg; males 6 kg) to the large Barbary macaque of the Atlas Mountains of Morocco and Japanese macaques of Japan (females 10 kg; males 12 kg). Some of the larger ground-living forms have very short tails, for example, stumptailed macaques of China, pigtail macaques of Southeast Asia, the Japanese macaques, and the Barbary macaques. Macaques are larger than most guenons (genus *Cercopithecus*) and smaller than most baboons (*Papio*).

The New World (*Platyrrhine*) monkeys live in Central and South America. They are comprised of two families, the marmosets (Callithrixidae) and all the other neotropical monkeys (Cebidae). The Cebidae include all the larger species of New World monkeys including spider (*Ateles*) and howler (*Alouatta*) monkeys as well as the tiny squirrel (*Siamiri*) monkeys and the cebus. Some of the Cebidae have prehensile tails, that is, tails that will support their full body weight. The New World monkeys branched off a common ancestor with the Old World monkeys about 35 million years ago after which they underwent an adaptive radiation. It seems likely that their common ancestor came across the Atlantic Ocean on floating vegetation at that time when South American and Africa were still closer together.

The cebus are a genus of Cebidae comprised of four species who live in Central America and the Amazon forest. They are all small animals (females 2 to 3 kg; males 3 to 4 kg) with little to slight sexual dimorphism in body size. They are larger than marmosets and squirrel monkeys, but smaller than spider and howler monkeys. At one time, they were widely used as organ grinders' monkeys. They inhabit a variety of forest types from tropical rainforest to dry deciduous forests, feeding primarily in the trees, but also foraging on the ground.

Members of each of these adaptive arrays, greater and less apes, macaques and other Old World monkeys, and cebus and other New World monkeys, are characterized by a different life history pattern.

PRIMATE LIFE HISTORIES

Reproductive maturity and the onset of reproduction is the keystone of what biologists call *life history*. By life history features, they mean gestation period, infancy, juvenility, puberty, onset of reproduction, litter size, birth interval, brain size, and life span. Like body form and behavior, these features have been shaped by selection within various developmental constraints (see chapter 8).

Prosimians, monkeys, lesser apes, great apes, and humans constitute a series displaying a stepwise increase in these so-called life history features. Although most primates bear single young, they differ in gestation period, life span, and age at sexual maturity. Increasing gestation period, immaturity, and life span correlate with increasing brain size and increasing number of periods of cognitive development (see chapter 12).

Typically, Old World monkey females have gestation periods of almost 6 months, first reproductive effort at 3 or 4 years, and a life span of approximately 20 to 30 years. Lesser ape females' life history is intermediate between that of the great apes and Old World monkeys in gestation length, brain size, and life span. Gestation in gibbons is about 7 months. Maximum life span is about 30 years. Females have their first offspring between 7 and 9 years of age.

Great ape females, in contrast, have gestation periods· of almost 8 months, puberty at about 8 to 10 years, first reproductive effort at 10 to 14 years, and life spans of more than 45 years. Humans have 9-month gestation periods, puberty at 11 to 16 years, first reproductive effort at about 19 years, and life span of more than 95 years. New World monkeys show much greater variation in life history features than Old World monkeys. Cebus are the slowest developing, and marmosets, the fastest developing species. Cebus females have their first offspring as late as 6 years of age.

Bogin (1997) argues that childhood and adolescence are new life history stages unique to humans. Childhood begins at the end of infancy, that is, with weaning. It is characterized by parental provisioning of special weaning foods. Childhood ends with the onset of juvenility, beginning with the eruption of the first permanent molars. Juvenility ends with a mid-growth spurt and adrenarche; adolescence begins with puberty; and adulthood begins with the onset of reproduction.

He contrasts these human life history stages with those of other primates who go directly from infancy to juvenility. He notes that across mammals weaning occurs when infants achieve between 3.2 and 4.9 times their birth weight. In chimpanzees this occurs at about 40 months. In humans, however, it occurs when infants have achieved less than 3 times their birth weight, at about 36 months, on the average. He notes that in both food-enhanced and food-limited societies, humans introduce some solid foods when infants have achieved about 2.1 times their birth weight. Table 7-1 summarizes primate life history features, which are depicted graphically in Figure 7-2.

TABLE 7-1   LIFE HISTORY FEATURES OF SELECTED ANTHROPOID
PRIMATES

Species	Gestation (months)	Eruption molar 1 (years)	Female age at first reproduction (years)	Maximum life span (years)
Humans	9	6	19	95
Great apes	8	3	10 to 15	45
Gibbons	7	1.75	8	
Macaques	5 to 6	1.4	3 to 4	30
Cebus	5	1.2	6	45

Note. From Harvey et al. (1987).

Bogin argues that the *childhood stage* is a unique human adaptation that facilitates a higher reproductive rate by allowing mothers to invest in subsequent offspring while their children are still immature. This is achieved through the combination of provisioning and kin aid in child care. He suggests that the human pattern arose in *Homo habilis* when brain size enlarged to 800 cm³ because of the heavy nutritional demands of brain development in children less than 5 years of age.[1]

Evidence suggests, however, that chimpanzees, at least, display some features of childhood. Indeed, wild chimpanzee mothers provide their weaned infants with some difficult-to-process foods (Silk, 1978) and aid them in processing such foods (Boesch, 1991b, 1993). The importance of this aid can be gauged by the fact that chimpanzees achieve proficiency in termiting and anting at about 5 years of age and in nut cracking, at about 8 years of age. They also show evidence of adolescence, that is, a period of adolescent sterility between puberty and the onset of reproduction (Watts, 1985).

Although similar, life histories of great apes do differ to some degree. Specifically, gorillas and bonobos seem to have earlier onset of reproduction than chimpanzees and orangutans. Gorillas also have a shorter birth interval than the other great apes and probably a shorter life span. Orangutans have the latest onset of reproduction and the longest birth intervals of all the great apes. These differences are consistent with apparent differences in rates of cognitive development within the great

---

1. Bogin (1997) also argues that the childhood stage was inserted into development and, therefore, according to Gould's (1977) definition of heterochrony, the evolution of human life history cannot be seen as involving what he calls "simple heterochrony." We discuss this issue in chapter 8.

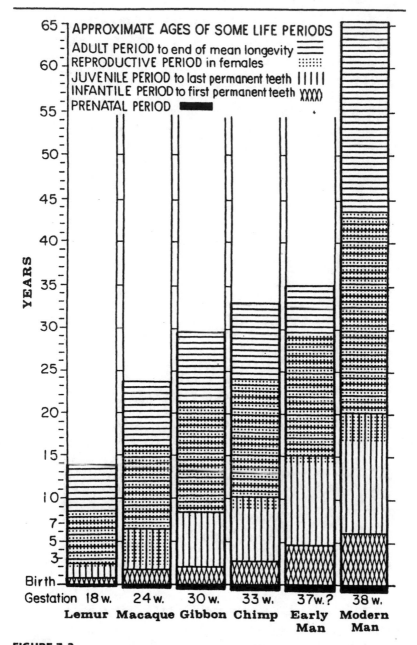

FIGURE 7-2
Graphic depiction of life history stages in selected primated (B. H. Smith, 1992). © 1992 John Wiley & Sons. Reprinted with permission of John Wiley & Sons, Inc.

apes. (They do not correlate with phyletic distances from humans among the great apes.)

Similarities in life history features reflect common ancestry. Great apes and humans share a common ancestor, A, who had life history features similar to those of living great apes. This common ancestor, A, which lived about 15 million years ago, evolved great apes' life history characteristics after it diverged from its common ancestor with lesser apes, B, which lived about 20 million years ago. B had life history features similar to those of living gibbons and siamangs. These characteristics evolved after B had diverged from its common ancestor with Old World monkeys, and so on (see chapter 9).

Each of these common ancestors, C, B, and A, gave rise to a new group of sister species, that is, the Old World monkeys (c), the lesser apes (b), and the great apes and hominids (a). A group of sister species that evolve from a common ancestor and fill different niches in a relatively brief time period is known as an *adaptive radiation*. Typically, the sister species form *an adaptive array* of species that share similar life history features but differ in diet, body size, and mating behaviors. The macaque species, the cebus species, the gibbon species, the African ape species, and the Australopithecine species are all examples of adaptive arrays (Fleagle, 1988). They all differ in body size, diet, and reproductive strategies.

Within mammalian species life history features are highly intercorrelated. Brain size correlates strongly with gestation length and life span. In fact, many biologists consider brain size to be the pacemaker of life history in mammals because large brains require a long period of development and delay the onset of reproduction. Because of their close relationships, these features tend to be evolutionarily conservative and hence similar among closely related species.

Large body size, long life span, few offspring, and prolonged development comprise a package of life history characteristics that tend to recur across various taxonomic groups. This package is often characterized as a *K life history strategy*. Conversely, small body size, short life span, many offspring, and rapid development is another such package that tends to recur. This package is often characterized as an *r life history strategy* (MacArthur & Wilson, 1967).

Among primates, the r life history strategy is exemplified in mouse lemurs, while the K life history strategy is exemplified in gorillas. Within a taxonomic group many life history characteristics correlate with body size and metabolic rate: smaller animals have higher metabolic rates and therefore live at a faster rate, while larger animals have lower rates and live at a slower rate.

> If a female mouse lemur (*Microcebus murinus*) and a female gorilla (*Gorilla gorilla*) were born at the same time, the mouse lemur could leave 10 million descendants before the gorilla became sexually mature. This astonishing difference results from the compounding difference in life-history patterns between the two species. Mouse lemurs can start reproducing when they are one year old and produce two litters a year thereafter with each litter containing up to three offspring. Gorillas are more like humans, not breeding until they are about ten years old, and even then they produce but a single young every four or five years. (Harvey, 1990, p. 81)

See Figure 7-3 for a graphic comparison of a gorilla and a mouse lemur.

As noted above, males and females of the same species may display differences in their body size (sexual dimorphism), especially in large-bodied species. This occurs as a consequence of sex differences in their life histories and growth patterns (*bimaturism*). When this occurs it is typically males who grow larger than females and take longer to do so (Shea, 1990). For this reason, female life histories are more similar than male life histories. Often in polygynous species such as in gorillas where some males have a higher reproductive potential than other males and than females, large body size and weaponry have been selected for male competition (Trivers, 1972).

Just as life history packages tend to recur, so certain packages of maternal-infant features tend to recur across avian and mammalian taxa: The first package, comprising small litters or single births, long gestation periods, higher neonatal body weight, slow growth rate, and the neonatal condition of sensory competence, is characterized as the *precocial condition*. The second package, comprising large litters, short gestation periods, low neonatal body weight, rapid growth rate, and the neonatal condition of hairlessness and sensory immaturity, is characterized as the *altricial condition* (Portmann, 1990). While the total energy investment of mothers in each offspring is greater in precocial species, the rate of energy investment of mothers in altricial species is greater (Martin & MacLarnon, 1990). As would be expected from their greater reproductive potential, altricial species tend to have **r** life history strategies while precocial species tend to have **K** life history strategies. See Figure 7-4.

In contrast to these conservative, intercorrelated **K** and **r** life history

FIGURE 7-3
Difference in body sizes of the mouse lemur and gorilla (Fleagle, 1988).

features, other features such as body size, diet, mating behavior, and social organization do not necessarily correlate within an adaptive array of closely related species. Among the African apes, for example, chimpanzees and bonobos live in large groups that break up into small foraging parties when foods are dispersed. Gorillas live in small, stable groups with one or two males. Chimpanzees are omnivorous, and gorillas eat herbivorous diets, while bonobos have an intermediate diet. Gorillas are larger than either of the chimpanzee species, reflecting their lower-grade diet. All three species display different mating patterns. These features seem to be less conservative than the preceding features, evolving rapidly in response to changing resources and social conditions.

Life history data reveal interesting patterns of similarities and differences within and across primate species. With few exceptions, primate females bear one infant at a time. (This pattern correlates with the two pectoral teats that are characteristic of primates.) Gestation periods show little variation within species or even among species within the same genera, while birth intervals show considerable variation within genera and even within species. Like gestation period, adult brain weight shows relatively little variation within genera and even families (Harvey, Martin, & Clutton-Brock, 1987).

The evolutionary conservatism of gestation periods and brain weight contrasts with the variation in adult female body weight, which often varies considerably among sister species within the same genus or family, by as much as a factor of 3. Even greater intrageneric and inter-familial variation occurs in adult male body weight. In many species, especially mid-size to large species (greater than 5 kg), males are as much as twice as large as females. Males of these sexually dimorphic species may achieve their greater body size through one of two forms of bimaturism, earlier female maturation or prolonged male growth, males maturing 1 to 3 years later than females. This pattern is especially pro-nounced in baboons and in the great apes (Shea, 1990).These species contrast with such taxa as the lesser apes, the gibbons and siamangs, and the marmosets, in which males and females are the same size.

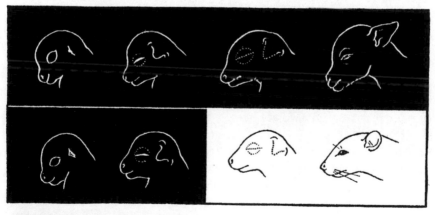

FIGURE 7-4
Comparison of the development of a primordial mammal having early birth as an altricial infant (*below*) with the development of the higher mammals with longer gestation periods and closure of the sensory organs during the early period (*above*). The black background indicates the time within the mother's uterus (Portmann, 1990). © 1990 by Columbia University Press. Reprinted with permission of the publisher.

Clearly, the most striking differences among groups within the primate order occur between humans and the great apes and between great apes and monkeys. Differences occur in body size, gestation length, age at sexual maturity, life span, and brain size. When fossil hominids are added to the picture, however, the gap between great apes and humans disappears, and the continuity from great apes to hominids to humans in all these features becomes clear. This must apply to cognitive abilities as well.

COGNITIVE DEVELOPMENT IN RELATION TO PHYSICAL DEVELOPMENT

Discussions in the preceding chapters have compared cognitive development in human and nonhuman primates in terms of chronological age. This chapter uses physiological indices of maturation rather than age for comparative purposes because they provide a relative rather than an absolute yardstick for comparing development. The eruption of molar teeth offers an ideal measure of relative maturation to compare with cognitive development.

DENTAL DEVELOPMENT

Data on molar tooth development is particularly useful for comparative studies of life history because it is highly correlated with brain size. Molars are doubly useful because they are accessible for both living and fossil species. Since molars fossilize more frequently than brains do, their eruption patterns offer valuable clues to brain size in early hominids. Best of all, they are more likely than any other tissue to fossilize. For these reasons, and because comprehensive data on molar tooth development are available for many primate species, this is the measure of relative maturation we use here.

Recent papers by Smith and her colleagues (B. H. Smith, 1993; B. H. Smith, Crummett, & Brandt, 1994; R. J. Smith, Gannon, & Smith, 1995) provide comprehensive data on dental development in monkeys, great apes, humans, and some early hominids. The eruption of the three molar teeth marks the end of infancy (and early childhood in humans), the end of childhood, and the achievement of full maturity (B. H. Smith, 1993). Table 7-2 compares the (mandibular) molar eruption sequences for macaque monkeys, gibbons, orangutans, gorillas, chimpanzees (*Pan*), Homo erectines, and humans.

It is interesting to note the following patterns: The molar development sequence in macaques begins about 2 years earlier than in great

TABLE 7-2 MOLAR ERUPTION SEQUENCE FOR SELECTED ANTHROPOID SPECIES IN YEARS OF AGE

Species	Molar 1	Molar 2	Molar 3
Macaques	1.4	3.2	5.5
Baboons	1.7	4	7
Orangutans	3.5	5.0	10.0
Gorillas	3.5	6.6	10.4
Chimpanzees	3.2	6.5	10.7
Homo erectines	4.5	9.5	14.5
Humans	5.4	12.5	18

Note. From B. H. Smith (1993), B. H. Smith et al. (1994).

TABLE 7-3 COGNITIVE DEVELOPMENT RELATIVE TO DENTAL DEVELOPMENT

Species	Highest level achieved	Occurrence of highest sensorimotor achievement	Occurrence of highest preoperational achievement
Monkeys	Sensorimotor up to fourth or fifth stage	After last deciduous tooth	Before first molar
Great apes	Preoperational up to symbolic	Long after last deciduous tooth	Coincident with third molar
Humans	Formal operational	Coincident with last deciduous tooth	Coincident with second molar

apes and occurs at approximately 2-year intervals. The sequence in great apes begins about 2 years later than in macaques and occurs at approximately 3-year intervals. The sequence in Homo erectines began about 1 year later than chimpanzees and 1 year earlier than in humans and occurred in approximately 5-year intervals. The sequence in humans begins about 2 years later than in great apes, and occurs in approximately 6-year intervals.

Humans, who begin the molar eruption sequence at 6 years and end at 18 years, traverse seven or eight subperiods of cognitive development. Homo erectines, who began molar eruption at 4½ years and ended it at 14½ years, apparently traversed five subperiods. Great apes, who begin molar eruption at 3 years and end at 10 years, traverse three cognitive subperiods. Macaques, who begin molar development at less then 1½ years and end at 5½ years, traverse only one cognitive subperiod.

Relating these cognitive developmental patterns back to life history patterns indicated by molar eruption, we see the following pattern. Macaques come close to completing their cognitive development by the end of infancy, great apes come close to completing their cognitive development by the end of childhood, and humans complete their cognitive development only at the onset of full maturity. (See Table 7-3 on molars and cognition.)

LOCOMOTOR DEVELOPMENT

Comparative data on locomotor development in wild bonobos, chimpanzees, and mountain gorillas reveal developmental changes in locomotion across three subperiods of infancy: class I (0 to 5 months); class II (6 to 23 months); and class III (2 to 5 years) in three species (Doran, 1992, 1997). Six forms of locomotor behavior in infants were described: quadrupedalism (palmigrade and knuckle walking); quadrumanous climbing and scrambling; suspensory behavior; bipedalism (supported and unsupported), leaping and diving, and somersaulting.

Class I infant chimpanzees and gorillas of both species begin to crawl at about 3½ to 4 months; palmigrade quadrupedalism in both begins at about 4½ to 5½ months (gorillas beginning slightly later). Forelimb-dominated behaviors are the norm in this age class. Chimpanzees in this age class are more independent than gorillas, climbing on lianas while gorillas climb on their mothers.

Class II infants (6 to 23 months of age) begin to differ significantly in their locomotor behaviors. In gorillas of this age quadrupedalism is already the predominant form of locomotion, accounting for more than 50% of locomotion, while in chimpanzees of this age it is an infrequent form of locomotion, accounting for less than 15% of their locomotion. Moreover, gorilla infants are already using the knuckle-walking form of quadupedalism about 50% of the time by 10 months while chimpanzee infants do not begin to use this form at that rate until about 30 months. This pattern reflects the fact that gorilla infants are moving on the ground most of the time while chimpanzee infants are rarely moving on the ground.

Class III infants (2 to 5 years) continue to show these differences. Gorilla infants continue to engage in more quadrupedal behaviors on the ground. Chimpanzee infants continue to engage in more climbing and suspensory behaviors above ground though they have begun to knuckle walk more frequently in their quadrupedal locomotion. By 3 years of age gorilla infants travel largely on their own, by 4 years they are using an essentially adult form of locomotion. In contrast, chimpanzees do not move independently of their mothers until they are about 5 years old. Doran (1997) notes that studies of captive gorillas report earlier ages for various locomotor landmarks (e.g., Maple & Hoff, 1982).

Gorilla and chimpanzee infants show similar locomotor behaviors at similar body sizes, which they achieve at different ages. Owing to the more rapid growth of gorillas, 1-year-old gorillas are similar in size to 2-year-old chimpanzees (and similar to juvenile bonobos). At this body size of roughly 9 kg, both species engage predominantly in quadrupedal knuckle-walking locomotion. In addition to differences in body size, differences in body proportions—broader scapulae and shorter fingers— render gorillas less suited to suspensory locomotion than chimpanzees (Doran, 1997).

Doran (1992) was unable to classify bonobo development into the foregoing categories because of lack of precise age data. Instead, she classified bonobos into infants, juveniles, adolescents, and adults. Bonobos of all ages engage more frequently in quadrupedalism and less frequently in quadrumanous climbing and scrambling than do chimpanzees. Bonobo juveniles and adolescents engage in more suspensory behaviors than do chimpanzees. Although adult bonobos and chimpanzees are roughly the same size, bonobos develop body size more slowly than chimpanzees do. Three-year-old bonobos achieve approximately 2.3 times their birth weight as compared to about 4.5 times for chimpanzees (Kuroda, 1989).

According to Doran's account the primary differences in locomotor development among African apes seem to follow from differences in body size in relation to age and from differences in degree of terrestrial versus arboreal locomotion.

Comparable figures for human infants are 8 months for crawling and standing, 12 months for unsteady walking, and 2½ years for competent walking (Gesell, 1945). For macaques, competent quadrupedal locomotion occurs before 1 month of age. For cebus, it occurs by 2 months of age.

PREHENSIVE DEVELOPMENT

The development of visually directed grasping is one of the hallmarks of Piaget's third sensorimotor-period stage. In human infants this occurs at about 3 or 4 months; in great apes, at 2½ or 3 months; in macaques, at about 1¼ months; and in cebus, at about 2 or 2½ months.

Mapping these locomotoric and prehensive developmental patterns onto cognitive development reveals the following patterns: (1) in macaques, prehension coincides roughly with competent walking; (2) in great apes, prehension coincides with crawling and precedes competent walking by about 6 months; (3) in human infants, prehension precedes crawling by about 6 months and competent walking by more than 12 months; and (4) in cebus, prehension slightly precedes competent walking. In other words, unlike other primates, human infants experience an extended immobility during the sensorimotor period. Several investigators have suggested that this enforced immobility may canalize human infants' interaction with objects (Antinucci, 1989; Parker, 1977; Vauclair & Bard, 1983).

BRAIN DEVELOPMENT

The course of cognitive development should parallel to some degree brain maturation. Unfortunately, few developmental data on brain maturation are available for monkeys and apes. Moreover, finding appropriate measures of brain maturation for primates is a complex issue (see chapter 12). If we use gross size increase as a measure of brain maturation, we see that for chimpanzees and humans, respectively, brain maturation parallels development. Humans are born when their brains are only about 25% of their adult weight, whereas chimpanzees are born when their brains are about 45% of theirs. Both species attain about 83% of their brain growth by the time their first molar erupts and about 98% by the time their third molar erupts (extrapolating from Smith's data). Although the relative rates of brain maturation are similar, chimpanzees mature almost twice as fast as humans (Passingham, 1975). By 1C years, they have traversed their full potential of three cognitive periods

PHYSICAL DEVELOPMENT IN HOMO ERECTINES

Recent work on the dentition of fossil hominids reveals that Homo erectine molar development apparently began at about 1 year later than in chimpanzees and 1 year earlier than in humans and occurred at approximately 5-year intervals (B. H. Smith, 1993). Preliminary analysis suggests that Homo erectines, who began molar eruption at about 4½ years and ended at 14½ years, traversed five subperiods of cognitive development.

If concrete operations is the highest attainment of Homo erectines, we can conclude that they completed the development of this terminal stage by the time their third molar erupted, that is, at about 14½ years of age. This compares with an age of about 12 years for the transition from concrete to formal operations in modern humans.

Extrapolating from data on chimpanzee and human development, we reconstruct the following parameters of cognitive development in *Homo erectus:* (1) sensorimotor-period development was completed before 3 years of age—about midway between the 2 years that human children require and the 4 years that chimpanzees seem to require; (2) the symbolic and intuitive subperiods were completed between about 3 and 8 years of age; and (3) the concrete operations period was completed by about 14 years of age at the onset of reproductive life. In other words, Homo erectines stood approximately midway between humans and great apes in their developmental course.

Two lines of circumstantial evidence suggests that Homo erectine infants completed the sensorimotor period more rapidly than living great ape infants do. First, Homo erectines may have been the first hominids who were exclusively terrestrial. If they had lost their ancestors' tree-climbing abilities, their infants would no longer be able to cling with their hands as Australopithecine infants did. Second, adult Homo erectines had substantially larger brains than Australopithecines or even Homo habilines, suggesting that their neonates also had larger brains. Selection would therefore have favored birth of neonates at an earlier developmental stage than occurred in their ancestors. Selection would therefore have favored greater maternal investment (Stanley, 1996) and greater mother-infant interaction in the face-to-face mode. This, in turn, would have favored early development of facial and vocal imitation and circular reactions (Parker, 1993).

In summary, comparing the development of monkeys, great apes, Homo erectines and humans reveals that molar eruption, locomotor developmental, brain maturation, and sexual maturity have slowed down, while early cognitive development has sped up and extended into new developmental periods in humans as compared to apes. Life history theory provides some insight into the relationships among these variables.

THE EVOLUTION OF DEVELOPMENT AND THE QUESTION OF INNATENESS

Use of the comparative approach implies a genetic basis for the cognitive character states under study. *Innateness* refers to highly probable species-typical outcomes contingent on normal developmental experiences (Elman et al., 1996). There are several lines of evidence that support the innateness of primate cognitive abilities: (1) the species-typical patterns of terminal-level abilities; (2) species-typical patterns of cognitive development; (3) species-typical associations between life history parameters and cognitive developmental patterns; (4) species-typical associations between brain size and terminal levels of cognitive ability; and (5) species-typical associations between brain development and cognitive development.

Innateness has been conceived in a variety of ways. Ethologists have conceived it in terms of an "innate schoolmarm" in the developing nervous system that orchestrates both classical and operant conditioning (Lorenz, 1965). Cognitive theorists have proposed two alternative models of innateness: representational versus architectural and temporal innateness. According to the representational innateness hypothesis of many neoinnatists, mental representations are hardwired into the nervous system (Carey & Spelke, 1994).

According to the alternative architectural and temporal hypothesis of the neoconstructivists, developmental changes in responding to small biases in architectural constraints in the cortex generate large-scale differences in information processing. Likewise, relatively small differences in developmental timing among various neurological systems generate large-scale changes. Finally, external regularities in the "problem space" can also favor particular behavioral or cognitive solutions.

The neoconstructionist or *emergent* view of innateness (Elman et al., 1996) has several advantages. First, it is more consistent with relevant data across a variety of fields than either the neoinnatist view or the strictly constructivist view. The neoconstructionist view is based on an integration of data from developmental neurology, psycholinguistics, and neural network simulation studies. Nonlinear dynamics modeling emphasizes the frequent disjunction between the visible behavioral outcomes and the invisible mechanisms underlying these outcomes (Elman et al., 1996).

Second, this view generates specific hypotheses concerning the mechanisms underlying innate systems. It vitiates the argument that concepts of innateness ignore the proximate mechanisms subserving particular adaptations. Finally, it is consistent with Piagetian and neo-Piagetian models of the epigenetic construction of succeeding stage of cognitive development up to the terminal level.

Adult human beings share many features with infant apes . . . like Peter Pan, we never grow up.

GRIBBIN, 1988

A philosophy of human growth and development that emphasizes the progressive appearance of new biological and behavioral traits is more satisfying empirically, and intellectually, than a view of development that emphasizes growth retardation and permanency of childhood.

BOGIN, 1988

CHAPTER 8
DEVELOPMENT AND EVOLUTION:
A PRIMER

After decades of focusing on genes and adults, evolutionary biologists are once again beginning to acknowledge that a complete view of evolution must include individual development (i.e., ontogeny). Evolution is produced by modifying ontogenies in descendant generations. Viewing evolution as adaptation through altered ontogeny provides many insights, as revealed in a rapidly growing literature sometimes called *evolutionary developmental biology* (e.g., Hall, 1992; McKinney & McNamara, 1991; McNamara, 1997; Raff, 1996). A major finding of this renewed interest is to document how developmental constraints on evolution have led to recapitulatory patterns of ontogeny in modern humans.

This chapter briefly reviews the role of development in evolution. This will explain why the human mind is not a product of an unpredictable sequence of "quirky" random evolutionary events, as is sometimes argued (e.g., Gould, 1995). Instead, increasing developmental constraint and increasing developmental complexity provide nonrandom evolutionary "arrows of time" (McKinney, in press-a, in press-b). In our discussion, we outline some of the basic ways that development evolves. We especially emphasize changes in developmental timing as a way of overcoming developmental constraints. We describe a classification scheme for analyzing timing changes and examine how such changes occur.

There is increasing appreciation in developmental psychology (Langer, in press), evolutionary psychology (Allman, 1994), and related disciplines of the key role of timing changes in mental ontogeny and evolution. A recent text on developmental psychobiology (Michel & Moore, 1995) discusses this growing interest in timing changes. Unfortunately, this renewed interest has also exposed the continued misunderstandings that have historically plagued discussions of mental development and evolution. Of particular note is the long-standing and very widespread idea that humans are underdeveloped or "juvenilized" apes, produced by a slowing of development (e.g., Gould, 1977; Montagu, 1989). This idea of the juvenilized ape has become basic dogma in dozens of popularized accounts of human evolution (e.g., Gribbin, 1988; Wesson, 1991), including popularized accounts of the emerging field of evolutionary psychology (Allman, 1994). Even "neural Darwinism" has attempted to incorporate this view (Edelman, 1987).

In contrast to those who hold this view are a number of workers, including us, who hold the opposite view, that humans evolved as overdeveloped or "adultified" apes by extending development (e.g., Gibson, 1991b; McKinney, in press-a, in press-b; McKinney & McNamara, 1991; Parker, 1996b; Shea, 1989). This view also contrasts with those of workers (e.g., Bogin, 1997) who argue that human development has evolved by inserting traits such as childhood and adolescence into the ontogenetic sequence. Instead, we will argue that these and other traits evolved by extension of stages within the ancestral human ontogeny (chapter 9).

DEVELOPMENT AS A CONSTRAINT ON EVOLUTION

Evolutionary change is clearly "constrained" by development (Maynard-Smith et al., 1985). Developmental constraint means that a species may not freely adapt to an environmental change because development limits available morphological and behavioral variation. In an ideal world, species could freely alter their morphology and behavior in order to adapt to unexpected change. But instead, to use the famous metaphor of Jacob Monod, natural selection is restricted to "tinkering" with existing patterns in order to produce new ones. Developmental constraints are a major category of constraints because natural selection can only tinker with the limited variety of developmental trajectories that are in existence at any given time. Environmental change produces evolution only to the degree that developmental variation is available for selection to act upon. As a result, development strongly influences both the rate and direction of evolution (e.g., Hall, 1992; Levinton, 1988; McKinney & McNamara, 1991; McNamara, 1997; Raff, 1996; Wray, 1992).

Furthermore, the constraining influence of development has in-

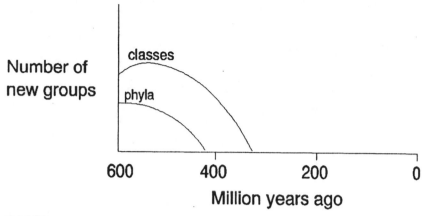

FIGURE 8-1

Concentration of origination of new major body plans such as phyla and classes in the early part of the Paleozoic era.

creased over evolutionary time as developmental processes have become less flexible. McKinney and McNamara (1991), Hall (1992), and Valentine et al. (1996) are among those who review the considerable developmental and paleontological evidence that the defining traits found in the basic body plans of all life on earth have gradually congealed after an initial period of relative plasticity. Figure 8-1 shows that origins of higher taxonomic groups, representing the most novel ontogenetic and evolutionary variants, are overwhelmingly clustered in the early Paleozoic era. This asymmetrical pattern of origination of basic morphologies is attributed to less constrained ontogenies in early Paleozoic multicellular life (Valentine, Erwin, & Jablonski, 1996). Furthermore, this pattern is also found within individual taxonomic groups. Detailed studies of fossil marine gastropods exemplify this pattern of increasing developmental constraint since the early Paleozoic era (Wagner, 1995).

Data from living organisms reveal similar findings. The defining traits that are characteristic of higher taxonomic groups (e.g., phyla and classes) alive today tend to appear early in ontogeny and show reduced levels of variation compared to traits that appear later in ontogeny (Hall, 1992). Stearns (1992) views modern constraints on evolution as the result of early adaptations that have become embedded within developmental pathways. Similarly, phylogenetic analysis documents how more closely related taxa tend to have more similar traits and more similar ontogenies (Harvey & Pagel, 1991). These follow the predicted pattern of nested evolutionary change arising from nested developmental modifications (McKinney & Gittleman, 1995).

Ontogenetic conservativism is also becoming evident at the genetic and cell level. A recent review of homeotic genes (Carroll, 1995) shows a striking degree of conservatism throughout such disparate groups as arthropods and vertebrates. These early acting genes sculpt the basic body plan of these animals. Recent evidence shows that the entire course of evolution in arthropods and vertebrates has largely involved regulatory changes in expression of highly conserved arrays of homeotic genes and the many developmental genes that they regulate (Carroll, 1995).

CAUSES OF INCREASING CONSTRAINT

Explanations for ontogenetic congealing or "hardening" have emphasized that, because development is a highly orchestrated process of interacting parts, the earlier interactions are less amenable to alteration. Alteration of early interactions would tend to have cascading effects on too many later interactions. As a result, developmental evolution has tended to involve mainly alterations of late ontogeny whereas early developmental interactions have become progressively "entrenched," or canalized, through stabilizing selection (Hall, 1992). Raff (1996) uses the term *phylotypic stage* to describe the early embryonic stage when the basic body plan traits that characterize an organism's phylum are produced.

Very general theoretical substantiation for this pattern of increasing constraint include models by Kaufman (1993) that show that even randomly interacting parts will quickly congeal into patterns of nonrandomness. More specific explanations have focused on the various kinds of interactions that occur in early ontogeny. Levinton (1988) reviews the concepts of genetic and epigenetic burdens that originate from the interdependency of genetic and tissue interactions, respectively, during early ontogeny. Hall (1992) reviews these and still other constraints that can limit variation in early development, including structural, cellular, and functional constraints. The limitations that such ontogenetic constraints place on evolutionary change have received a variety of names, for example, *burden, epigenetic traps, epigenetic cascades, generative entrenchment,* and *gene nets* at the genetic level (reviews in McKinney & McNamara, 1991; Hall, 1992; Raff, 1996; Wills, Briggs, & Fortney, 1994).

EVOLUTION BY TERMINAL DEVELOPMENTAL CHANGE

These increasing constraints on alteration of early ontogeny have meant that mutations altering later development have been favored. This does not mean that terminal addition to development is the only way that evolution has occurred. *Terminal addition* is the term historically applied to the addition of traits at the end of development. We emphasize that many other kind of developmental changes can also occur. For example, *terminal subtraction,* which removes traits at the end of development, is relatively common (Fong, Kane, & Culver, 1995; Hall, 1992). Also, nonterminal changes can occur, including early changes in development (Wray, 1992). Furthermore, different changes can affect different traits in different ways in an evolving lineage. This is often called *mosaic evolution* and is well documented in fossils (Levinton, 1988). One trait may thus exhibit terminal subtraction during evolution whereas another may show terminal addition.

The key point is that universal terminal addition, whereby all traits always evolve by addition to their development, is simply wrong. This of course also means that universal recapitulationism, wherein all aspects of ontogeny perfectly repeat all aspects of the individual's evolution, is wrong.

But we caution against "throwing the baby out with the bathwater." The many lines of evidence noted above in favor of increasing developmental constraint over evolutionary time implies that terminal developmental changes (terminal subtractions, additions, and substitutions) are likely the most common type of evolution (Hall, 1992; Levinton, 1988; McKinney & McNamara, 1991). Direct evidence for this includes Mabee's (1993) thorough phylogenetic study of living species, which finds that terminal addition may account for up to 51.9% of character state evolution in centrarchid fishes. She also shows that, when other terminal changes (such as terminal subtractions and substitutions) are included, up to 75% of change is accounted for. Similarly, Sordino and others (Sordino, Hoeven, & Duboule, 1995) have shown how terminal changes in limb appendages (via minor regulatory gene change) can explain the evolutionary origin of fingers and toes from fish fins, by unequal proliferation of cells.

We emphasize that this is a statistical and not a deterministic process: early changes in ontogeny do still occur, especially before the philotypic stage (Wray, 1992), but are less likely than late-acting changes. This also implies that recapitulatory patterns may be expected as a common statistical tendency in that more closely related species will share more similar ontogenies. The more closely related are species, the more likely that their ontogenies will not differ until later (terminal) stages of development. This agrees with Mayr (1994), who, in citing agreement with Gould (1977), has noted that "recapitulation, properly understood, is simply a fact."

Finally, terminal changes leading to an enlarged brain are, in many ways, the "ultimate" evolutionary response to increasing developmental constraint. As discussed in the chapter 13, increasing plasticity of behavior is a highly adaptive solution to the problem of reduced plasticity of morphology. This has traditionally been implicitly acknowledged by the recognition, by many evolutionists, that increased behavioral plasticity is correlated with higher grades of evolution (Bonner, 1988).

DEVELOPMENTAL TIMING AND EVOLUTION

Given that terminal changes are the commonest way to modify development, what kind of terminal changes can occur? The most commonly observed terminal changes involve alterations of developmental timing (McKinney & McNamara, 1991). These can produce either: (1) "overdevelopment," in which the developmental trajectory is extended, or (2) "underdevelopment," in which the trajectory is truncated. For example, adults of a later descendant species may closely resemble juveniles of its ancestral species. In such a case we might infer that the new species arose through some kind of underdevelopment, or "juvenilization," process. Such processes are not only easy to describe, but are also the most common types of evolutionary patterns by far. The fossil record is rampant with gradational evolutionary sequences that document evolution by minor alterations of late ontogeny. The most widely discussed schemes for describing gradational developmental-evolutionary sequences have sought to describe developmental changes in terms of the timing alterations that produced them.

BASIC HETEROCHRONIC PATTERNS

Heterochrony (literally "different time") refers to changes in the timing of development. A certain developmental stage could be "sped up" or it could be terminated prematurely, to cite just two examples. But whose development are we comparing? Sped up relative to what? Evolutionary biologists have most often used this term in reference to timing changes between species, such as between humans and our ancestral primate

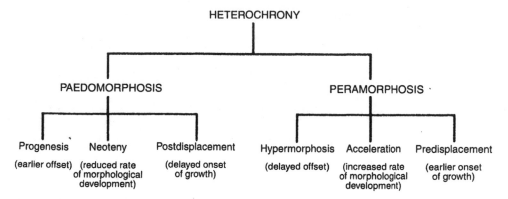

FIGURE 8-2
Classification of heterochronic patterns. Overdevelopment (peramorphosis) and underdevelopment (paedomorphosis) can each occur by three basic timing changes. (McKinney & McNamara, 1991).

species. However, heterochronic terms can also be employed when comparing developmental patterns between individuals of the same species. For either genetic or environmental reasons, such developmental variations occur within nearly all species. These terms can be applied to development of behaviors as well as morphological traits.

A casual perusal of developmental variations, even within humans, initially presents a seemingly bewildering array of timing changes. Some individuals are accelerated in some stages or traits but delayed in other stages or traits. Other individuals may show early onset of some traits but late onset of other traits. A main advantage of the heterochronic approach outlined here is the way it can hierarchically categorize all these confusing patterns into just a few basic groupings.

The two most basic categories reflect underdevelopment versus overdevelopment. We use these terms loosely to refer to developmental trajectories that are ultimately truncated or extended. This can refer to developmental trajectories of the whole individual or also to developmental trajectories of organs and behaviors within an individual. As shown in Figure 8-2, the heterochronic category that has historically described underdevelopment is *paedomorphosis* (literally, "child formation"). If all aspects of the individual's development are paedomorphic, then the result is an adult that looks and behaves like a juvenile of the ancestor. Paedomorphosis, also called juvenilization or *terminal truncation*, thus produces individuals that are adults in being sexually mature, but otherwise retain the ancestral juvenile's morphology and behavior.

Many domesticated pets, such as cats and dogs, are paedomorphic relative to their wild ancestors (McNamara, 1997). This is because humans have selectively favored smaller, cuter (more juvenilized) morphologies and more playful behaviors (Morey, 1994).

The three timing processes that produce paedomorphosis are:

1. *Progenesis*, or earlier offset. Offset refers to termination of development. By terminating development sooner, an organism (or trait or stage) may become underdeveloped.

2. *Postdisplacement*, or late onset. By starting development later, the organism (or trait or stage) may become underdeveloped if growth rate and offset time remain unchanged.

3. *Neoteny*, or slower rate of development. By retarding developmental rate, the organism (or trait or stage) may be underdeveloped if onset and offset time are unchanged.

These three processes are graphically illustrated in Figure 8-3. This more clearly shows how development of a trait can be altered in just these three ways. Note a key distinction: Only neoteny actually alters the rate, or velocity, of development throughout the duration of development. Progenesis and postdisplacement result from changes at a single point in time without direct influence on the developmental rate per se. This is not trivial because people often confuse these, such as by saying that progenesis is retarded development, or neoteny is delayed development. In fact, changes in rate are qualitatively distinct from changes in onset and offset times that development begins and ends. This leads to the complex process discussed below whereby a descendant ontogeny shows a combination of changes, such as showing slower rate combined with earlier or even delayed offset. Note also that the trait (y-axis) in Figure 8-3 can refer to developmental cognitive stages, such as those of Piaget (discussed further below).

As shown in Figure 8-2, the heterochronic category that has historically described overdevelopment is *peramorphosis* (literally, "beyond formation"). If all aspects of the individual's development are terminally extended (peramorphic), then the result is a juvenile that looks and behaves like an adult of the ancestor. Conversely, the juvenile continues to develop "beyond" the developmental point at which the ancestral adult stopped so that the descendant peramorphic adult is "overdeveloped."

For simplicity, we will usually refer to peramorphosis as *terminal*

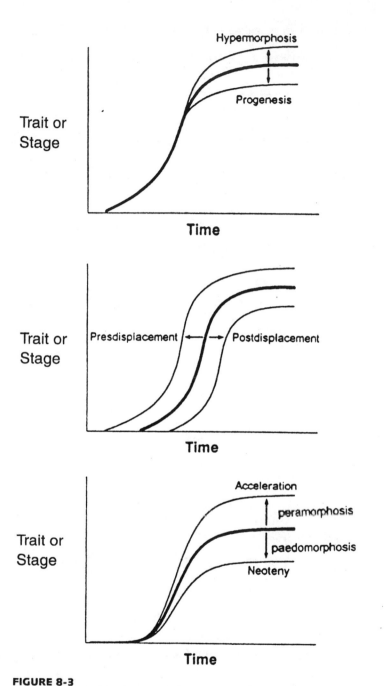

FIGURE 8-3
Six basic patterns of heterochrony when seen as minor modifications of simple growth curves. (McKinney & McNamara, 1991). Used with permission of the publisher.

extension. We prefer the phrase *terminal extension* over the historical phrase *terminal addition* because extended development is often not an additive process. Development is a highly complex process with many kinds of cellular interactions. Terminal extension of development may therefore involve much more than simply adding new cells or tissues. Terminal extension of the same trajectory can produce morphological, behavioral, or cognitive traits that are not simple linear extrapolations of ancestral traits. For example, terminal extension of the complex neuronal interactions during brain growth has produced neuronal interconnections in the human brain that are much more complex than would be produced from strictly additive processes (see chapter 12). This, in turn, has resulted in a rich repertoire of behavioral and cognitive skills that are sometimes not readily predictable from observing ancestral primate behaviors and cognitions.

Among the most common examples of terminal extension are species with large body sizes. Such size is often a result of delayed maturation (prolonged growth as embryos and juveniles) so that the organism follows the same developmental trajectory as an ancestor but for a longer period. This also causes many individual morphological and often behavioral traits to "overdevelop" too. One of the most famous examples is that of the "Irish Elk," the largest known deer. This organism had the largest known set of antlers because they, along with other organs, extended their growth trajectories as the elk grew to larger overall body size (McNamara, 1997).

Three timing processes produce peramorphosis (terminal extension) (see also Figures 8-2 and 8-3):

1. *Hypermorphosis,* or later offset. By terminating development later, an organism (or trait or stage) may become overdeveloped.

2. *Predisplacement,* or early onset. By starting development sooner, the organism (or trait or stage) may become overdeveloped if growth rate and offset time remain unchanged.

3. *Acceleration,* or faster rate of development. By accelerating developmental rate, the organism (or trait or stage) may become overdeveloped if onset and offset time are unchanged.

All the timing changes above can also apply to behaviors as well as cognitive traits and stages. As we discuss shortly, humans are in many ways overdeveloped apes in this respect.

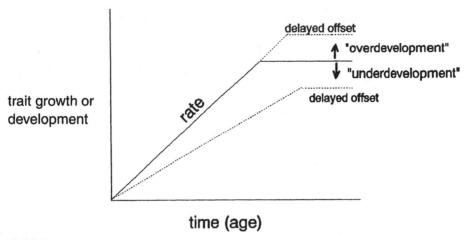

trait growth or
development

rate

delayed offset

↑ "overdevelopment"

↓ "underdevelopment"

delayed offset

time (age)

FIGURE 8-4
Changes in development or growth of a trait during ontogeny, where rate is slope of the line. Offset, or termination, of growth occurs where the line levels off. Note how overdevelopment can occur if delayed offset in the descendant is accompanied by increased rate of growth.

COMPLEX HETEROCHRONIC PATTERNS

Many, if not most, ontogenetic changes in evolution apparently occur via combinations of the simple patterns shown in Figure 8-3. Evolution often shows heterochronic combinations that involve alterations of both rate and offset time, (e.g., McKinney & Gittleman, 1995; McKinney & McNamara, 1991). This has contributed to the confusion in heterochronic studies because a descendant species may have a slowed growth rate (neoteny) yet become overdeveloped (peramorphic) because it has a later offset of growth (hypermorphosis), as shown in Figure 8-4.

The evolutionary tendency toward heterochronic combinations is especially relevant for human evolution because, as discussed below, humans seem to have a generally slowed rate of morphological growth but grow for a much longer period relative to ancestral ontogenies. This results in an overdeveloped (peramorphic) morphology, not an underdeveloped one that would occur from slowed growth rate alone.

GLOBAL HETEROCHRONY

All of the changes in developmental timing discussed above can affect the development of most or all of the entire organism. Such cases have been called *global heterochrony*, because they are "global" in terms of their effects on the individual (McKinney & McNamara, 1991). If, for example, sexual maturation is delayed by a few years, then many tissues and organ systems may continue to grow during that time. Specifically, this case could be called global hypermorphosis because it results from delayed offset of the juvenile phase. This would lead to terminal extension of growth and development of the entire organism. This, in turn, could cause larger body and brain size. Similarly, global progenesis via early maturation can occur; global acceleration via rapid body growth and other changes are all well documented in evolution (McKinney & McNamara, 1991; McNamara, 1997). Miniaturization in salamanders and other groups is a common example of change via global progenesis that simultaneously affects many aspects of growth (Hanken & Wake, 1993).

Global heterochrony is a common process because it is a developmentally easy way to produce an organism with organs and traits that are already coadapted to each other and to their environment. The proportional changes that are extended are thus preadapted to produce interacting organs or tissues that natural selection has shown to be functionally integrated and efficient. Thus, in the Irish Elk there was a proportional extension of growth in body and antler dimensions that, in theory, permitted the individual to maintain morphological dimensions that were physiologically and ecologically adaptive (McNamara, 1997).

Global heterochronies are often associated with the evolution of life histories (see chapter 7). Life history traits include body size at birth, age at maturity, number of offspring, life span, and a large number of other traits that affect reproductive output (Stearns, 1992). Life history theory has attempted to produce general principles that interrelate these traits to each other and explain their evolution in terms of natural selection. This has proved to be a very complex task because we have discovered that many factors influence the evolution of life stages (Stearns, 1992). However, as explored by McKinney and Gittleman (1995), global heterochrony is one of the major mechanisms by which life history evolves.

Of special importance for this book is that human (and indeed higher primate) evolution has largely involved the sequential delay of the major life history stages, ranging from the embryonic and infancy, through adolescence and old age. As discussed below, this has not only caused the extended growth of body and brain size, but it has allowed for prolonged learning periods. Such sequential delays in the termination of developmental stages is called *sequential hypermorphosis,* a term originally described in McNamara (McKinney, in press-a, in press-b; also see McKinney & McNamara, 1991; McNamara, 1997).

DISSOCIATED HETEROCHRONY

Developmental timing changes need not affect all, or even most, of the individual's development. Instead, most heterochronic changes in evolution affect only certain tissues or organs. These have been called *dissociated heterochrony* because the developmental timing changes of the affected tissues are separated or dissociated from the development of other tissues. In humans, for example, dental and facial growth is generally retarded whereas body and brain is terminally extended (Shea, 1989). Such changes have traditionally been called *mosaic evolution*. Within the brain, a well-known example is a disproportionate increase of the prefrontal cortex in hominoid evolution (see chapter 12).

Dissociated heterochronies are enormously important mechanisms for reducing developmental constraints. Whereas patterns of developmental covariation clearly constrain evolutionary change, dissociated heterochronies indicate that these covariant patterns are not unbreakable. Species clearly differ in the amount of dissociation, or developmental flexibility, that they exhibit (Hall, 1992). The more often such dissociations occur, the more variation is provided for natural selection to act upon. As noted by Levinton (1988), successful adaptation to change often depends on the ability to decouple traits that were formerly integrated into a functional whole.

Furthermore, dissociated heterochronies are a major mechanism of *evolutionary innovations*, referring to exceptionally novel traits that permit major new adaptations. Hall (1992) and McKinney and McNamara (1991) describe, for example, how changes in the rate and timing of cell migration during early embryogenesis have led to major new adaptations in jaw, teeth, and facial morphology.

Cognitive and behavioral evolutionary innovations are also produced in this way. Dissociation in the rate and timing of cognitive and behavioral development permits "decoupling" to produce major cognitive and behavioral novelties. Michel and Moore (1995) argue that "delaying or accelerating the development of grammatical skills in humans will create differences in intelligence, social skills, emotional range, and other aspects of personality" (p. 118). Langer (1996) discusses how the synchronous alignment in humans of physical cognition with logico-mathematical (LM) cognition was a major evolutionary step that per-

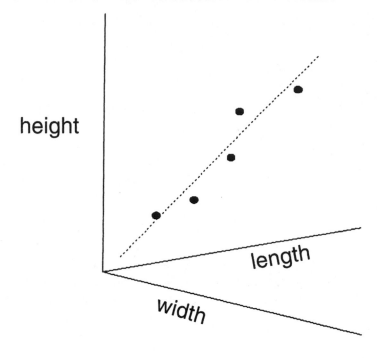

FIGURE 8-5
Three-dimensional view of plotting three variables that change during the ontogeny of a species. Points represent individuals at different ontogenetic stages.

mitted full information flow between these two key domains. This dissociated heterochronic shift toward early onset of LM cognition allowed a "logic of experimentation" and many other cascading cognitive interactions, not available in other primates where these two domains development asynchronously. Such *cognitive dissociations* are common in primate and human evolution, as discussed in chapter 9 and in Parker (1996b). These dissociations occur across and within a variety of different domains and serics. Similarly, one may identify behavioral dissociations in evolution as a consequence of decoupled shifts in rate or timing of development of suites of behaviors (McKinney & Gittleman, 1995; McKinney & McNamara, 1991).

ALLOMETRY

The anatomical shape changes caused by heterochrony are often called *allometry* (literally, "different measure"). Allometric extensions are often expressed in three dimensions. Figure 8-5 shows such an example; the three axes represent length, width, and height of an organ, such as the brain. Note that shape change does not occur if the slope of the line is equal to 1, which is called *isometry*. With isometry, the trait will increase in size, but shape will not change, if terminal extension occurs and the slope is equal to 1. But such a slope is very rare in complex (multicellular) organisms.

In the vast majority of cases, the slope is not equal to 1. Global heterochrony, including terminal extension, usually causes shape change. Simply by extension of growth, by one of the three processes above, the shape of an organ and, usually, the entire organism will change. This has crucial implications for behavior because terminal extension of brain development is the evolutionary process that produced human cognitive abilities. Indeed, the traits on the axes of Figure 8-5 can be behaviors so that terminal extension in growth of one or both traits can lead to behavioral allometry (McKinney & Gittleman, 1995). Similarly, by plotting cognitive developmental changes on Figure 8-5, we can depict cognitive allometry to describe the global or dissociated heterochronic rate and temporal shifts noted above in the evolution of various domains and series.

OVERDEVELOPMENT IN HUMAN EVOLUTION

A major theme of this book is that humans generally evolved by progressive overdevelopment (peramorphosis). This overdevelopment was not entirely global because many fine-scale dissociated heterochronies in morphology, brain, and cognition have also occurred. However, the general underlying pattern was one of extended development of life history stages, brain growth, and cognition (see chapter 7). The later offset (delay) in the termination of each stage of life history, somatic and brain growth, from embryogenesis to maturation to death is sequential hypermorphosis, noted above.

These sequential extensions of life history, somatic and brain growth, have permitted cognitive development to be extended well beyond that of ancestral cognition. These sequential delays have often been mistaken for the paedomorphic process of underdevelopment, but as in Figure 8-4, they are more properly viewed as producing overdevelopment. By pro-

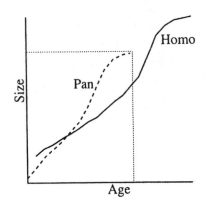

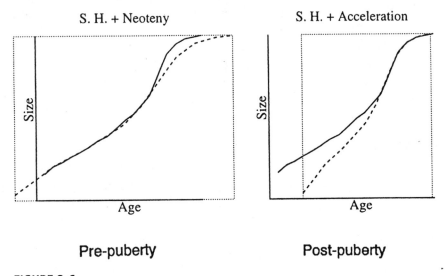

S. H. + Neoteny

S. H. + Acceleration

Pre-puberty

Post-puberty

FIGURE 8-6

Transformation of chimpanzee (*Pan*) ontogeny into modern human ontogeny, requiring sequential hypermorphosis (S. H.) (extending segments). Before sexual maturity, reduced growth rate is also required; after maturity, accelerated rate is also required. (Sean Rice, personal communication).

ducing a larger, more complex brain accompanied by longer life spans and learning periods, these progressive sequential extensions have progressively terminally extended the cognitive development of the human lineage. McNamara (1997, p. 293) thus depicts the evolution of humans as a *peramorphocline*, a common evolutionary pattern whereby descendant species become progressively overdeveloped.

Rice (1997) has recently improved upon past descriptions of human sequential hypermorphosis, depicted in Figures 8-4 and 8-6, with computer models that more accurately compare the complex growth curves of most organisms. Figure 8-6 shows how chimpanzee somatic (body) growth can be converted into a modern human growth curve by extending segments (e.g., stages) of the chimpanzee curve as in sequential hypermorphosis. Note, however, that a good fit is obtained only if the rate of growth is slowed (neoteny) in the chimpanzee before the growth spurt of sexual maturation and accelerated after the growth spurt. If chimpanzee growth is similar to that of ancestral humans of its adult size (e.g., B. H. Smith, 1992), then human overdevelopment has involved both sequential extensions of stages and change in growth rate within each stage.

Let us examine four types of sequential extensions in human evolution: life history, somatic, brain, and cognitive development.

Sequential Extensions of Life History Events. Sequential delays in nearly all life history events (e.g., fetal growth stage, sexual maturation, and death) are perhaps the most basic developmental changes that make us "human" (i.e., that affect our cognitive skills as discussed below). Sequential delays of stages permitted a longer time for growth (size) and development (complexity) of the brain and many somatic features. This led to increases in both organizational complexity and size of brain and body. Such sequential heterochronies are very common modes of evolution in the history of life (McKinney & McNamara, 1991).

The fossil evidence for such sequential ontogenetic delays in human evolution is quite extensive and growing rapidly. B. H. Smith (1992 and many references therein) has gathered considerable evidence on the evolution of human life history, especially maturation in fossil hominids. Comparison of early human tooth eruption and other dental patterns to those of living primates indicates a progressive delay in age of maturation from Australopithecines to modern humans. Life history patterns (e.g., timing of many life history events) of *Homo erectus,* for example, are intermediate between those of the great apes and modern humans

(Parker, 1996b; B. H. Smith, 1992). Bogin (1997) discusses evidence for prolongation of stages, especially childhood, in terms of modern patterns of human growth.

Sequential Extensions of Somatic Growth. Shea (1989, 1992) and many references therein) reviews morphological evidence against human evolution via juvenilization. This evidence indicates that the most prominent pattern in the morphological traits that make us most human is overdevelopment via terminal extension. For instance, human leg bones, which are crucial to the bipedal complex, are overdeveloped relative to apes and ancestral hominids, owing to longer growth. Body size, cranial dimensions, and other traits are also peramorphic (Shea, 1989, 1992).

Sequential Extensions of Brain Development. Gibson (1990, 1991b, and many references therein) discusses how the sequential delays in human developmental stages have been responsible for our larger, more complex brain (see also McKinney, in press-a, in press-b). Delayed offset of fetal growth, when all cortical brain cells originate, leads to a bigger brain (Deacon, 1997). Delay in offset of infancy and juvenile stages of development leads to a more complex brain as these stages of dendritic growth are extended. Most regions of the brain show a pattern in which dendrite and glial cell growth, synaptogenesis, and myelination have a longer time to occur, and thus go through—and extend beyond—ancestral developmental patterns.

Delayed myelination and synaptic function are well-studied examples of such terminal extension because they play key roles in maturing memory, intelligence, and language skills (Gibson, 1990; Konner, 1991). Myelination is highly canalized, following a specific sequence that is similar in many species. In rats, monkeys, and humans, brain stem areas myelinate before cortical areas, and within the neocortex, primary motor and sensory areas myelinate before association areas. But the rate and timing of this sequence vary among species; rhesus monkeys show myelination until at least 3½ years, but it persists in humans until after 12 years (Gibson, 1991b). Dendritic growth shows similar delays, persisting in humans until at least 20 years. Association (especially prefrontal) areas of the brain myelinate at different rates and times that correlate with development of cognitive and behavioral complexity (Deacon, 1997). From such evidence, Gibson (1991b) concludes that "on neurological grounds . . . there is nothing paedomorphic about the adult human brain." In direct contrast to the infant primate, adult human brains are "large, highly fissurated and myelinated with complex synaptic morphology." Deacon (1990a, 1990b) offers a similar inference.

Sequential Extensions of Cognitive Development. As discussed in earlier chapters, comparison of cognitive stages between humans and other primates indicates that monkeys and apes follow the same general sequence of cognitive development as humans. Humans, however, extend cognitive development and ultimately go beyond the development of modern apes and ancestral humans. The highest levels achieved by monkeys is about equal to that of 1-year-old children in the Piagetian sensorimotor period.

Apes do complete the sensorimotor period at about 3 years of age and go on to complete the first subperiod of preoperations, terminating their cognitive development at 8 or 10 years of age at a level approximating that of 3-or 4-year-old children in the preoperations period. Human cognitive development goes "beyond" that of other primates by development of additional modalities and increased number and complexity of schemes.

MECHANISMS OF DEVELOPMENTAL TIMING CHANGE

Developmental timing changes have traditionally been attributed to so-called regulatory genes (e.g., Gould, 1977). Small mutations in such genes were thought to produce large morphological changes by altering their regulation of developmental processes. In the case of human evolution, for instance, it is often noted (e.g., Weiss, 1987, Wills, 1993) that we share over 98% of our genome with chimpanzees and gorillas. Given this small difference, it is inferred that broadly acting "regulatory" genes must account for much, if not all, of our distinctiveness (Gribbin, 1988).

But it is now generally recognized that the phrase *regulatory gene* represents a great oversimplification (Levinton, 1988). There are many kinds of genes that regulate development, and these often act in different ways (Raff, 1996). Carroll (1995) discusses how timing (heterochronic) changes in homeotic gene expression at many levels in the gene network can produce morphogenetic variation. A useful concept is the *gene net,* which refers to the network of genes and their products that act to integrate yet "modularize" development (Bonner, 1988). This would explain the mechanisms by which dissociated heterochronies discussed above could occur. Slatkin (1987) has discussed how the quantitative genetics of heterochrony can be applied to understand how rate and timing of tissue growth are modularized and inherited (also see Hall, 1992).

Specific mechanisms of this modular growth regulation often seem to involve the diffusion across cell membranes of *morphogens*. These are complex biochemical signals that control cell multiplication and differentiation, often via diffusion gradients (Wolpert, 1991). Localized rate and timing changes in development of tissues and organs are thus often mediated by changes in rate and timing of morphogen production and quantity (McKinney & McNamara, 1991). On the other hand, global heterochronies are often due to alterations in rate and timing of production of growth hormones that are broadly acting throughout the body of the developing individual (McNamara, 1997). A good example in humans (and mammals in general) is the insulinlike growth factor I (IGF-I), which stimulates cell multiplication in many tissues and is circulated throughout the bloodstream (McKinney & McNamara, 1991).

MECHANISMS OF COGNITIVE TERMINAL EXTENSION

Brain development is thought to be influenced by about one-third of the human genome, or about 30,000 genes (Wills, 1993). Given that less than 2% of our entire genome differs from our closest primate relatives', mutations in only a small proportion of these genes must account for our cognitive skills. Although much work remains to be done, there is significant evidence that genes controlling the timing of the production of polypeptide growth factors, such as IGF-I noted above, are of great importance in determining tissue development, including the brain. These growth factors play a key role in initiating (onset) and terminating (offset) cell multiplication, cell migration, and cell differentiation (McKinney & McNamara, 1991; McNamara, 1997). Hormones are especially important as heterochronic regulators of neural development. They can act via a variety of mechanisms: directly on neuron growth and number, on connectivity, and on neural pruning (Michel & Moore, 1995).

An example of an important growth factor influencing timing of brain development is the insulinlike growth factor IGF-II, which is active during the fetal stage and stimulates brain tissue growth (McKinney & McNamara, 1991). Prolonged production of neurons is followed by prolonged stages (terminal extension) of axonal and dendritic growth (Finlay & Darlington, 1995; Gibson, 1991b). Similarly, there is an extension in the time at which neural competition begins to produce regressive loss of neurons. This allows an increase in neural connectivity before neural pruning begins (Edelman, 1987).

Later aspects of brain development follow a similar sequence of delays. As described above, different areas of the brain myelinate at different times, and in humans each of these areas tends to myelinate at a later time than in our living primate relatives (and immediate ancestors). These delays underlie the behavioral and cognitive terminal extensions discussed in the next chapter. In particular, areas within the advanced association (especially prefrontal) parts of the brain myelinate later and at different rates and times that correlate with development of cognitive and behavioral complexity (Deacon, 1997). It seems likely such myelination patterns occur from delayed production of a variety of biochemical mediators that govern them.

Such sequential prolongations of neuron multiplication, connectivity, and myelination mean that humans initially develop more slowly, which is one reason why the idea that humans evolved via juvenilization has been so popular (e.g., Gould, 1977; Montagu, 1989). Early sensorimotor development in humans, for example, is slower than in monkeys. But the pace of cognitive development accelerates so that early preoperational abilities are superseded in humans by about 4 years of age, whereas in great apes early preoperational abilities continue to develop until 7 or 8 years old. As McNamara (1997, p. 299) notes, whereas true juvenilization results in an underdeveloped adult, humans are neurologically and cognitively overdeveloped as adults.

ANTECEDENT VERSUS CONSEQUENT HETEROCHRONY

Determining the exact mechanisms underlying heterochronic changes is complicated by their hierarchical and sequential nature. This is not a major problem as long as we are specific about our level of explanation. We have just seen, for example, that genetic causes can be related to biochemical and thus to tissue-level changes. These, in turn, can be related to cognitive changes by describing the underlying sequence of changes in brain development.

But even when we focus solely on one spatial level in the hierarchy of change, such as the tissue or behavioral level, the sequential (temporal) nature of heterochrony can produce complexities that we must acknowledge in attempting to explain heterochronic mechanisms. Later changes in rate or timing are often produced by earlier changes.

An important example of this is the marked acceleration of the late sensorimotor and early preoperational stages of cognitive development in humans as compared with apes. Although this can certainly be justifiably described as acceleration, it is important to realize that this acceleration is a product of earlier heterochronic changes that may not be produced by acceleration per se. In particular, we noted that delay in the termination of early brain growth was a fundamental cause of the enhanced cognitive construction abilities that promote extended and accelerated cognitive development. In this case, we suggest the term *antecedent heterochrony* to refer to the earlier rate or timing change that produces later changes, which may or may not be of the same type. In this case, the delayed offset of brain growth is the antecedent heterochrony. In contrast, the later extension of cognitive development is a consequent heterochrony that results from the earlier change.

This distinction does not necessarily mean that antecedent heterochrony is more important. Rather, it seeks to identify which processes precede others and thereby produce a more complete explanation of developmental change. Other examples of this would be that, in sequential hypermorphosis, the late offset of one stage will automatically produce a late onset of the succeeding developmental stage. In this case, the antecedent heterochrony is the late offset of the earlier stage producing the consequent late onset of the next stage.

SUMMARY AND CONCLUSIONS

Diversity in the evolution of life has often been produced by modifications of developmental trajectories, or ontogenies. Early multicellular life, in the early Paleozoic era, apparently possessed developmental programs that were less constrained than modern organisms'. But as life has evolved, the basic body plans of all major groups have become progressively more constrained as interactions of early ontogeny became more entrenched. As a result of this increasing constraint, evolutionary change has tended to occur by modifying later parts of the developmental trajectory. Examples of such late-acting modifications include terminal additions and subtractions. Universal recapitulationism does not occur because early developmental changes are still possible, as are terminal subtractions and other nonadditive terminal changes.

When late-acting developmental modifications occur, they usually involve developmental timing changes (heterochrony). The two basic categories of heterochronic change are: (1) underdevelopment, also called paedomorphosis, juvenilization, or terminal truncation, and (2) over-development, also called peramorphosis or terminal extension. *Termi-*

nal extension, which has been especially important in human evolution, is a much better description than the traditional phrase of *terminal addition* because development is often not an additive process. When development is prolonged, such as in extended human brain growth and maturation, the behavioral and cognitive extrapolations may be more than additive. Cognitive abilities, for instance, may multiply rapidly or become enhanced in ways that are perceived as qualitative advances.

Heterochronic changes can be simple or complex, global or dissociated. Global heterochronies affect most of all of the individual, as in a miniatured descendant, which can be underdeveloped (juvenilized) in many aspects of morphology and behavior. Global heterochronies are often associated with changes in life history traits. A good example is the sequential hypermorphosis (sequentially delayed offset) of much human development.

Dissociated heterochronic changes affect only some of the individual's morphological, behavioral, or cognitive traits. Such changes are very important as mechanisms for morphological, behavioral, and cognitive evolutionary innovations because they reduce developmental constraints on those traits. Advances in human cognition may well have occurred via such cognitive dissociations allowing the flow of information among formerly temporally isolated domains. The shape changes produced by heterochronic shifts are called *allometry. Cognitive allometry* refers to the relative alterations of cognitive domains relative to one another.

Sequential terminal extension has been an underlying theme in human evolution. Four key aspects of human evolution display terminal extension via sequential hypermorphosis (sequential delays of all developmental stages): life history, somatic morphology, brain development, and cognitive development.

Such developmental timing changes are ultimately caused by alterations in complex genetic interactions called gene nets and their biochemical by-products such as morphogens. A key example of such an alteration in human evolution has been change in timing of the production of certain growth factors, such as IGF. In identifying these underlying heterochronic mechanisms we must often account for the hierarchical and sequential nature of developmental evolution.

A faithful application of the traditional idea of recapitulation to psycho-physical behavior would mean that as the individual develops it will rehearse in a chronological way the chief stages of psycho-physical evolution in the phylogeny, except as this has been deranged by direct or cenogenic modifications. There rests upon the advocate of this view, then, the necessity of showing that the sequence of developmental stages is really representative of the racial sequence, and that the constituent phases in the progression are truly homologous with the ancient stages and not merely analogous or convergent phenomena.

DAVIDSON, 1914, P. 78

CHAPTER 9
THE EVOLUTION OF
HUMAN MENTAL DEVELOPMENT

This chapter focuses on the evolutionary origins of patterns of cognitive development displayed by living hominoids. It uses evolutionary techniques of examining the distribution of these patterns among closely related living primate species to see when and in whom they originated. This analysis demonstrates that hominoid cognitive development evolved primarily through terminal extension and acceleration resulting in developmental recapitulation.

THE CLADISTIC TECHNIQUE OF CHARACTER RECONSTRUCTION

The cladistic approach to evolutionary reconstruction described below differs significantly from the classical approach of comparative psychology. Because classical comparative psychologists worked in the laboratory to identify universal laws of learning, they focused on *model organisms* that were convenient for controlled research. Over the years, this resulted in the use of a few species that were hardy and easy to reproduce and house—the pigeon, the white rat, and the rhesus monkey (Beach, 1965). In contrast, ethologists and other evolutionary biologists focus primarily on reconstructing the evolution of new character states. Because identification of differences and similarities within a group of closely related species is the first step in this process, model organisms have little to offer in this regard (Raff, 1996).

The cladistic techniques involved in evolutionary reconstruction can be illustrated with a simple example. We can reconstruct the evolutionary origins of the form of primate locomotion known as brachiation, by mapping various locomotor patterns of primate species onto a family tree of primates. First, note that all of the living apes display the *brachiation complex* (the features that facilitate two-handed suspensory hanging and locomotion under branches) and none of the primates in the next most distant branch, the Old World monkeys, display this complex. Then, draw a line under the common ancestor of all the apes, which represents the origin of the brachiation complex. This process is called *character mapping.* Figure 9-1 presents a diagram for character mapping of brachiation and other modes of primate locomotion.

Mapping characteristics onto an family tree to reconstruct their common ancestry reverses the procedure phylogenetic taxonomists use to construct the tree in the first place. This other process, which we call *tree construction,* is the primary function of cladistic methodology, that is, determining phylogenetic relations among species. It involves mapping character states of hypothetically related species onto alternative branching diagrams (*cladograms*) until the most parsimonious tree is found. (The most parsimonious tree is the one that requires the fewest evolutionary transformations.) Based on the most likely evolutionary tree, phylogenetic taxonomists identify sister species on adjacent branches and classify them into higher taxonomic groups such as genera and families based on common ancestry (Hennig, 1979; Ridley, 1986; Wiley, 1981).[1]

Hennig (1979) argued that evolutionary trees (cladograms) must be based on the distinction between (1) homologous characters that are uniquely shared (*shared derived character states*) by a given group of sister species owing to their origin in a recent common ancestor and (2) homologous characters that are broadly shared by a larger group of species (*shared character states*) owing to their ancient origin in a distant common ancestor. So, for example, the character state of brachiation, which occurs in all the living apes but none of the Old World monkeys, is a shared derived character state among the apes (the ingroup) relative to the Old World monkeys (the outgroup). The same characteristic may be shared, shared derived, or uniquely derived depending upon the frame of reference. Brachiation, for example, was a uniquely derived character

1. The procedures described here are those developed by the school of evolutionary taxonomists known as *cladists.* The branching diagrams they construct—cladograms —differ from conventional family trees in that all species are represented at the ends of branches. None are represented at the nodes as they sometimes are on phylogenies.

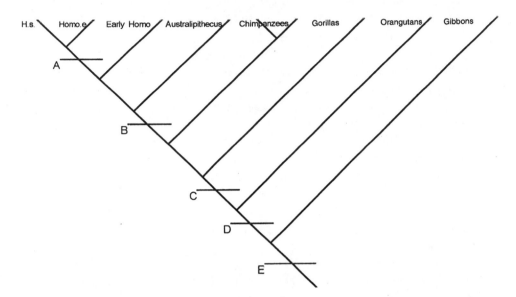

FIGURE 9-1
Character mapping for brachiation and other forms of locomotion in primates:
A = fully terrestrial bipedalism, B = semiarboreal bipedalism, C = knuckle
walking, D = fist walking, E = brachiation.

in the common ancestor of all the apes, it is a derived character state
among all living apes, and a shared character state among living great
apes because it is also present in lesser apes.

Homologous characteristics contrast with characters that have arisen
in only one species (*uniquely derived character state*). They also contrast with characters that have evolved through convergent evolution
in distantly related species who did not inherit them from their common ancestor (*homoplasies*) (Wiley, 1981). Bipedal locomotion in birds
and hominids is a homoplasic character because it was not present in
the common ancestor of birds and mammals (and is only present in one
derived group of mammals).

Biologists distinguish between shared and shared derived characters
by comparing character states of closely related species with those of
more distantly related species. So, for example, the character state of
brachiation, which occurs in all the living apes but none of the Old
World monkeys, is a shared derived character state among the apes (the
ingroup) relative to the Old World monkeys (the outgroup). The same

characteristic may be shared or shared derived or uniquely derived depending upon the frame of reference. Brachiation is a shared character state among great apes because it is also present in lesser apes.

Determining whether character states shared by various species are homoplastic, shared, or shared derived is the key to constructing evolutionary trees. Shared derived character states indicate closeness of phyletic relationship, whereas shared or homoplastic character states do not. Evolutionary trees are constructed on the basis of shared derived character states because only shared derived character states indicate common ancestry.

Classically, the characters that evolutionary biologists have used to construct trees have been anatomical features. More recently they have used comparative molecular data, especially from DNA, for the same purpose. Whatever character states evolutionary biologists use to construct trees are assumed to have a genetic basis. Since before Darwin's time, evolutionary biologists have recognized that developmental patterns as well as anatomical structures are species-typical characters that have a hereditary base. Modern evolutionary biologists call these developmental patterns *life history patterns.*

Once tree construction is complete, the cladistic method can be reversed to do character mapping, that is, to discover the ancestry and sequence of evolution of character states. This can be done by *mapping character states onto a previously constructed tree to see which common ancestor gave rise to shared derived characters.* In order to avoid circularity, only characteristics other than those used to construct the tree can be mapped on it to reconstruct common ancestry (Brooks & McLennan, 1991). It is important to note, however, that improbable evolutionary sequences revealed by character mapping may stimulate reexamination of the process of tree construction (McLennan, Brooks, & McPhail, 1988; Wiley, 1981).[2]

Classically, ethologists have used instinctive behaviors, particularly among insects and birds, as character states to infer common ancestry and reconstruct the evolution of specific behaviors in related species

---

2. Cladistic procedures for reconstructing common ancestry of a character are similar to ethological analyses of behaviors in related species. Indeed ethology arose as a technique for using behavioral characters for assessing phylogenetic relationships (Lorenz, 1950; McLennan et al., 1988). The two approaches offer much to each other: Cladism offers a systematic approach to reconstructing phylogeny. Ethology offers knowledge of characters that would seem to bring considerable insight into phylogenetic relationships (McLennan, Brooks, & McPhail, 1988).

(Lorenz, 1950). Like tree construction, character mapping begins with identification of differing versions of homologous structures in closely related species (contrasting character states). The versions of a characteristic displayed by closely related species are compared to versions in more distantly related species: If molar teeth is the character, then different molar numbers or shapes could represent different moral character states. If locomotion is the character, then quadrupedalism, brachiating, knuckle walking, and bipedalism represent different locomotor character states in anthropoid primates. These are all examples of *simultaneous character states*, that is, character states that can be measured in a particular instant of time.

Comparative data can be used in two ways to reconstruct common ancestry. First, it can be used to reconstruct the evolution of terminal stages of development by mapping the highest developmental stage achieved by each species onto our family tree. This process is called *simultaneous character mapping*, because it uses simultaneous character states.

Second, it can be used to reconstruct the evolution of development by mapping the ontogenetic sequences (e.g., all the stages of development and their timing) of each species onto our family tree. This process is called *ontogenetic character mapping*. These character states are called *ontogenetic character states* because they cannot be measured in a particular instant of time, but only over a span of time (de Queiroz, 1985). Because we are interested in reconstructing the evolution of cognitive development, we will use the second approach.

The developmental stages of cognition in monkeys, great apes, and humans described in preceding chapters constitute the ontogenetic character states in our character mapping. Cognitive development is an appropriate character because it is a life history feature that is solidly grounded in maturation of brain function. It is important to note, however, that brain maturation depends upon feedback generated by the behavior of the developing organism and the behavior of conspecifics during sensitive periods of development (see chapter 12 for discussion).

RECONSTRUCTING THE EVOLUTION OF COGNITIVE DEVELOPMENT IN HOMINOIDS USING ONTOGENETIC CHARACTER MAPPING

The first step in reconstructing the common ancestry of various patterns of cognitive development (character mapping) is to select a previously derived family tree of the focal species. In this case this is a phylogeny

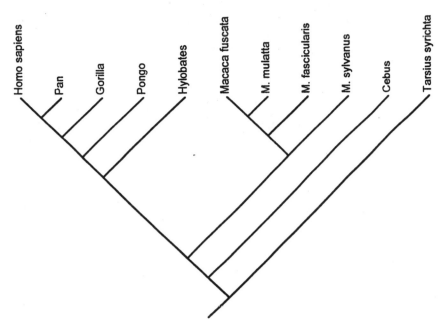

FIGURE 9-2
Primate phylogeny (D. Maddison & Maddison, 1992).

of living primates and fossil hominids on which characters states can be mapped (see Figures 9-2 and 9-3).

The second step is to identify the diagnostic character states in the focal species. in the case of our developmental reconstructions, *the entire cognitive developmental sequence in each group rather than its terminal stage is its ontogenetic character state.* Ontogenetic transformations, the timing of stages, and the temporal relationship among domains are the most significant parts of the character state. Instantaneous character states are incomplete characters that are only parts of ontogenetic character states (de Queiroz, 1985).[3]

3. Henning (1979) recognized several criteria in addition to outgroup comparisons for distinguishing derived from shared characters. These included (1) geological precedence (i.e., earlier appearance in the fossil record of a monophyletic group), (2) "chorological progression" (i.e., geographic/spatial parallels with branching sequences), and (3) ontogenetic character preference" (i.e., earlier appearance in ontogeny). De Queiroz (1985) rejects the ontogenetic character preference criterion for determining polarities, relying solely on phylogenetic polarities determined by the outgroup method. This approach prevents circularity in arguing from ontogeny to phylogeny and then back to ontogeny.

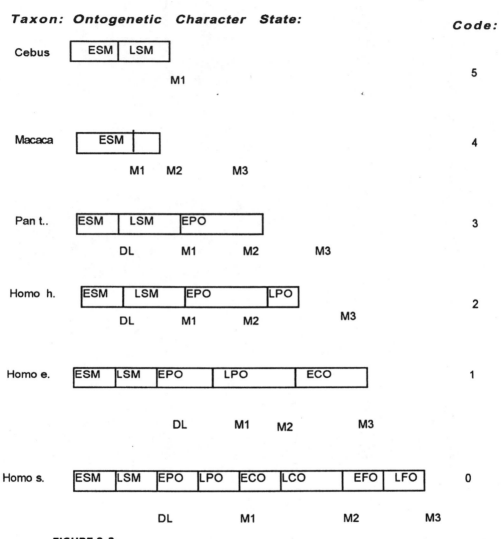

FIGURE 9-3

Ontogenetic character states for cognitive patterns in selected primates.

Cognitive development occurs in the context of dental and brain maturation. Therefore, we use *the pattern of cognitive development relative to molar development* in humans, hominids, great apes, macaques, and cebus monkeys as our ontogenetic characters. Figure 9-4 depicts these characters.

Figure 9-4 reveals the following pattern: First, macaques complete their maximum development at the third-, fourth-, or fifth-stage sensorimotor period, depending on the domain, before the eruption of the second molar. Second, great apes complete their maximum development at the symbolic/preoperational subperiod shortly after the eruption of their third molar. Third, Homo erectines probably completed their maximum development at the early concrete operational period before the eruption of the third molar. Fourth, humans complete their maximum de-

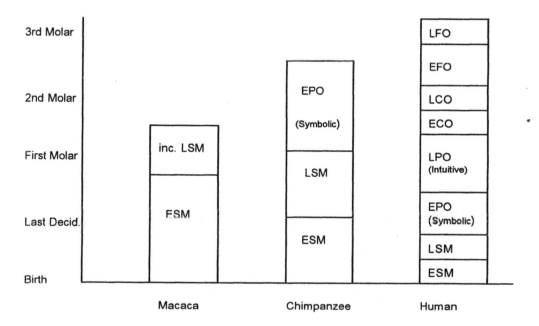

ESM= Early Sensorimotor; LSM= Late Sensorimotor,
EPO= Early Preoperations, LPO= Late Preoperations,
ECO= Early Concrete Operations, LCO= Late CO
EFO= Early Formal Operations, LFO= Late Formal Operations

FIGURE 9-4
Cognitive development relative to molar maturation in macaques, chimpanzees, and humans.

velopment at the formal operational period before the eruption of the third molar.

Eruption of the first molar occurs at approximately 1⅓ years in macaques, 3 years in chimpanzees, and 5½ in humans. Eruption of the second molar occurs at approximately 3 years in macaques, 6½ years in chimpanzees, and 12½ years in humans, and eruption of the third molar occurs, respectively, at 6 years, 10½ years, and 18 years (B. H. Smith, 1992).

Figure 9-4 illustrates the following developmental pattern: The longer the interval between eruptions and hence the longer the period of molar development, the greater the number of subperiods of cognitive development traversed. Macaques, whose molars erupt in a 1-, 3-, 6-year sequence, traverse only one and a half developmental subperiods. Chimpanzees, whose molars erupt in a 3-, 6-, 10-year sequence, traverse three developmental subperiods. Homo erectines, whose molars erupted in a 4-, 9-, 14-year sequence, apparently traversed four and a half or five developmental subperiods. Humans, whose molars erupt in a 6-, 12-, 18-year series, traverse seven or eight developmental subperiods (see chapter 7 for more detailed descriptions).

Molar eruption sequences provide an age-independent standard for comparing development. As the foregoing data indicate, however, molar eruption sequences can be translated into ages of development in the various species. Age data give important clues to understanding the evolution of the pace of development.

Macaques apparently complete stages 3, 4, and 5 of the various sensorimotor-period domains by about 2 years of age, that is, about 2 years before sexual maturity. Great apes complete all six stages of the sensorimotor period in all domains by 3 or 4 years of age. They traverse the first few stages (first through fourth) of most series at about the same rate as human infants, but they traverse the later (fifth and sixth) stages at markedly slower rate than human infants in the space and causality series in the physical domain, but at about the same rate in the object concept domain.

They also develop more slowly than human infants in classification and logic in the logical-mathematical domain. They develop in the domain of gestural imitation in the social domain at almost the same rate as human infants but more slowly in pretend play. They go beyond the sensorimotor-period stages and traverse the first subperiod of the preoperations period, the symbolic subperiod. They probably have completed their cognitive development by the beginning of adolescence at 7 or 8 years.[4]

Humans, typically, complete the sensorimotor period at about 2 years, about 2 years earlier than great apes do. They complete the symbolic sub-period at about 4 years, approximately 4 years before great apes do. They go on to traverse four or five more subperiods of cognitive development by the age of 16 or 18 years, at the end of adolescence. The higher the developmental subperiod, the greater the variation in age at completion.

Comparative data on the cognitive abilities and life history of fossil hominids suggest that Homo habilines may have achieved late preoperational thinking and Homo erectines achieved early concrete operational thinking in the spatial domain (see chapter 6). Their dental development suggests that they had completed their highest level of cognitive development by 14 years. Comparative data from great apes suggest that they must have completed the late sensorimotor period by age 3.

Mapping ontogenetic patterns (as represented by the ontogenetic character states) onto the primate phylogeny reveals the following patterns. First, the macaque ontogenetic pattern (ontogenetic node 4) originated sometime before the common ancestor of macaques. Lacking comprehensive data on New World monkeys and other Old World monkeys, we cannot be sure where it originated. Given the report that langur monkeys display a pattern similar to the macaques' (Chevalier-Skolnikoff, 1983), we suspect that this pattern originated in the common ancestor of all Old World monkeys.[5]

---

4. It is important to note that great apes show various lacunae and limitations in their sensorimotor and symbolic abilities relative to those of human infants and children. These lacunae are most notable in three areas: First, although they imitate gestures and some actions on objects, great apes lack the capacity for vocal imitation. Second, although they show secondary circular reactions typical of the third sensorimotor stage and tool use and other behaviors characteristic of the fifth stage, they show incomplete tertiary circular reactions or testing the properties of objects in relation to other objects and physical forces. Third, and related to this, they show relatively little ability to detach objects from their actions on those objects and hence to create relations among objects. This is reflected in their tendency to make mobile rather than stable constructions with objects and to use their own bodies rather than substrates as a base for objects.

5. Although there are several reports of tool use in captive Old World monkeys, for example, hamadryas baboons and lion-tailed macaques, these performances are highly scaffolded and probably do not involve fifth-stage understanding of the relationship between the tool and the goal and the dynamics of movement of the tool relative to the goal. These performances seem rather to involve fourth-stage instrumental activities based on an attempt to contact the goal with the stick.

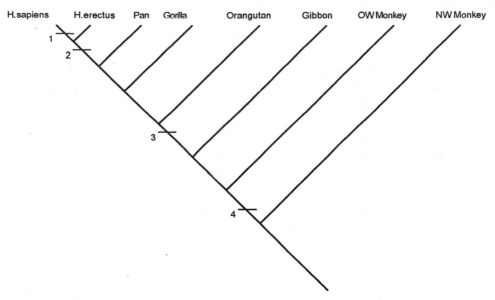

FIGURE 9-5
Mapping phyletic origins of cognitive patterns.

The ontogenetic pattern of cebus monkeys is atypical of New World monkeys as well as Old World monkeys. For this reason we do not map it onto the common ancestor New World monkeys. Anecdotal reports on other New World monkeys suggest that, like Old World monkeys, they lack intelligent (fifth-stage) tool use. This ability seems to be unique to cebus monkeys. Confirmation of this prediction must await cognitive studies of other New World monkey species.

Second, the great ape ontogenetic pattern (node 3) apparently originated in the common ancestor of the great apes. (More data on the lesser apes are needed to confirm this.) Third, the Homo erectine ontogenetic pattern (node 2) originated in the common ancestor of that species and *Homo sapiens.* Fourth, the human pattern (node 1) originated in a common ancestor after the common ancestor of *Homo erectus* and *Homo sapiens.* Figure 9-5 depicts these phyletic events.

Discovering the ancestry of patterns of cognitive development is important for two reasons. First, it illuminates the nature of heterochronic changes entailed in the evolution of cognitive development in hominoids. Second, it helps us generate and test alternative hypotheses concerning the adaptive significance of various cognitive abilities (see chapter 10).

Examination of Figure 9-5 implies several points. First, and most obviously, it shows that the addition of new subperiods of cognitive development (i.e., early and late preoperations, early and late concrete operations, and formal operations) occurred through *extensions* of previous subperiods during hominid evolution. Within these subperiods, it shows *addition of new stages*, for example, the evolutionary *extension* of primary circular reactions (stage 2) present in Old World monkeys, into secondary circular reactions (stage 3) and incipient tertiary circular reactions (stage 5) in great apes. It also shows *consequent preterminal extension and elaboration* of all three kinds of circular reactions in hominids.

Second, it implies *earlier onset* and *acceleration* in the rate of development of the late sensorimotor and early preoperational subperiods in humans as compared to great apes. These changes apparently occurred concomitant with the addition of later stages of cognitive development in hominid ancestors. This explains why human infants complete the sensorimotor period by 2 years of age and the symbolic subperiod of preoperations by 4 years of age while great apes complete the sensorimotor period by 4 years of age and the symbolic subperiod by about 6 or 7 years of age.

Third (not represented in Figure 9-5), various domains changed their developmental pace relative to one another, *synchronizing* during hominid evolution. The fifth and sixth stages of the causality, space, and imitation were *realigned* relative to the object concept series so that all the series develop synchronously. Late sensorimotor period logical development also synchronized with other sensorimotor domains. These changes occurred through *acceleration* of development in the causality, spatial, and logical domains relative to the object concept, and earlier onset of logical development.

Fourth, also apparent from comparative data, evolutionary changes in rates of development were associated with increased richness and breadth of expression of particular abilities. These *preterminal elaborations* included (1) extension of imitation from the facial and gestural modalities to the auditory/ vocal modality, (2) increases in the number and kinds of object manipulation schemes, (3) increases in the numbers and kinds of each class of circular reactions, and (4) increases in the spatial and temporal stability of constructions with objects.

TABLE 9-1   CHARACTER STATE STEP MATRIX OF OBSERVED AND INFERRED ONTOGENETIC CHARACTERS OF PRIMATE TAXA

Ontogenetic characters	Homo sapiens	Archaic H. sapiens	Homo erectus	Homo habilis	Pan troglodytes	Hylobates	Macaca
1234567	1234567	123456	12345	1234	123	12	1
1234567		−1	−2	−3	−4	−5	−6
123456	+1		−1	−2	−3	−4	−5
12345	+2	+1		−1	−2	−3	−4
1234	+3	+2	+1		−1	−2	−3
123	+4	+3	+2	+1		−1	−2
12	+5	+4	+3	+2	+1		−1
1	+6	+5	+4	+3	+2	+1	

Notes. Pairwise distances indicate hypothetical distances and directions of transformation during ontogenies. Key: ESM-LSM-EPO-LPO-ECO-LCO-EFO = 1234567; ESM-LSM-EPO-LPO-ECO-LCO = 123456; ESM-LSM-EPO-LPO-ECO = 12345; ESM-LSM-EPO-LPO = 1234; ESM-LSM-EPO = 123; ESM-LSM = 12; ESM = 1. ESM = Early sensorimotor, LSM = Late sensorimotor, EPO = Early preoperations, LPO = Late preoperations, ECO = Early concrete operations, LCO = Late concrete operations, EFO = Early formal operations.

Realignment among domains apparently generated more powerful cognitive constructions, which fueled the extension of cognitive development. It may have been a driving force in the elaboration of imitative and circular schemes. These cognitive constructions created new relationships among objects, which stimulated the capacity to reflect on these relationships.

Given the definition of ontogenetic character states (see Figure 9-4) and the reconstruction of their common ancestry (Figure 9-5), we can create a diagram of the hypothetical steps involved in the evolution of the ontogenetic character states. Table 9-1 presents a step matrix representing hypotheses about the order and distances among these ontogenetic character states as recommended by Mabee and Humphries (1993).

If each stage in the evolution of ontogenies entailed terminal addition of a new subperiod of ontogenetic development as Table 9-1 suggests, then a certain number of phylogenetic events are implied. Specifically, this model implies the following scenario:

1. The common ancestor of Old World monkeys had evolved an ontogeny that extended into early sensorimotor period in most domains and into the fifth stage in a few domains.

2. The common ancestor of lesser apes and great apes probably had

evolved an ontogeny that extended into the late sensorimotor period in most domains.

3. The common ancestor of the great apes evolved an ontogeny that extended into the sixth stage of the sensorimotor period and into the early preoperations period.

4. The common ancestor of hominids probably had evolved an ontogeny that consolidated the early preoperations-period achievements.

5. The species preceding Homo erectus (H. habilis or H. rudolfensis) probably had an ontogeny that extended into late preoperations.

6. Homo erectus had apparently evolved an ontogeny that extended into early concrete operations, archaic Homo sapiens had evolved an ontogeny that extended into late concrete operations, and finally modern Homo sapiens evolved an ontogeny that had the potential to develop into formal operations.

Although it is unclear if each evolutionary step involved one such transition, this model generates testable hypotheses. Specifically, it suggests that (1) archaic Homo sapiens should have displayed development to late concrete abilities; (2) living lesser apes, gibbons and siamangs, should display development to late sensorimotor abilities (fifth and perhaps sixth stage); and (3) New World monkeys (with the exception of cebus) should have a less extended ontogeny than New World monkeys. Alternatively, the common ancestor of both groups may have evolved the pattern we see in macaques. (Variation among species within taxonomic groups is likely. We know, for example, that cebus monkeys display development to later sensorimotor abilities in more domains than macaques do. It also seems likely that baboons display development to later stages than macaques do.)

CONCLUSIONS ABOUT HETEROCHRONY IN THE EVOLUTION OF HOMINOID LIFE HISTORY AND COGNITIVE DEVELOPMENT

The foregoing comparative evolutionary analysis provides clues to the nature of heterochronic changes in cognitive development that must have occurred between the origins of hominoids and the appearance of modern humans. (Refer to chapter 8 for a primer on the categories of heterochronic change.)

The evolutionary changes in molar development entailed progressive

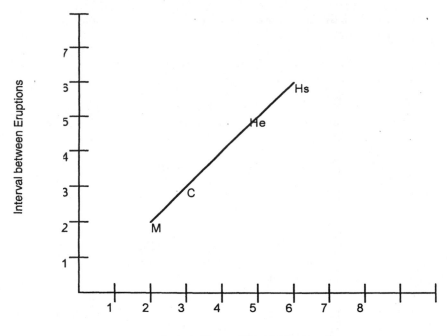

Ages in Years at First Molar Eruption

FIGURE 9-6
Rate of deceleration in molar development in macaques, chimpanzees, Homo erectines, and *Homo sapiens.*

delays in the onset of molar eruption and progressive increases in the intervals between molar eruptions. This pattern of heterochronic change is known as *sequential hypermorphosis.* It can be illustrated by plotting age of molar eruption against intervals between eruptions in years (see Figure 9-6).

Evolutionary changes in anthropoid brain development also involved progressive delays in the onset of various aspects of cortical development and progressive increases in the intervals between them (Gibson, 1990, 1991a, 1991b). In other words, the evolution of hominid brain development also reveals the pattern of *sequential hypermorphosis* (Gibson, 1995) (see chapters 8 and 12). Because of differences in neural and behavioral development, cognitive development showed a different pattern.

Evolutionary changes in cognitive development entailed sequential addition of new stages at the end of development through terminal extension and elaboration of preceding stages. It also entailed earlier onset and offset of each stage preceding concrete operations. This change entailed the following categories of heterochronic change:

1. *terminal extension* (of the late preoperations, concrete, and formal operations periods of cognitive development [time hypermorphosis])

2. *predisplacement*, that is, earlier onset and offset (of late sensorimotor and early preoperations subperiods of development

3. *acceleration* of development (of the sensorimotor and early preoperations periods [condensation or rate hypermorphosis])

4. *dissociated heterochrony* or realignment of particular series and domains relative to one another, resulting in synchronous development across series and domains in modern humans

In addition to these heterochronic changes, there was *consequent preterminal extension and elaboration* of schemes and operations and their breadth and application within series and domains.

In other words, the evidence favors the hypothesis that the evolution of cognitive development in hominids occurred through *overdevelopment* (adultification or peramorphosis) rather than *underdevelopment* (juvenilization or paedomorphosis). Because peramorphosis (overdevelopment through terminal extension) leads to ontogenetic recapitulation, this supports the old contention that human cognitive development recapitulates the stages of its evolution in ancestral forms (Baldwin, 1897, 1902, 1906; McKinney & McNamara, 1991; Parker, 1996b; Parker & Gibson, 1979).

It is important to note that the reasoning behind the neorecapitulation model for human mental development is not circular. It is not based on the sequence of ontogenetic stages in human infants and children. (It does not depend upon ontogenetic precedence.) Rather, as de Queiroz (1985) recommends, it is derived exclusively from mapping of comparative data on development onto a primate tree. In other words, it is based on outgroup comparison rather than the ontogenetic criterion. Significantly, Mithen (1996) has reached a similar conclusion using a different developmental model for the stages of the evolution of hominid cognition.

THE OVERDEVELOPMENT MODEL'S NONIMPLICATION OF PERMANENT CHILDLIKE BEHAVIOR IN GREAT APE ADULTS

Skeptics may object that adult great apes do not behave like 3-year-old children nor do monkeys behave like year-old children (as if that were the implication of the model). As counter examples, they might point to such adult activities as mating, maternal care, foraging, dominance, and territorial contests. Not only great apes and monkeys, but all other mammals—and indeed all adult vertebrates—engage in such basic life functions as maintenance, reproduction, and defense.

Vertebrates employ a wide variety of behavioral strategies in these basic activities. In most cases these strategies rely on species-typical instinctive behaviors and innately canalized learning (Lorenz's [1965] "innate schoolmarm"). Learning spans an enormous range of phenomena, from highly specific one-time learning through imprinting, to individual trial-and-error learning, to highly flexible forms of learning over prolonged periods through such sophisticated means as social imitation. All vertebrates depend upon learning, but the kind of learning varies enormously across taxa. Intelligent species are adept at learning a variety of means for solving problems, whereas unintelligent species learn a very narrow range of specific skills and associations. In other words, learning is not equivalent to intelligence (e.g., Parker & Baars, 1990).

It is important to emphasize that whereas some species have evolved intelligence as an adaptive strategy, most have not. Successful adaptation does not imply intelligence. Relatively few vertebrate species display intelligence as an adaptive strategy. There are several reasons for the rarity of intelligent strategies. The first is that intelligence depends upon a sizable repertoire of behavioral options (a fairly large repertoire of motor patterns) that can be flexibly recombined for new purposes. This in turn requires a fairly large and sophisticated nervous system. In addition to reliable metabolic energy, such nervous systems require a good and reliable supply of oxygen. Finally, intelligent nervous systems require prolonged protected periods for assimilating and integrating the body of data they require to operate effectively. These prerequisites can evolve only as sequels to earlier, simpler adaptations (see chapter 13).

Whether or not recapitulation of the mind is correct, it provides a means to establish the framework of hypothetical architectural phases which is needed to continue with my study. Indeed it would seem a missed opportunity verging on academic negligence if I were to ignore the idea of recapitulation.

MITHEN, 1996, P. 63

CHAPTER 10
COGNITIVE ADAPTATIONS OF APES AND HUMANS

This chapter models selection pressures involved in the evolution of ape and human cognitive development. This approach assumes that cognitive abilities are adaptations rather than beneficial side effects of other adaptations (contra Gould & Lewontin, 1979). The primary reason for rejecting the alternative hypothesis that higher cognitive abilities arose as side effects of noncognitive abilities is the enormous energetic expense of the enlarged brain that underpins higher cognition (Parker, 1990b).

Building models for the evolution of human intelligence has become a popular endeavor for anthropologists and psychologists in recent years. Like ours, most of these scenarios are based on comparisons of cognition in monkeys, apes, and modern humans supplemented by interpretations of archeological data. Also like ours, most recent models focus on at least three succeeding levels of complexity: great ape cognition, *Homo erectus* cognition, and modern human cognition. The latter focus is due to the growing consensus that the earliest hominids, the Australopithecines, were apelike in their cognition and that the capacity for full-fledged language and culture emerged with the appearance of modern *Homo sapiens*. Donald's (1991) *Origins of the Modern Mind* and Mithen's (1996) *The Prehistory of the Mind* offer two such scenarios that complement and contrast with ours.

Donald, a neuropsychologist, proposes successive levels of mental adaptation (all of which persist in humans): (1) the episodic culture of monkeys and apes, (2) the mimetic culture of *Homo erectus*, (3) the mythic culture of modern *Homo sapiens*, and (4) the theoretic culture of literate humans. He defines mimesis as "conscious, self-initiated, representational acts that are intentional but not linguistic" (Donald, 1991, p. 168). He distinguishes mimesis from imitation and mimicry by its invention of intentional representations in a variety of modalities. Mimesis is a general ability that does not fit the modular model of mental capacities. Donald suggests that mimetic culture is well adapted for diffusing social knowledge, particularly social roles and rudimentary skills and believes it was first manifested in tool making by *Homo erectus*. Donald argues that monkeys and apes share episodic culture based on procedural memory. Although he believes that great apes display the highest level of episodic culture, he follows Terrace and Bever in believing their mentality is "restricted to situations in which the eliciting stimulus, and the reward, are clearly specified and present, or at least very close" (Donald, 1991, p. 152).

Comparative data support much of Donald's scenario but suggest that another level of cognitive ability needs to be added between monkeys and *Homo erectus*. We emphasize the greater cognitive abilities of apes as compared to monkeys and believe that great apes display some rudimentary symbolic capacities.

Mithen (1996), an archeologist, describes hominoid evolution in four acts: Act 1 opens 6 million years ago with ancestral apes; Act 2 opens 4.5 million years ago with ancestral hominids; Act 3 opens 1.8 million years ago with *Homo erectus*; and Act 4 opens 100,000 years ago with modern humans. He uses Karmiloff-Smith's (1995) model of the stages of cognitive development in children to describe three phases of evolution of the human mind. Phase 1 was dominated by the domain of general intelligence; phase 2 was enriched by the addition of multiple specialized intelligences (social, natural history, and technical), each working in isolation from the others; and phase 3 was characterized by the flexible flow of knowledge among the multiple intelligences.

Mithen argues that in Act 1, ancestral great apes, modeled on chimpanzees, had a powerful general intelligence plus several specialized mental modules, including one for social intelligence (which first evolved in monkeys). He argues that in Act 2, early hominids continued this basic adaptation. In Act 3, a language module was added between the time of *Homo erectus* and Neanderthals. This new module communicated with the social module, but remained isolated from the technological and natural history modules, which also remained isolated from each other. Only in Act 4, with the origin of modern humans, did the barriers among the modules break down, resulting in flexible, general purpose intelligence. Despite his hesitancy to adopt a recapitulation model for the evolution and development of the human mind, he acknowledges that it provides a useful heuristic for explicating his architectural model of the evolution of the human mind.

Mithen's use of a comparative developmental approach parallels ours (though we emphasize different developmental models). His neorecapitulation model of intellectual development in modern humans also parallels ours, as does his conclusion that humans display greater fluidity in their cognition than other primates do. We differ with Mithen, however, regarding the origins and onset of that greater fluidity and regarding the validity of modular models of cognitive abilities.

Our three-part scenario elaborates on our idea that problem solving is a complexly orchestrated activity situated in natural contexts (see chapter 6). We present our favorite scenarios for the evolution of three sequentially developing cognitive levels: (1) the emergence of symbolic abilities in the great apes, (2) the emergence of concrete operational abilities in Homo erectines, and (3) the emergence of formal operational thought in modern *Homo sapiens*. We call the first the *Apprenticeship Scenario*, the second, the *Joint Attention Scenario*, and the third, the *Declarative Planning Scenario*.

THE APPRENTICESHIP MODEL FOR THE EVOLUTION OF GREAT APE COGNITION

Great apes (and humans) share the following derived characteristics: fifth- and sixth-stage sensorimotor intelligence across physical (causal and spatial), logical-mathematical, and social (imitation, self-awareness) domains; symbolic-level preoperational intelligence across these same domains (including symbolic play); and perhaps even late intuitive–early concrete operational intelligence in the logical-mathematical domain under human tutelage. Evolutionary reconstruction through comparison of the ingroup (great apes) and the nearest outgroup (Old World monkeys) reveals that this new combination of abilities must have arisen in the common ancestor of living great apes (see chapter 9).

Several scenarios have been proposed to explain the evolution of primate intelligence. These include social models, ecological models, and socialization models. (See chapter 11 for a comparison of alternative models.) Whereas many primate species engage in the kinds of social competition that favor Machiavellian intelligence, only great apes have developed symbolic-level Machiavellian strategies. Likewise, many primates engage in the kinds of foraging that favors spatial and causal understanding, but only great apes have developed symbolic-level understanding of physical relations. These facts imply that any model that purports to explain the evolution of great ape intelligence must show why it differs from monkey intelligence. The social and the ecological hypotheses fail to do this; that is, they fail to explain the taxonomic distribution of these abilities (see chapters 9 and 11).

Whereas the primacy of the social intelligence model is widely accepted, it rests on two poorly documented assumptions. The first assumption is that Machiavellian social competition in monkeys is intelligent. The second is that monkeys and apes have similar intellectual abilities (Donald, 1991; Tomasello et al., 1993) and only humans display cognitive fluidity (Mithen, 1996). Both assumptions are probably wrong. It is questionable, given research on social intelligence (Cheney & Seyfarth, 1990, 1991), that monkeys display any of the symbolic capacities that underlie the complex social strategies displayed by apes (e.g., de Waal, 1983).

Associational learning, for which there is ample contextual support during socialization, can explain most of the Machiavellian strategies in monkeys. Also, given the extensive documentation of imitation in great apes (Custance et al., 1995; K. J. Hayes & Hayes, 1951; Miles et al., 1996; Russon, 1996; Russon & Galdikas, 1993, 1995), it is unlikely that they lack this cognitive capacity. If these two assumptions underlying the social hypothesis are wrong—if monkeys lack social intelligence and great apes display it, and great apes also display technological and logical intelligence—this undermines the argument that social intelligence was the primary force behind the evolution of intelligence. It implies that these three domains co-evolved. If we begin with this premise, we are compelled to explain the unusual coincidence of these three kinds of intelligence.

We propose that the coincident cognitive abilities of great apes are best explained in terms of a rare coincidence of ecological, logical, and social selection pressures for apprenticeship for extractive foraging. We envision these abilities as having emerged in the following sequence: (1) simple intelligent tool use, (2) complex intelligent tool use (symbolic level), (3) imitation of tool use (late sensorimotor level), (4) pretend play and teaching of tool use (symbolic level), and (5) self-awareness (late sensorimotor level).

First, simple forms of intelligent tool use (characteristic of the fifth and sixth sensorimotor stages) arose in early great apes as an adaptation for foraging on a variety of nutritious seasonal foods that require extraction. These probably included hard-shelled nuts and fruits, small fossorial animals, honey, and possibly roots and tubers, all of which are difficult to extract from the matrices in which they occur without some kind of tool. Such foods would have been at a premium during dry seasons or drying periods (Parker & Gibson, 1977). The variety of techniques involved in such extraction (pounding, probing, digging, and hitting) favored flexible extraction strategies as opposed to specialized anatomical manipulators for stereotyped tool use (Gibson, 1986).

Second, these simple abilities became elaborated in later great apes into more complex forms, including use of tool sets, anvils, and wedges. These more elaborated forms of intelligent tool use favored such logical-mathematical skills as classification and seriation. Seriation, for example, is implied in the selection of a series of tools of different sizes and diameters for excavating honey. Classification, perhaps multiple classification, is involved in seeing that the same object can simultaneously serve two or more functions, as in using a rock as an anvil, a hammer, or a wedge or using a stick as a probe or a weapon, or for that matter, seeing a branch on a tree as a potential probe.

Third, probably coincident with more advanced tool use, imitation arose as an adaptation for learning tool-mediated extractive foraging on nutritious but difficult-to-process foods. We suspect that imitation was also favored as a means for communicating referential information about the location and nature of such foods through various forms of mimicry of actions involved in extraction of those foods (Parker & Gibson, 1979). Given the long apprenticeship required for social learning of efficient tool use (Boesch, 1991b; Teleki, 1974), it seems unlikely that this ability could have been acquired through individual learning alone. Imitation serves as a means for reducing infant dependency on mothers who otherwise must share such foods with their postweaning offspring (Silk, 1978).

Fourth, teaching through demonstration arose as an adaptation for hastening the development of tool-mediated extractive foraging to offspring. As indicated earlier, teaching seems to depend upon the capacity for symbolic play, which develops from imitation at the beginning of the symbolic subperiod of preoperations. Pretend play is crucial for teaching because it involves the incipient capacity to play reciprocal roles and to exchange roles, and hence to anticipate the behavior of others.

Fifth, self-awareness of the kind revealed by MSR (mirror self-recognition) evolved in concert with imitation and teaching. Like imi-

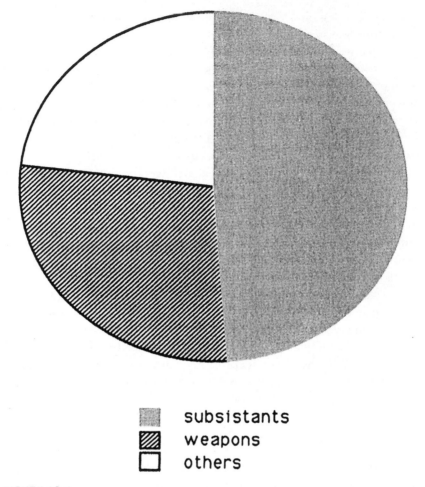

subsistants
weapons
others

FIGURE 10-1
Relative frequency of wild chimpanzee use of tools for various functions across populations (McGrew, 1992). Reprinted with the permission of Cambridge University Press.

tation, MSR depends upon kinesthetic-visual matching (Mitchell, 1993) of the model's actions by the imitator. Likewise, teaching by demonstration involves the teacher's ability to anticipate matching of her actions by the imitator. It also involves the teacher's ability to identify the pupil's level of ability to imitate, to identify inadequacies in the pupil's imitative actions, and to identify the segments of her actions that the pupil needs emphasized and repeated. In other words, imitation and teaching both entail awareness of the relationship between one's actions and those of another (Parker, 1996a).

If intelligent tool use was originally favored as an adaptation for omnivorous extractive foraging, then we should expect convergent adaptations for this ability. New World cebus monkeys, who also specialize in extractive foraging on embedded foods, show such convergent adaptations (Parker & Gibson, 1977). Lion-tailed macaques and baboons, who are also omnivorous extractive foragers, are also capable of tool use (Westergaard, 1988, 1992). The absence of tool use abilities in other New World monkeys and in most macaques and other Old World monkeys suggests that these abilities arose two or three times independently in primate evolution.

If the extractive-foraging hypothesis is correct, we should also expect tool use to occur more frequently in the context of extractive foraging than in any other context, such as self-grooming, at least in great apes. (Humans continue to use tools in extractive foraging, but the range of functions of tool use has expanded in our species.) This seems to be true in wild chimpanzees and orangutans. Figure 10-1 compares chimpanzee tool use in various contexts.

Similarly, if the apprenticeship hypothesis is correct and imitation arose as an adaptation for learning tool use on a variety of embedded foods, we should expect that it should occur more frequently in that context than in any other in great apes. There are no data on this. Likewise, if demonstration teaching arose as an adaptation for teaching tool-mediated extraction of embedded foods, it should be occur most frequently in that context. The two clear cases of demonstration teaching in wild chimpanzees occur in this context (Boesch, 1991b).

Whereas intelligent tool use was apparently favored by natural selection for efficiency in feeding on embedded foods, imitation, teaching, and self-awareness were probably favored by social selection (Trivers, 1985; West-Eberhard, 1983) for imitative learning and demonstration teaching. Specifically, they were probably favored by parental selection or parental manipulation favoring offspring who were capable of learning tool use through imitation.[1]

1. *Social selection* refers to selection by other members of the same social group or species as compared to *natural selection*, which refers to selection by members of other species (e.g., predators, diseases) and by abiotic factors (e.g., climate) (West-Eberhard, 1983). Social selection encompasses sexual selection, kin selection, parental manipulation, and reciprocal altruism (see Trivers, 1971, 1972, 1974, 1985). Parental

An Atavistic Model. If the apprenticeship hypothesis is based on the behavior of chimpanzees and humans, why not postulate that tool use and associated abilities evolved in the common ancestor of chimpanzees and hominids? The problem with this interpretation is that intelligent tool use, imitation, pretend play, and MSR co-occur in all the living great apes; in other words, these are shared derived character states. This fact lead us to propose the atavistic hypothesis that this complex of cognitive abilities arose in the common ancestor of these sister species (Parker, 1996a; see Parker & Gibson, 1979, for a similar hypothesis about intelligent tool use).

The atavistic hypothesis is the most parsimonious explanation for the co-occurrence of this complex of abilities in all five living sister species. It is supported by the discovery of widespread tool use for extractive foraging in two populations of wild orangutans (Fox & van Schaik, in press; van Schaik et al., 1996), who were the first great apes to diverge. Further support for the apprenticeship hypothesis comes from the convergent evolution of intelligence tool use, imitation, and self-awareness in dolphins (Martens & Psarakos, 1994; Tayler & Saayman, 1973). Although we know that dolphins engage in sophisticated alliances (Connor, Smolker, & Richards, 1992), they may also use their cognitive abilities in their hunting strategies and in apprenticeship for these strategies (see chapter 11).

A Transitional Model. Finally, the apprenticeship model for the origin of late sensorimotor and early preoperational intelligence in great apes is convincing because it provides a transitional model for the evolution of hominid intelligence. According to this transitional model, the fluidity across cognitive domains that is characteristic of the human mind (Karmiloff-Smith, 1995; Langer, 1996; Mithen, 1996) had its origins in the apprenticeship adaptation of the common ancestor of the great apes. Although modest in degree, the establishment of the connection between the social and technological domains was crucial to the subsequent evolution of hominid cognition. *Imitation and teaching were the keys to this connection, and they first evolved in the common ancestor of great apes.* Tomasello et al. (1993) and others are right that imitation and teaching were crucial to the capacity for culture. They are wrong,

---

manipulation of offspring is differential parental investment in offspring according to their potential contribution to the parent's total fitness, for example, favoritism toward male offspring.

however, in placing its origins at the base of the hominid radiation. An evolutionary analysis places the origins of these abilities at the base of the great ape radiation in the Miocene.

An evolutionary analysis tells us that early hominids must have had cognitive abilities at least equivalent to those of the common ancestor with great apes. These abilities must have served as preadaptations for continued technological and social intelligence in the earliest hominids. The most parsimonious model is that the Australopithecines continued to rely on extractive foraging for nutritious foods, probably expanding their dependence on such foods. Extractive foraging provides a natural bridge into a scavenging niche. Therefore we hypothesize that early *Homo* simply extended extractive foraging with tools to extraction of bone marrow with hammer stones and extraction of meat from tough hides by cutting with sharp-edged flakes (Parker & Gibson, 1979).

THE JOINT ATTENTION MODEL FOR THE EVOLUTION OF HOMO ERECTINE COGNITION

Most investigators agree that after the origins of symbolic intelligence in great apes, the next major transition in the evolution of cognitive abilities of hominoids probably occurred with the emergence of Homo erectines about 1.6 million years ago. As in the case of great apes, this transition probably involved interacting technological and social components. This transition is evidenced in the larger body and brain sizes of these creatures as well as their more developed technology and their greater geographic spread.

Recent evidence suggests that the earlier hominids, Australopithecines and early *Homo*, were small-bodied, relatively small-brained hominids who, like living apes, relied on trees for protection from predators (Stanley, 1992). They were bipedal but not exclusively so (Abitbol, 1995; McHenry, 1986). Although they undoubtedly used tools, their technological intelligence probably did not exceed that of modern chimpanzees significantly (Wynn & McGrew, 1989).

Evidence from an almost complete skeleton of an immature *Homo erectus* (Turkana boy) suggests that this species was the first to display essentially modern body proportions and striding locomotion. It was also probably the first species to display a life history pattern markedly different from that of living great apes (B. H. Smith, 1993). Life history analysis based on dental evidence suggests that Turkana boy matured at a rate midway between that of great apes and modern humans. Evidence also suggests that this species had a brain size about two-thirds that of modern humans, as compared with Australopithecines and great apes, whose brain size was about one-third that of modern humans (Begun & Walker, 1993).

Several scholars have suggested that the emergence of Homo erectines coincided with a significant change in maternal behavior attendant upon the loss of infants' ability to cling, which must have occurred with the loss of tree climbing (Stanley, 1992). The evolution of a more modern pelvis for bipedal striding and the loss of tree climbing combined with the larger brains of neonates resulted in the birth of neonates who required more maternal care. The loss of clinging by infants put greater pressure on mothers to attend to their infants and conversely greater pressure on infants to elicit appropriate maternal care. This pressure likely favored the evolution of more distal communicative signals between mother and infant, including greater face-to-face gestural and vocal interaction as well as greater emphasis on mutual imitation (Parker, 1993).

Consistent with this scenario is the greater facility for vocal and facial imitation in human infants as compared with great ape infants. Also consistent is the greater emphasis on mutual imitation in our species. These are all elements of a major episode of preterminal extension and elaboration of the modalities and rates of imitation and circular reactions (see chapter 9).

Greater emphasis on distal communication would have favored the development of the capacity for joint maternal-infant attention to important environmental events as a means for signaling the significance of such events to the child in the absence of direct tactile communication. The symbolic capacities that grow out of imitation would have provided the ontogenetic and phylogenetic foundations for symbolic communication of a kind seen in event representation capacities of human children during the symbolic subperiod of preoperations (Nelson & Gruendel, 1986; Nelson & Seidman, 1984).

Greater dependency may also have favored greater paternal investment in offspring in the form of provisioning both mother and offspring, at least during critical reproductive and developmental periods. Greater specialization of males and females attendant upon greater infant and juvenile dependency would have favored gender-specific apprenticeship in different productive and reproductive tasks. This, in turn, would have favored increased emphasis on demonstration teaching by parents and pretend play rehearsals of gender-specific roles by juveniles. Such prolonged apprenticeship would have favored the representation of certain recurrent activities, which provides a bridge into the development of grammar (Nelson & Gruendel, 1986; Nelson & Seidman, 1984).

Another line of evidence for a significant shift in Homo erectine adaptation is found in the fact that this species or its contemporary sister species were the first to spread out of Africa. They had spread throughout Asia and Europe by 1 million or even 2 million years ago. They made more elaborate tools, apparently relied more on meat, scavenged or hunted, and were able to survive in nontropical climates, perhaps owing to shelter construction or clothing. It is unclear whether they controlled fire.

Clearly, the greater reliance on technology must have favored the evolution of more complex cognitive abilities in both technological and social domains. Particularly, it would have favored increased ability for imitation and pretend play and teaching skills for the social transmission of technological skills. Wynn's (1989) analysis of Acheulian stone tools suggests that their flaking pattern implies understanding of such Euclidean spatial concepts as angles, which are characteristic of children in early concrete operations. Achievement of competence in such activities must have required an extended period of immaturity for prolonged apprenticeship.

Given a life history that was midway between that of great apes and modern humans, it is tempting to speculate that the cognitive development of these creatures was midway, probably in the range of early concrete operations. It is also tempting to speculate that the pace of cognitive development peaked midway between that of great apes and modern humans, that is, that it coincided with sexual maturity at about 14 to 16 years.

The joint attention model for the emergence of elaborated social imitation and pretend play adaptation for increased infant dependency provides a bridge to the succeeding model for the emergence of declarative planning in modern humans. Specifically, it provides a bridge to the evolution of language through event representation and a bridge to the dominance of social selection in the evolution of modern humans.

THE DECLARATIVE PLANNING SCENARIO FOR THE EVOLUTION OF HIGHER INTELLIGENCE AND LANGUAGE IN MODERN HUMANS

The last major transition in the evolution of hominoid cognition probably occurred quite recently, probably in the last 100,000 years with the emergence of anatomically modern *Homo sapiens*. If Homo erectines stand roughly midway between great apes and modern humans in their life histories, the first anatomically modern humans were probably equivalent to modern humans in both their life histories and their cognitive abilities. The debate concerns when and where exactly these creatures first appeared and when and where their characteristic artifacts first appeared. Current evidence suggests that they first emerged in Africa approximately 100,000 years ago and that their characteristic technologies first appeared at about 80,000 years ago, also in Africa. This evidence also suggests that these newly evolving humans migrated out of Africa, partially or fully replacing other populations of *Homo* in other parts of the Old World (Mellors & Stringer, 1989).

The emergence of anatomically modern humans was followed by several major social and technological innovations. First, tool cultures become more standardized and complex, varying regionally and, on a historic scale, changing rapidly. Generally they are characterized by blades, microliths, and hafting. Second, both stationary and mobiliary art emerge for the first time in the form of cave paintings, carvings, statues, and ornaments. Third, elaborated burials of bodies first appears. Fourth, large-scale migrations, apparently in pursuit of seasonal herds, becomes apparent. Finally, occupation of new, more marginal habitats begins to occur (Mellors & Stringer, 1989).

These innovations suggest that these hominids were the first to display fully modern capacities for cognition, language, and *culture*. The apparent absence of all of these behaviors in earlier species of *Homo* argues that these abilities first appeared in modern *Homo sapiens* (Noble & Davidson, 1996; Parker & Milbrath, 1993). The argument in favor of this is that propositional language and culture are necessary conditions for the appearance of the foregoing innovations. The common element underpinning these innovations is declarative planning.

We favor the hypotheses that modern forms of cognition and language arose as adaptations for declarative planning and mental testing of alternative technological and social scenarios (Parker & Milbrath, 1993). Declarative planning, unlike procedural planning which precedes it developmentally, involves symbolic representation of and hence anticipation of sequences of actions and their likely consequences. It also involves flexible addition, deletion, and substitution of elements of the plan. This, in turn, allows modification of plans before execution. Early aspects of this kind of planning emerge developmentally in conjunction with concrete operations. Full development of declarative planning depends upon the achievement of hypothetical-deductive reasoning characteristic of formal operations (DiLisi, 1987).

Declarative planning depends upon full-blown propositional language, which emerges developmentally during concrete and formal operations. Socially shared and engaged planning is the essence of culture in Goodenough's (1981) sense of plans, recipes, procedures, rules, and values. Although it is easy to imagine myriad advantages for declarative planning, we favor the notion that it arose in the Upper Paleolithic as an adaptation for expansion into highly variable and seasonal habitats characterized by unstable resources and perhaps extreme climatological shifts.

Not only would this expansion have placed a premium on new technologies and routines and social complementation (Reynolds, 1993) for employing them, it also would have favored experimentation with new forms of social exchange (Gamble, 1976). Specifically, social selection may have favored marriage as a form of social alliance between families. Marriage alliances would have facilitated exchange of resources between regions during periods of scarcity in a given region. In addition, kinship ties would have favored defensive or offensive coalitions against other groups (Gamble, 1976).

The declarative planning model for higher cognition is appealing because it focuses on common functions crosscutting various domains of cognition—technology, logic, and social manipulation. It thereby emphasizes the "cognitive fluidity," the integrative, flexible cross-domain aspect of human cognition that apparently distinguishes it from that of other primates (Mithen, 1996).

ONTOGENETIC NICHES AND THE EVOLUTION OF HOMINID COGNITIVE DEVELOPMENT

The preceding section models the adaptive significance of simultaneous character states (see chapter 9). This section models the adaptive significance of *ontogenetic characters*, which are the focus of this book. Specifically, it hypothesizes that stages of cognitive development within each domain recapitulate the stages of their evolution. This hypothesis raises questions about the origin and maintenance of ontogenetic characters (Mayr, 1994). Have they been maintained by selection or by other evolutionary forces such as phylogenetic inertia and developmental constraints? Are recapitulated stages adaptive in current environments?

In a paper on these issues, Mayr (1994) reiterates earlier views that recapitulated stages are maintained either by selection for their direct functions or by developmental constraints imposed by their indirect functions. He describes these indirect functions in terms of what he calls the *somatic program:* "The development of a structure is controlled both by a genetic and a somatic program. The inducing capacities of the surrounding embryonic tissues form the somatic program" (Mayr, 1994, p. 230). He gives the example of archaic embryonic stages serving as temporary but indispensable templates for organ formation. (Developmental constraints are phenotypic biases arising from the interactions of developmental systems [Alberch, 1990; J. M. Smith et al., 1985].)

In accord with Piaget's (1970) and Trevarthan's (1973) view of psychogenesis as an extension of embryogenesis, we extend the concept of the somatic program into postnatal development, thereby extending Mayr's paradigm to cognitive development. We therefore rephrase the question about origins and maintenance of recapitulated stages: To what degree do these developmental stages serve direct functions maintained by current selection, and to what degree do they serve indirect epigenetic functions maintained by developmental constraints or phylogenetic inertia?

The distinction developmental biologists make between internal and external factors in evolution (e.g., Alberch, 1990; McKinney & McNamara, 1991) is useful for addressing these questions. By external factors they mean natural selective forces from the environment, and by internal factors they mean epigenetic constraints on developing biotic systems. In the case of hominoid cognitive development, the environment is social, and therefore the selection is social. The concept of *ontogenetic niches*, a phrase originally coined to refer to differential resource use by conspecific animals at different developmental stages (and sizes) (Werner & Gilliam, 1984), is also useful for thinking about the adaptive significance of developmental stages.

The adaptive significance of developmental stages needs to be addressed through specifically designed research programs. At this point, we can only suggest certain approaches to this research. In order to address the question of direct adaptations, we need to know what kinds of environments developing organisms adapt to. Then, we need to know how they adapt to these environments.

Creation of a descriptive taxonomy of ontogenetic social niches of hominoid development is an initial step toward studying developmental environments. By *ontogenetic social niches*, we refer to the stage-typical social environments in which immature individuals develop and on which they depend for their subsistence, survival, and training. These are the niches that are relevant to ontogenetic adaptations (Borchert & Zihlman, 1990), and we suggest the following:

1. the maternal dependency niche of infancy (from birth to weaning)

2. the maternal surveillance and modeling niche of early childhood (from weaning to locomotor independence)

3. the apprenticeship niche (from locomotor independence to adolescence)

4. the courtship and occupational niche (from adolescence to onset of reproduction)

A taxonomy of ontogenetic social niches is prerequisite to describing the parameters of those niches. We suggest the following parameters of the earliest ontogenetic niches in hominoids. In the maternal dependency niche, those parameters would include the nature and frequency of maternal handling, carrying, and nursing, and the nature of mother-infant communication, as well as the nature and frequency of maternal interactions with other conspecifics. These and others are the features to which hominoid infants must have adapted through parental and kin selection (forms of social selection).

As indicated above, we suggest that vocal and facial imitation and reciprocal secondary circular reactions between mother and infant, "the game" (J. S. Watson, 1972), arose as coadaptations for maternal-infant bonding and establishment of joint attention in this context (Parker, 1993). Specifically, we suggest that these adaptations arose in association with the emergence of dedicated terrestrial locomotion and the attendant loss of tree climbing in adults and children and partial clinging in infants that occurred with the evolution of Homo erectines (Stanley, 1992).

In the surveillance and modeling niche of early childhood, those parameters would include maternal weaning and food-sharing strategies. It would also include maternal encouragement of play and other forms of practice, as well as intervention strategies to protect offspring from siblings, playmates, and other conspecifics. Finally, it would include modeling of such basic activities as finding and preparing food. Selection would have favored these maternal abilities as an adaptation for hastening independence of one offspring preparatory to investment in another and hence re-establishment of sexual relations by the mother. These were the features to which young early hominoid children must have adapted. Later in human evolution, probably in the evolution of

Homo sapiens, symbolic games played an important role in the development of grammar (Bates et al., 1979).

According to this scenario, pretend play arose as an adaptation by young children for imitating subsistence and social roles. This capacity probably first arose in the common ancestor of the great apes as an adaptation for imitative learning of tool-mediated extractive foraging (Parker, 1996a). In the apprenticeship niches of later childhood, selection would have included maternal encouragement to establish independent sleeping sites, feeding strategies, and social networks. At this point in development, children would have shifted their attachment to peers and mentors, whereas mothers were primarily preoccupied with a new infant. Mothers probably encouraged older offspring to help care for siblings and do simple subsistence tasks. These are features to which older hominoid children must have adapted.

Extended apprenticeship became increasingly important as hominids depended more on technological modes of subsistence. Teaching became increasingly important in early hominids as an adaptation for modeling key subsistence activities and technologies. Gender-specific modeling of subsistence activities would have been favored by the increasing division of labor between male and females in later *Homo* species. This would have favored father-son and uncle-nephew apprenticeships as well as other forms of mentorship.

We suggest that late preoperational and early concrete operational capacities for causal and spatial understanding, for example, understanding of whole-part relations and intermediate transmission of forces, arose as adaptations for learning subsistence technologies. These technologies were associated with manufacture and use of stone tools, butchery of animal foods, procurement and processing of vegetable foods, and construction of clothing and shelters in early *Homo* species (Parker & Gibson, 1979; Wynn, 1989).

In the courtship and occupational niches of adolescence, individuals were faced with the task of demonstrating their competence in subsistence and political tasks. This demonstration was both a practical matter of survival at the level of individual and kin competition, and a means for attracting and competing for mates. Throughout hominid evolution, procuring and sharing meat has been a mating strategy for males (Parker, 1987). Likewise competence in subsistence tasks is crucial to maternal competence. In later stages of human evolution, engaging in political alliances and reciprocal networks became increasingly important (Gamble, 1976).

We suggest that the late concrete and early formal operational capacity for declarative planning based on logical-mathematical reasoning and propositional language arose in anatomically modern *Homo sapiens* as an adaptation for planning political and economic alliances through reciprocal networks based on affinal kinship ties. Planning of this kind may have also relied upon specialized information-processing capacities for evaluating the costs and benefits of social exchanges (Cosmides & Toobey, 1991). Propositional language and hypothetical deductive reasoning allowed individuals and social groups to predict and evaluate probable outcomes of alternative scenarios before deciding which course to follow. It also allowed them to make contracts, notably marriage contracts which established reciprocal obligations between kin groups. (Parker & Milbrath, 1993). This in turn favored parental manipulation of mate choices of offspring.

According to our terminal extension model, a series of new stages of cognitive development evolved at the end of the developmental period in a series of ape and hominid ancestors. In addition, at least one major episode of preterminal extension and elaboration of imitative abilities occurred during the evolution of *Homo* species. Sequential evolution of these new stages culminated in the modern forms of cognitive development from the sensorimotor through the late concrete operations period.

According to our ontogenetic niche model, these forms of cognitive development were favored by natural and social selection. Social selection in the form of parental and kin manipulation and reciprocal altruism favored infants, children, and adolescents who were able to motivate and respond appropriately to parental investment (Trivers, 1985). Likewise, social selection favored increasing parental investment and manipulation through symbolic communication and teaching. Finally, reproductive competition has favored increasing cognitive and linguistic capacities for planning.

A thorough assessment of ontogenetic social adaptations would require a systematic investigation of all the ontogenetic niches, developmental stages and periods, and social selection pressures in each domain in each species.

Even if such an analysis were to reveal that each developmental stage in every domain is adaptive today, we could not conclude that selection alone was maintaining ontogenetic recapitulation.

According to our epigenetic model, each new stage or subperiod in cognitive development is constructed out of the raw materials of the preceding stage. The developing brain is the construction site for these changes, which begin prenatally and continue throughout psychogenesis (see chapter 12). Insofar as the epigenetic hypothesis is confirmed by continuing research, we must conclude that ontogenetic recapitulation is maintained by developmental constraints. This preliminary analysis then suggests that hominoid stages of cognitive development are maintained by both selection (natural and social) and developmental constraints.

CONCLUSIONS

In this chapter, we have approached the evolution of hominoid cognition from two perspectives, First, we have proposed hypotheses to explain the evolution and adaptive significance of ape and hominid cognition viewed as simultaneous character states. Second, we have proposed hypotheses to explain the evolution and adaptive significance of cognition viewed as ontogenetic character states. This latter approach involved identifying specific ontogenetic social niches and hypothesizing the nature of age-specific adaptations to these niches. This approach also allowed us to distinguish current adaptation from maintenance of previous adaptations. This social niche approach is consistent with our emphasis on the increasing role of various forms of social selection in the evolution of the genus *Homo*.

Model makers have opposing views regarding the specificity of selection for cognitive abilities in human and nonhuman primates. One group proposes that selection has favored such specific cognitive modules as a language module, an exchange module, and an intuitive mechanics module (Cosmides & Toobey, 1991; Pinker, 1994; Tooby & Cosmides, 1992). Others propose that selection has favored, first, domain-specific and, then, fluid cognitive abilities. Perhaps the most popular view is that selection has favored general purpose problem solving and learning abilities.

As our evolutionary scenarios suggest, we imagine selection focusing on responses to related sets of adaptive challenges and opportunities. In the case of ancestral great apes, we argue that a major challenge and opportunity was social transmission of tool-mediated strategies for foraging on a variety of high-energy embedded foods. We propose that apprenticeship in these foraging strategies favored a suite of characteristics including imitation, self-awareness, and pretend play as well as at least sixth-sensorimotor-stage understanding of physical causality and spatial relations.

Our orchestral model envisions adaptive behaviors in everyday life as sets of cognitive abilities composed in response to the demands of problem situations rather than as pure expressions of cognitive domains (see chapter 6). This model implies that selection operates on component abilities insofar as they participate in combinations that contribute to adaptive outcomes. It also implies that selection operates on mechanisms that assemble and recombine these abilities. Various cognitive domains are traditional psychological constructs derived from analysis of test data and spontaneous behaviors, rather than naturalistic units. The degree to which specific domains reflect unified organizational features of the brain is unclear (see chapter 12).

Model making has become a popular endeavor among cognitive theorists. The relative merits of competing models have been difficult to evaluate. In the following chapter, we introduce cladistic methods for constructing and evaluating adaptive scenarios. We illustrate the heuristic value of this method by comparing various scenarios for the evolution of primate cognition.

The meaning of adaptation can be clearly stated in cladistic terms: adaptation is apomorphic function due to natural selection. . . . An evolutionary definition of adaptation must have an historical component specifying selection as the evolutionary agent responsible for the appearance of the feature.

CODDINGTON, 1988, P. 5

CHAPTER 11
COMPARING ADAPTIVE SCENARIOS FOR PRIMATE COGNITION

Some biologists refer to adaptive scenarios as "Just So Stories" because they can seem as fanciful as Rudyard Kipling's tales of the origins of such characters as the leopard's spots. The fault lies not in the generation of fanciful tales, but in the failure to use criteria for diagnosing adaptation. This chapter introduces cladistic criteria for evaluating evolutionary scenarios and illustrates their application to competing models of cognitive evolution in primates.

EVOLUTIONARY CONCEPTS OF ADAPTATION

Biologists use the term *adaptation* in various ways: (1) to refer to physiological adjustments during an individual's life span, (2) to refer to a physiological or behavioral function that varies within species, and (3) to refer to functions or features that distinguish species in a genus or family or higher category (Coddington, 1988). Evolutionary biologists have been particularly concerned with the third meaning of adaptation. The classic definition of species-typical adaptation is a means or mechanism that has been shaped by selection for a specific function or purpose (Williams, 1966). Adaptation is thereby distinguished from beneficial effects of a feature that may have arisen as serendipitous side products of adaptation (Williams, 1966).

Evolutionary biologists use a variety of approaches to distinguish adaptations from beneficial side effects. The classic approach is the *argument by design* wherein evidence for purported function is made by analogy with design for function by human engineers. Dawkins (1986), for example, compares the sonar of bats to mechanical sonar devices designed by naval engineers. Another standard approach is to study differences in fitness associated with the presence or absence of features. Another approach is to study the correlation across species between a feature and a particular life style. The most definitive approach is to demonstrate increased survival or reproductive success (fitness) associated with the presence of the feature (Williams, 1992).

Yet another approach is implicit in the definition of adaptation provided by cladistics. According to this definition, *adaptations are shared derived characters*, that is, characters shared by two or more sister species that inherited it from a common ancestor (Coddington, 1988). This cladistic meaning implies the designation of from whom, when, why, and where the character arose:

> Adaptational hypotheses usually take the form: trait *M0* is an adaptation in taxon *C* for function *Fx*. From a cladistic point of view, such a statement is incomplete. With respect to the phylogeny depicted in [Figure 11-1], the complete statement would be *the derived trait* M1 *arose at time (t) in the stem lineage of taxa* C, D, *and* E *via selection for the derived function* F1 *with respect to the primitive trait* M0 *with primitive function* F0 *in taxa* A *and* B. (Coddington, 1988, p. 5)

Hypothetically, for example, the derived trait of bipedalism arose about 4.5 million years ago in the common ancestor of *Homo* and *Australopithecus* via selection for habitually transporting objects. It arose from the primitive trait of knuckle walking, which was selected in the common ancestor of African apes and its descendants *Gorilla* and *Pan* for occasionally transporting objects during quadrupedal locomotion. Figure 11-1 graphically depicts a phylogeny.

Note that the cladistic definition of adaptation depends upon a clear specification of the derived trait and its function (its design) as well as of the primitive trait and its function in a group of related species. It also specifies the selection pressures that favored the derived trait.[1] Most significantly, it demands that the hypothesis address why the new trait is unique to a particular taxonomic group.

Appropriate definition of traits or character states is crucial to every

---

1. This definition of *adaptation* seems to sidestep the thorny issue of distinguishing evolved adaptations from side effects or benefits (Ghiselin, 1969; Williams, 1966) Certainly, however, the diagnosable presence of a character in several related species suggests that it has been directly or indirectly selected.

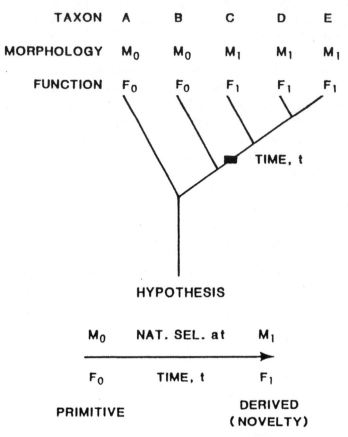

FIGURE 11-1
Depiction of cladistic method for discovering adaptions (from Coddington, 1988, p. 6).

approach to understanding adaptation. Relevant character states should be coherent, integral, and independent of one another. They also should be genetically patterned and species typical (Clark, 1964). Finally, they should be homologous. Cladistic methodologies help ensure homology.

CLADISTIC METHODOLOGY

THREE APPLICATIONS OF CLADISTIC METHODOLOGY

Cladistics is a school of evolutionary taxonomy known as *phylogenetic taxonomy*. Cladistic methodology has three distinct applications. First, it offers taxonomists a systematic methodology for determining phylogenetic relationships. Second, It offers ethologists a methodology for

TABLE 11-1 THREE CLADISTIC METHODOLOGIES

Method use	Proponents	Key concepts	Applications
Taxonomic classification	Henning (1979)	Shared derived character states	Identification of sister species
Reconstruction of character evolution	Brooks and McLennan (1991); Harvey and Pagel (1991).	Shared derived character states	Identification of common ancestor originating shared derived character
Definition of adaptations	Coddington (1988)	Shared derived character states	Identification of shared derived character states as adaptations

identifying the common ancestry of characters (see chapters 8 and 9). Finally, and most important to this chapter, it also offers evolutionary biologists criteria for identifying adaptations and a methodology for generating and testing adaptive hypotheses (see chapter 10). All of these applications rely on the concept of shared derived character states. This chapter emphasizes the third approach. Table 11-1 summarizes these three different applications of cladistic methodology.

As we indicated in chapter 9, the first application of cladistics involves constructing cladograms by mapping character states of hypothetically related species onto alternative branching diagrams (cladograms) until the most parsimonious tree is found. (The most parsimonious tree is the one that requires the fewest evolutionary transformations.) Based on the most likely evolutionary tree, evolutionary taxonomists identify sister species on adjacent branches and classify them into higher taxonomic groups, such as families, based on common ancestry (Hennig, 1979; Ridley, 1986; Wiley, 1981).[2]

Hennig (1979), who codified this methodology, argued that evolutionary trees (cladograms) must be based on the distinction between (1) homologous characters that are uniquely shared by a given group of sister species (shared derived character states or *synapomorphies*) owing to their origin in a recent common ancestor and (2) primitive homologous

2. The procedures described here are those developed by the school of evolutionary taxonomists known as *cladists*. The branching diagrams they construct—cladograms—differ from conventional family trees in that all species are represented at the ends of branches. None are represented at the nodes as they sometimes are on phylogenies. Phylogenies, in contrast, represent purported common ancestors at the nodes or branch points.

characters that are broadly shared by a larger group of species (shared character states or *symplesiomorphies*) owing to their ancient origin in a distant common ancestor.

Homologous characteristics contrast with analogous characters that have arisen in only one species (uniquely derived character state or *autapomorphies*). They also contrast with analogous characters that have evolved through convergent evolution in distantly related species who did not inherit them from their common ancestor (*homoplasies*) (Wiley, 1981).

Determining whether character states shared by various species are homoplastic, shared, or shared derived is the key to constructing evolutionary trees. *Shared derived character states indicate closeness of phyletic relationship, whereas shared or homoplastic character states do not.* Evolutionary trees (cladograms) are constructed on the basis of shared derived character states because only shared derived character states indicate common ancestry. These cladograms may then be transformed into phylogenies in which putative common ancestors of the various branches are identified (Skelton, McHenry, & Drawhorn, 1986)

As we saw in chapter 9, the second application of cladistic methodology begins once tree construction is complete. At this point, the cladistic method can be reversed to do *character mapping*, that is, to discover the ancestry and sequence of evolution of character states. This can be done by mapping character states onto a previously constructed tree to see which common ancestor gave rise to shared derived characters. In order to avoid circularity, only characteristics other than those used to construct the tree can be mapped on it to reconstruct common ancestry (Brooks & McLennan, 1991). Character mapping is a vital part of evolutionary modeling because it alone can identify the common ancestor in whom a given character originated. It therefore plays a role in both the construction and the testing of adaptive hypotheses, as elaborated below.

The third application of cladistic methodology, on which we will focus, is the identification of adaptations mentioned at the beginning of this chapter, and generation and testing of adaptive scenarios. Both rely upon identification of shared derived character states and the construction of cladograms.

USE OF CLADISTIC METHODOLOGY FOR EVALUATING ADAPTIVE SCENARIOS

During the past century and a half, many anthropologists and a few psychologists have tried their hands at modeling the adaptive significance of primate, and especially hominid, cognition and behavior. In response to the fanciful nature of these scenarios, various anthropologists have proposed guidelines for constructing evolutionary models. In one of the best-known guides, Tooby and DeVore (1987) urged modelers to keep their scenarios within the bounds of evolutionary biological concepts. They listed 25 evolutionary guidelines for "strategic modeling," by which they meant modeling based on evolutionary concepts. They did not use the cladistic definition of *adaptation* or the cladistic approach to modeling.[3]

The cladistic definition of *adaptation* is significant because it implies a standard for evaluating alternative adaptive scenarios (Coddington, 1988). Only adaptive scenarios that specify the phylogeny of the character as well as putative selection pressures can be considered potentially valid. We call this the *phyletic criterion*.

Specification of all of these features in one taxonomic group meets the minimum phyletic criterion for an adaptive hypothesis. Specification of these features in two distantly related taxonomic groups that show convergent functions strengthens the hypothesis.

> The observation of a functional trait in a single taxon clearly can tell us nothing about how and why it arose and, indeed, adaptation as current utility makes no claims about origin. It makes only the weak claim that if the trait functions one way in one case, then maybe similar traits function the same way. On the other hand, if for each instance of convergence one has elucidated the function of the derived trait as well as the primitive homologue and the derivation is consistent with selection for the derived function, then one has essentially performed the kind of analysis advocated here. (Coddington, 1988, p. 17)

We call this the *convergence criterion*.

Only those adaptive scenarios that can be made to meet the phyletic criterion and the convergence criterion achieve credibility within the

---

3. Use of the cladistic methodology obviates Tooby and Devore's (1987) distinction between conceptual and referential models by focusing on reconstructing the common ancestry of specific character states. This approach is inherently conceptual and referential. The conceptual part lies in the methodology; the referential aspect is determined by application of the procedure. This approach obviates the problem with most models of hominid evolution, which have begun with a reference species rather than a set of character states. Cladistics also obviates the problem of misuse of such evolutionary concepts as analogy and reconstruction. Cladistic character mapping focuses specifically on shared derived homologous characters. It is true, rather than metaphorical or analogous, evolutionary reconstruction (Moore, 1996).

TABLE 11-2   CLADISTIC CRITERIA FOR ADAPTIVE HYPOTHESES

Phyletic criterion	Convergence criterion
Description of character states	Description of character states
Demonstration that character states are synapomorphic (shared derived)	Demonstration that character states are synapomorphic (shared derived)
Specification of common ancestor who originated the synapomorphic character states	Specification of convergent characters in a different lineage whose members evolved them independently
Demonstration of common functions in sister species	Demonstration of common functions in two lineages

cladistic frame of reference. Unfortunately, this excludes scenarios that address derived character states that are unique to a particular species. (Other approaches may be useful for evaluating such scenarios [McPeek, 1995].). (See Table 11-2 on cladistic criteria for adaptive scenarios.)

Insofar as alternative adaptive scenarios are mutually exclusive in their formulations and predictions, only one—at most—can be correct. Therefore, insofar as conformance to the aforementioned criteria indicates potential validity of adaptive scenarios, such competing scenarios should vary in their conformance. Consequently, we should expect competing scenarios to display different degrees of conformity to cladistic criteria. Indeed demonstration of conformance or nonconformance is the rationale for such criteria.

ADAPTIVE HYPOTHESES CONCERNING THE ORIGINS OF INTELLIGENCE IN ANTHROPOIDS

The remainder of this chapter illustrates the use of the preceding cladistic criteria for evaluating various alternative adaptive scenarios for the evolution of primate cognition. Global evaluations by category of hypothesis are difficult because different scenarios address different traits in different taxa. Moreover, scenarios often develop through a series of transformations and incarnations. This phenomenon as well as other ambiguities make clear comparative analysis difficult.

The following sections review some hypotheses that focus on the phylogeny of cognitive character states and the selection pressures on the evolution of cognitive abilities in monkeys and great apes. These hypotheses are evaluated in terms of their adherence to the phyletic and convergence criteria outlined above. This exercise is intended to illustrate the proposed approach to constructing and evaluating adaptive hypotheses. It is also intended as an argument in favor of more explicit formulation of evolutionary hypotheses in cladistic terms.

In his 1976 essay on the subject, Humphrey (1976) noted that the literature on animal cognition contains few hypotheses about the adaptive significance of animal intelligence. Despite the burgeoning literature on primate cognition, this is still true 20 years later. Until recently, most adaptive models have been proposed by biological anthropologists. The rarity of adaptive models may reflect the greater number of comparative psychologists relative to biologists and biological anthropologists engaged in studies of primate cognition.

Our discussion focuses on recent hypotheses concerning the evolution of cognition in monkeys and apes. These hypotheses fall into four categories: social, ecological, socialization, and locomotor.

SOCIAL HYPOTHESES

The idea that primates evolved intelligence primarily as an adaptation for social interaction and especially for deception has been proposed by several investigators. Frans de Waal (1983) documented long-term political strategies of competing male chimpanzees in a colony in Arnheim, Holland, in his book *Chimpanzee Politics*. Robert W. Mitchell (1986) classified various levels of deception that have evolved in primates and other animals. Whiten and Byrne (1988) reprinted various articles purposing the adaptive value of social manipulation in their book *Machiavellian Intelligence*.

Whiten and Byrne particularly credited Nicholas Humphrey's (1976) article "The Social Function of Intellect" with stimulating interest in this model. They also noted the influence of Allison Jolly's (1966) comments on social intelligence in lemurs, Chance and Mead's (1953) ideas about social behavior and primate evolution, and Hans Kummer's (1967) concept of tripartite interactions among baboons.

There are several models of Machiavellian intelligence among primates. Recently, de Waal (1996b) has proposed that the capacity for social reciprocity and some sort of accounting of favors given and received has been selected in monkeys and apes. Cheney and Seyfarth (1990) have suggested that vervet monkeys have evolved the capacity to recognize and categorize conspecifics by family membership as an adaptation for social competition. Harcourt (1988) has provided the most explicitly evolutionary formulation of alliance hypotheses.

Harcourt (1988) argues that tripartite alliances may have provided advantages in contests over scarce high-quality clumped resources and that this would have provided selection pressures responsible for the evolution of primate intelligence. He argues that the information processing required for participation in three-animal contests is both quantitatively and qualitatively greater than that required for two-animal interactions. Quantitatively, participants in two-animal contests must process information about the adversary's consanguinity and competitive ability. Participants in three-animal contests must also process information about the risk of injury to potential recipients who are not supported based on who else will attack them. Qualitatively, participants in two-animal contests must assess their abilities relative to potential adversaries; participants in three-animal contests must also assess the abilities of potential allies relative to one another. Harcourt suggests that three-party contests may be unique to primates, thereby implying that it is a shared derived characteristic of primates (though his examples are from Old World monkeys and apes). He does not, however, explain why such contests should be unique to primates.

In contrast, other early social hypotheses are somewhat amorphous. First of all, the concept of intelligence tended to be global in nature; the exact nature of social intelligence as a trait or traits was rarely specified. Second, even when specific abilities were specified, as in tripartite interactions, or levels of intentionality, the clades or groups of related species that share the trait or traits were rarely specified explicitly, nor were their derived statuses specified. More recently, however, some social theorists have turned to developmental models of social cognition that provide a consistent framework for identifying specific social character states and their phyletic distribution (Whiten, 1991).

Comparative studies using these approaches provide the data on cognitive differences between monkeys and great apes in imitation, pretend play, mirror self-recognition, and deception that are prerequisite to modeling common ancestry. In his book *The Thinking Ape*, Byrne (1995), for example, concludes that, among primates, imitation, teaching, and mental state attribution are restricted to great apes (or even to common chimpanzees). He notes that some of these abilities may also occur in dolphins (Byrne, 1995).

His formulation implicitly meets the phyletic criterion by specifying the clade that shares derived traits. He and other researchers are beginning to specify the exact nature of the shared derived traits and the primitive traits from which they were derived. Hence they are moving toward specifying the species who share the derived trait and the originating common ancestor. Only a few of these models, like Byrne's and

Harcourt's, mention analogous behaviors in distantly related species, much less discuss their adaptive functions. Research into distantly related taxa might reveal convergent adaptations, as, for example, among dolphins, that would meet the convergence criterion.

ECOLOGICAL HYPOTHESES

The hypothesis that primate intelligence evolved primarily as a means for exploiting environmental resources is generally represented as the alternative to the social manipulation hypothesis. There are several versions of this hypothesis. In one of the earliest, Clutton-Brock & Harvey (1980) proposed that greater comparative brain size (CBS) was favored by the demands of foraging on scarce food sources that are scattered in time and space. They reported that among primate species, frugivores (fruit eaters), who forage on such foods, display greater CBS than folivores (leaf eaters).

Taking a similar approach, Milton (1981, 1988), the only proponent of an ecological hypothesis represented in the *Machiavellian Intelligence* compilation, suggests that greater CBS and greater intelligence arose as an adaptation for efficient foraging for high-quality foods that are complexly distributed in time and space. Specifically, she contrasts the foraging strategies and brain indices of two sympatric New World monkeys, folivorous howlers and frugivorous spider monkeys, that she studied in Panama. She notes that the home range size of spider monkeys is roughly 25 times larger than that of howlers, while the brains of spider monkeys are 2 times larger than of those of howlers by a variety of measures. She suggests that the greater seasonality and patchiness of the fruits spider monkeys specialize on provide a far more complex problem in resource location than do the abundant leaves and figs howler monkeys specialize on. She also notes that spider monkey infants enjoy a more protracted period of dependency and learning than howler infants do (Milton, 1988).

Milton argues that greater CBS is generally associated with frugivory as opposed to folivory not only in primates but also in bats, rodents, insectivores, and lagomorphs. She also argues against social hypotheses for primate intelligence, on the grounds that "there is little evidence to suggest that social systems (or breeding systems) *per se* show a strong primary relationship to either primate brains or brain size in other mammalian orders" (Milton, 1988, p. 296). In a related, but more narrowly focused model, Gibson (1986) proposes that primates and nonprimates who engage in extractive foraging on embedded foods have higher encephalization and neocortical progression indices than related species who do not rely on this strategy (see the section below on the correlation method).

In an earlier model, Parker and Gibson (1977) proposed that intelligent tool use arose in the common ancestor of great apes and hominids as an adaptation for extractive foraging on a variety of embedded foods. Recently van Schaik, Fox, and associates (Fox & van Schaik, in press; van Schaik et al., 1996) have reported tool use in wild orangutans. On the basis of these data, they have also speculated that tool use arose in the common ancestor of the great apes.

Boesch and Boesch (1984) have proposed that chimpanzees have evolved levels of spatial knowledge comparable to that of 6- or 7-year-old human children. They argue that this knowledge arose as an adaptation for optimizing their foraging for tools and nuts in the Tai forest. Birute Galdikas (1995) has proposed that the intelligence of orangutans arose as an adaptation for efficient foraging on highly dispersed irregularly fruiting plants. She also proposed that imitation and observational learning in orangutans arose as an adaptation for socialization during the extended period of mother-infant association that precedes the more solitary existence of adults. Because both of these last two scenarios focus on single species, they fail to address cladistic criteria.

Parker and Gibson's (1977) extractive-foraging scenario and van Schaik et al.'s (1996) scenario meet the phyletic criterion. Parker and Gibson's scenario also meets the convergence criterion by specifying a convergent adaptation in cebus monkeys. The frugivory hypotheses of Clutton-Brock and Harvey (1980) and Milton (1988) and Gibson's (1986) extractive-foraging hypothesis fall into a different category than the other ecological hypotheses. They meet the correlation criterion for adaptation, which is discussed below.

SOCIALIZATION HYPOTHESES

In a third approach to the evolution of cognition, various investigators have proposed syntheses of the social and ecological hypotheses. We call these socialization hypotheses. Humphrey mentions this idea, saying, "One of the chief functions of society is to act as it were as a 'polytechnic school' for teaching subsistence technology" (Humphrey, 1976, p. 20). In his recent book, Byrne (1995) also acknowledges the plausibility of socialization hypotheses when he says, "perhaps it is not 'either/or' at all. Social and technological skills are not independent in practice" (Byrne, 1995, p. 209).

Building on their extractive-foraging hypothesis for hominoid evolution, Parker and Gibson (1979) suggested that imitation arose as in the common ancestor of great apes and hominids as an adaptation for learning tool-mediated extractive foraging and for transmitting information about the nature and location of distant food sources. The communication hypothesis was based on an analogy with communication of the location of food sources by bees.

Parker (1996a) recently amplified this hypothesis to suggest that teaching and self-awareness as well as imitation arose in the common ancestor of the great apes and hominids as an adaptation for teaching and learning tool-mediated extractive foraging. According to this apprenticeship hypothesis, teaching arose as a maternal adaptation for increasing the efficiency of imitative learning by segmenting out and repeating key elements the offspring needed to master. Both imitation and teaching presuppose an awareness of the orientation of the self and the self's actions relative to those of the other. The apprenticeship hypothesis was stimulated by the pattern of correlated abilities among the great apes. It partially meets the convergence criterion because it suggests that dolphins show an analogous adaptation.

Both formulations meet the phyletic criterion. Like the closely related extractive-foraging hypothesis, the apprenticeship model depends upon the supposition that the common ancestor of the great apes was an omnivorous extractive forager whose feeding strategy was similar to that of living chimpanzees. This atavistic hypothesis is proposed to explain why only chimpanzees and humans (and possibly orangutans) of the living descendants of the great ape radiation show the shared derived function in the wild (see below).

LOCOMOTOR HYPOTHESES

Yet another approach to the evolution of cognition might be called the locomotor approach. In the most explicit locomotor model, Daniel Povinelli and John G. Cant (1995) have proposed that self-awareness arose in the common ancestor of great apes as an adaptation for arboreal clambering. They define clambering as "a form of locomotion in which the body is orthograde, with various combinations of the four appendages grasping supports in different directions, both above and below the animals" (Povinelli & Cant, 1995, p. 401). They note that orangutan clambering is a highly flexible, unstereotyped locomotor compliance to complex demands that the arboreal substrates impose on a large-bodied animal.

They argue that the demands of this form of locomotion favored the evolution of mental representation of the body as a means for deciding appropriate movement strategies. They contrast clambering to the highly stereotyped locomotor patterns of sympatric long-tailed macaques and gibbons, which require only proprioceptive awareness. As the authors note, their hypothesis depends upon the notion that the common ancestor of the great apes depended upon a clambering form of locomotion similar to that of living orangutans. This, in turn, suggests that they were large-bodied creatures. This is unlikely, given Cope's "rule" that the common ancestor of reptilian and mammalian radiations is typically a small-bodied omnivore (Stanley, 1973).

The clambering hypothesis is similar to the hypothesis that complex locomotor adaptations favored the evolution of representational intelligence in orangutans (Chevalier-Skolnikoff et al., 1982). This earlier formulation, however, applied only to orangutans and hence failed to address any phyletic issues.

Note that the clambering hypothesis meets some elements of the phyletic criterion by specifying the selection pressures and the originating common ancestor. If clambering is unique to one living species, however, it does not meet the requirement for a shared derived character. Likewise, it does not meet the convergence criterion because no analogous adaptation is identified.

If, on the other hand, as recent work suggests, clambering is also seen in other living primates, it is possible to look at these species for possible analogies. A 1995 article defines a locomotor category of cautious climbing characterized by the following features: "slow deliberate movements, a toe to heel foot grasp, hook postures of the foot, extension of the hip and knee joint, and frequent bipedal behaviors when terrestrial" (Sarmiento, 1995). According to this model, this locomotor category is shared by all the great apes and was present in their common ancestor. It is also present in colobine monkeys (colobus and langurs) in the Old World and ateline (woolly and spider) monkeys from the New World. If clambering is a form of cautious climbing, we can test the model on the other primates who have independently evolved this form of locomotion. The colobine and ateline monkeys, however, show no signs of mirror self-recognition. In the absence of data on convergent functions, it seems more likely that use of self-awareness or insight is a side benefit of a character with another adaptive function.

SUMMARY EVALUATION OF ALTERNATIVE ADAPTIVE HYPOTHESES

Preliminary analysis of various hypotheses about the evolution of intelligence in primates reveals that relatively few are couched in the explicitly evolutionary terms required by the cladistic approach to adaptation. Nevertheless, several hypotheses about the adaptive significance of intelligence in primates meet at least one of the criteria. An increasing number of scenarios meet the phyletic criterion by specifying the shared derived character state (e.g., Byrne, 1995; Harcourt, 1988). Several scenarios (e.g., Byrne, 1995; Parker & Gibson, 1979; Povinelli & Cant, 1995) meet the criterion of specifying the taxonomic group or common ancestor with that character state. Only the extractive-foraging scenario (Parker & Gibson, 1977, 1979) explicitly meets the convergence criterion by specifying the convergent function as well as the convergent character state. The closely related apprenticeship scenario partially meets the convergence criterion by specifying the convergent character states.

According to Coddington's (1988) formulation outlined above, the phyletic criterion is necessary but insufficient to reconstruct the evolution of adaptation, while the convergence criterion, which depends on the phyletic criterion, is the additional requirement for doing so. Given the central role of convergence in this cladistic analysis, it is useful to discuss this phenomenon specifically in the case of cebus monkeys in relation to primate cognitive evolution.

WHY CEBUS MONKEYS ARE A BOON TO ADAPTIVE SCENARIOS

Cebus monkeys, also known as "the poor man's chimpanzee" because of their extraordinary manipulative abilities, have been popular subjects of cognitive studies. When interpreting the evolutionary significance of their performances, however, two things should be borne in mind. First, cebus and great apes are more distantly related than are macaques and great apes. So, if cebus display abilities that are not present in macaques and other Old World monkeys, these abilities must have evolved after the divergence of New World and Old World monkeys. Second, other New World monkeys do not share the abilities of cebus either. In other words, the cognitive abilities of cebus and great apes are not present in other New World or Old World monkeys. Therefore, these abilities could not have been present in the common ancestor of New World and Old World monkeys and great apes. These abilities must have arisen through convergent evolution.

Convergent abilities are a boon to evolutionary model building. If we find that cebus display an ability analogous to that of great apes and we find that it serves similar functions in the two groups in the wild, then we have the basis for an adaptive hypotheses. Specifically, the fact that both chimpanzees and cebus monkeys display tool use and prototool use in the context of extractive foraging on a variety of seasonal embedded food sources suggests that intelligent tool use arose as an adaptation for this kind of feeding strategy (Parker & Gibson, 1977). Recent reports of tool use in extractive foraging in wild orangutans supports the interpretation that this feeding strategy was characteristic of the common ancestor of great apes and hominids (van Schaik et al., 1996).

Contrasting patterns of convergent and nonconvergent abilities are also a boon to evolutionary model building. It is equally significant that cebus monkeys display differences from as well as similarities to great apes. Most notably, they fail to achieve the symbolic subperiod of preoperations. They also fail to imitate novel actions (Visalberghi & Fragaszy, 1996) and to recognize their mirror images.[4] This contrasting pattern of abilities in cebus and great apes, for example, suggests the hypothesis that imitation and mirror self-recognition are causally related, while tool use is not (Parker, 1991).

Differences among great ape species in the apparent significance of cognitive abilities in the wild raises another problem for evolutionary reconstruction. Persistence of cognitive character states in the absence of apparent adaptive advantage in the wild (e.g., tool use in gorillas) raises the issue of the difference between the origin and maintenance of adaptations (Coddington, 1988).

ORIGIN AND MAINTENANCE OF ADAPTATIONS: WHY GORILLAS ARE A FLY IN THE OINTMENT OF ADAPTIVE SCENARIOS

In his discussion of adaptation above the species level, Coddington (1988) emphasizes the distinction between the origination and the maintenance of adaptations. He notes that some evolutionary hypotheses imply both origin and maintenance of the shared derived trait. These hypotheses argue that the trait conserves its original function in all the taxa that share it. These hypotheses are highly testable because the trait can be investigated in current time.

In contrast, other evolutionary hypotheses imply origin but not maintenance of the shared derived trait. Such hypotheses argue that a trait

---

4. Although cebus monkeys successfully perform some simple tool-using tasks, the quality of their actions is very different from that of great apes. They show none of the characteristic deliberation of great apes that suggested insight to Köhler (1927). The qualitative and quantitative differences between the cognition of apes and cebus, as well as their independent evolution argue against Tomasello and Call's (1997) claim that great apes are no more intelligent than monkeys. The significant differences between great apes and cebus in brain size and life history also render this conclusion doubtful.

that was once adaptive is now maintained by phylogenetic inertia or developmental constraints in some species. These hypotheses, which generally apply to higher taxonomic groups, are generally untestable.

Note that *mosaic evolution* can result in mixed cases. Some species in an adaptive array might maintain the original function of a shared derived trait, while others do not. This seems to be the case in tool use in great apes. Gorillas, for example, do not show the shared derived trait of tool use in the wild, though they do in captivity (Parker, Mitchell, & Miles, in press)). According to some measures, gorillas develop more rapidly than other great apes. There is no doubt, however, that gorillas, having diverged after orangutans and before chimpanzees, are great apes (Ruvolo, 1994). This situation is inconvenient for scenario builders.

Generally speaking, primatologists and anthropologists have had one of several reactions. First, they have ignored gorillas and concentrated on chimpanzees or orangutans. Second, they have denied that gorillas have cognitive abilities comparable to those of other great apes (Gallup, 1977; McGrew, 1992). Third, they have suggested current functions for the mental abilities of gorillas (Byrne & Byrne, 1991; Byrne & Whiten, 1992). Fourth, they have explained the absence of cognitive abilities as secondary losses of shared derived abilities (Povinelli, 1994).

All but the last of these responses to apparent anomalies of gorilla cognition betray nonevolutionary thinking. The phyletic criterion provides a corrective to such nonevolutionary approaches. Specifically, it suggests that intelligent tool use, imitation, and self-awareness are shared derived character states that gorillas, orangutans, chimpanzees, bonobos, and humans derived from their common ancestor (Parker, 1996b).

AN ALTERNATIVE APPROACH TO TESTING ADAPTIVE HYPOTHESES: THE CORRELATION CRITERION

In contrast to the cladistic-derived phyletic and convergence criteria, the correlation approach addresses the adaptive significance of characters that recur sporadically within and across a wide range of taxa. According to this approach, correlations between characters across taxa is evidence for adaptive significance. If, for example, nuptial plumage and polygyny are correlated, and absence of nuptial plumage and absence of polygyny are correlated, this suggests that nuptial plumage is an adaptation for male competition in courtship (Brooks & McLennan, 1991).

The correlation approach seems particularly well suited to analysis of characters that have evolved repeatedly through parallel or convergent evolution and therefore are widely distributed among both closely and distantly related taxa. Nuptial plumage and other sexually selected characters are one such class of characters. Dietary specializations such as frugivory or folivory are another such class of taxonomically widespread, recurrently evolving characters. We call this test for adaptation the *correlation criterion*.

This approach is valid if the correlation is based on the number of times the association has evolved independently rather than the number of species displaying the correlation (Harvey & Pagel, 1991). In the interests of controlling for phylogeny, this approach parses out the effects of shared derived characters. Consequently, it does not meet the phyletic criterion. Because it often includes closely related species, many of the correlations must have arisen from parallel evolution while others may have arisen from convergent evolution.

Some of the ecological hypotheses proposed to explain the larger brain size of primates have been based on such correlations. The first of these is Clutton-Brock and Harvey's (1980) demonstration that frugivores have larger CBS than folivores. Their category of frugivory is broad and includes all primate species in whose diet fruit represents the largest component (therefore it encompasses omnivores and many extractive foragers). CBS is a genus-specific measure they propose to control for effects of phylogenetic affinity. As indicated above, they suggest that increased CBS follows from the greater information-processing demands of foraging on resources that are complexly arrayed in time and space. They also note that CBS correlates with home range size. Milton's (1988) model (discussed above) is very similar to that of Clutton-Brock and Harvey. These are both adaptive scenarios that meet the correlation criterion.

A second ecological scenario in this category is Gibson's (1986) argument that primates who engage in extractive foraging on embedded foods have bigger brains than nonextractors as measured by both Jerison's encephalization index and Passingham's neocortical progression index. Gibson distinguishes between primate species who use anatomical manipulators and those who use tools to extract embedded foods. She finds that primate species in both categories have bigger brains by both measures. This is an adaptive argument that meets the correlation criterion.

In a third scenario, Dunbar (1992, 1993) proposes that neocortical volume in primates correlates with group size rather than extractive foraging. He hypothesizes that neocortical volume limits the information-processing capacity of organisms, which, in turn, limits the number of social relationships they can simultaneously monitor. Strictly speaking, this correlation argument is about constraints on evolution rather than adaptation. Dunbar tries to counter Gibson's extractive-foraging hypothesis using different primate species than she does. (He omits large-brained nonsocial orangutans from his sample.) The significance of Dunbar's correlation is further undermined by other intervening variables (e.g., the fact that group size correlates with home range size, which correlates with frugivory [Milton, 1988]).

These investigators have used the correlation criterion to argue for three different models for the evolution of brain size in primates and other mammals. (Unfortunately for purposes of comparison, however, they have used different measures of relative brain size, different measures of dietary adaptation, and different samples of primate species.) Milton and Gibson's ecological models provide adaptive scenarios whereas Dunbar's provides an evolutionary constraint scenario. Both the adaptive scenarios meet the correlation criterion. Neither of these meet the phyletic criterion (because they cut across taxonomic groups) or the convergence criterion (because in these statistical calculations convergence is conflated with parallel evolution).

CONCLUSIONS

This chapter has reviewed a cladistic methodology for constructing and evaluating adaptive scenarios. This method is useful for explaining relatively rare adaptations that are characteristic of a few taxonomic groups, as, for example, cognitive abilities of the great apes. Drawing on the cladistic definition of adaptation and its associated methodology, we have distinguished two criteria for assessing the validity of such scenarios: the phyletic criterion and the convergence criterion.

Coddington (1988) characterizes convergence as the capstone methodology for reconstructing adaptation. Based on this, we have discussed the importance of recognizing convergent adaptations of cebus monkeys and great apes. We have also emphasized the heuristic value of making evolutionary models explicit through specification of taxonomic groups (clades) that share derived characteristics as required by the phyletic criterion.

We have also considered the alternative correlation approach to testing adaptive scenarios. In contrast to the cladistic approach, this approach is useful for explaining adaptations that are relatively widespread but sporadically distributed among taxonomic groups. Finally, in an effort to illustrate their heuristic power, we have suggested how cladistic criteria apply to some well-known alternative hypotheses—ecological, social, socialization, and locomotor—about the evolution of cognitive abilities in various primate groups.

If adult human behavior were really neotenous, it would consist primarily of reflexes and modal action patterns rather than of highly intelligent constructed acts. Similarly, if the adult brain were neotenous, it would be dominated by brain stem rather than cortical structures. The view that the human brain is neotenous is held by scholars . . . who have never studied brain development. Developmental neurobiologists often have claimed just the opposite, that neural ontogeny parallels neural phylogeny.

GIBSON, 1990, P. 120

CHAPTER 12
THE EVOLUTION AND DEVELOPMENT OF THE BRAIN

This chapter briefly reviews comparative data on the nature of primate brains, their development, their relation to cognition, and their evolution. It also discusses evolutionary implications of brain development. Following this, it discusses two competing models for the evolution of the human brain and argues that overdevelopment (peramorphosis) rather than underdevelopment (paedomorphosis) has shaped the evolution of the quintessential human characteristic.

COMPARATIVE BRAIN ANATOMY AND FUNCTION

The *neuron* is the nearly universal unit of information processing in metazoan nervous systems. Likewise, the miniature nervous system composed of a connected loop of sensory and motor neurons is the nearly universal unit of organization of metazoan nervous systems (Jerison, 1976). Vertebrate brains are organized into the hindbrain, the midbrain, and the forebrain. See Figure 12-1 for diagram of vertebrate brains.

The *hindbrain* is composed of the medulla, pons, and cerebellum, which control the visceral processes (breathing, swallowing, and circulation) and coordination of movement. In lower vertebrates the midbrain is composed of the tectum, which coordinates sensory and motor pathways. In mammals the *midbrain* (mesencephalon) is composed of the cranial nerves and the anterior and posterior colliculi, which process and relay information from the eyes and ears. The *forebrain* is composed of the diencephalon and the telencephalon (the end brain). The dien-

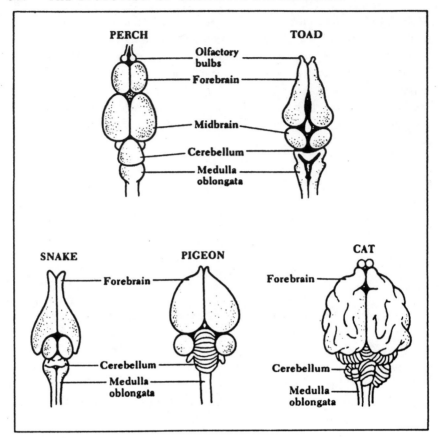

FIGURE 12-1
Brain parts in vertebrates (Sagain. 1977).

cephalon is composed of the thalamus and the hypothalamus, which, respectively, relay messages to the cortex (in mammals) and control appetites and emotions. The telencephalon is composed of the olfactory lobe, the corpus striatum (the basal ganglia), the limbic lobe, and the cerebral hemispheres. The cortex is designed for the reception, analysis, and storage of information (Butler & Hodos, 1996).

The *cerebral cortex* is composed of two hemispheres that are connected by a broad band of fibers, the corpus callosum, and by various commissures. In humans, and probably most mammals, the right and left sides of the cortex are specialized for slightly different functions. The cortex has six layers, with layer 1 the outermost and layer 6 the innermost, each layer receiving or sending its axons to specific structures inside or outside the cortex. The cortex is also organized in vertical columns that include these six layers (Gibson, 1991a).

The cortex is a bilaterally symmetrical rind separated into four lobes: the frontal lobes (over the eyes), the temporal lobes (over the ears), the parietal lobes (above the temporal lobes), and the occipital lobes (in the back). The occipital lobes are specialized for the analysis and synthesis of visual information coming from the retina. The temporal lobes are specialized for analysis and synthesis of auditory information coming from the cochlea of the ear. The parietal lobes are specialized for analysis and synthesis of tactile information from the skin and for association among sensory modalities (Butler & Hodos, 1996). The frontal lobes are specialized for planning, decision making, and motor control (Luria, 1973). The Sylvian fissure separates the temporal lobe from the frontal and parietal lobes; the Central fissure separates the frontal lobes from the parietal lobes. Likewise, the Lunate sulcus separates the visual cortex from the parietal cortex. See Figure 12-2 for a depiction of the major cortical sulci and motor homunculi of human brains (bearing in mind that the relative size of parts reflects their relative functional importance).

The *frontal lobes* in humans and other anthropoid primates are composed of two sectors, the posterior sector containing the motor cortices and the anterior sector containing the prefrontal association cortices. The motor cortices are composed of the primary motor cortices and the association cortices. The motor cortex contains a motor homunculus of the contralateral side of the body, which controls movements. (In humans this includes Broca's area, which is involved in speech production.) The motor cortex receives and integrates information from the sensory cortices in the parietal lobe—which contains a sensory homunculus—and subcortical structures. The association cortices contain the limbic association areas, which are involved in social behavior.

The prefrontal association areas are involved in decision making, planning, emotional regulation, and creativity (Semendeferi, in press). The enlarged frontal lobe of humans is implicated in human language, advanced planning, and hypothetical-deductive reasoning (Deacon, 1997; Noack, 1995).

Each cortical region includes both primary or projection areas and secondary zones. The projection areas, which are in afferent (incoming) layer 4, are highly specialized by modality. They are surrounded by secondary areas in layers 2 and 3 that contain more interneurons subserving associative and synthetic functions (Luria, 1973).

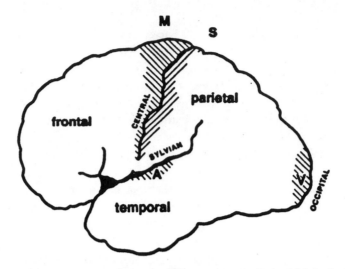

The primary cortical areas of the human brain (left side) include: M, motor; S, somatosensory; A, auditory; and V, visual, with the lobes of the brain and Sylvian and central sulci indicated.

A homunculus that illustrates the order and scale in which body parts are represented on the primary somatosensory and motor areas of the brain (left hemisphere).

FIGURE 12-2
Human cortical sulci and motor homunculi (Falk, 1992). Used with permission.

COMPARATIVE NEUROGENESIS

Increases in the size of these three divisions of the brain in Tertiary mammals occurred roughly proportionately with one another and with the brain as a whole. Hence the trend in vertebrate brain evolution from the Eocene to the present has not been merely corticalization (Jerison, 1976).

The size of various parts of the brain has been explained by the principle of proper mass according to which the relative masses of neural tissue associated with different functions reflect the relative importance of those functions within and among species (Jerison, 1976). Therefore, it is interesting to note that the brains of monkeys and apes are similar to those of other mammals in their structure and function, in the proportions of their parts to the whole (Finlay & Darlington, 1995), and in the order of maturation of the various parts (Gibson, 1991b). This similarity is quite astounding given that they vary by about 1000-fold in size, that is, from 1 to 1000 grams.

Deacon (in press) notes that, although mammalian brains differ by many orders of magnitude in size and neuron numbers, they vary little in the volume of their neuronal cell bodies. Larger brains require longer axons, which require larger neuronal cell bodies and increased thickness of *myelin* sheaths, which increase the ratio of white to gray matter. They also require more supporting glial cells. Larger brains take longer to develop than smaller brains. Although the proportions of major brain parts are relatively constant across mammals, the parts and the whole differ considerably in size among species. Generally speaking, the timing and duration of neurogenesis determine the size of adult brain structures in such a way that the later neurogenesis begins, the larger the resulting structure (Finlay & Darlington, 1995). The later the onset time, the greater the number of stem cells that have proliferated and hence the greater the number of neurons. The later the development of a neuron within a brain, the smaller its size (smaller cell bodies have shorter axons) (Deacon, in press).

Primate brains differ significantly from those of most other mammals in their larger size, particularly their larger cortices, and the higher proportion of the basal metabolism devoted to brain function (Armstrong, 1983). The brains of anthropoid fetuses are twice as large for a given body weight as those of the fetuses of other mammals (Martin, 1983). These differences apparently follow from the early segmentation of the primate embryo, which gives a greater (as compared to other mammals) proportion of the embryo to brain as opposed to body mass (Deacon, in press; Sacher, 1982). A unique pattern of human brain size and growth arises early in *embryogenesis* when there is an increased production of cortical neurons versus noncortical neurons during neurogenesis (Deacon, 1990a).

Recent discoveries reveal that embryo development, including nervous system development, is similarly patterned across phyla by the action of homeotic genes, which are expressed in serial order along their chromosomes (Raff, 1996). The fact that the products of these genes act via chemical processes in cells imposes an upper limit on the size of cell populations controlled by genetically orchestrated processes. This is indicated by the fact that both the smallest and the largest animals go through the majority of the steps in embryonic fate determination while they are still less than 1-cm long. These size-limited molecular mechanisms cause any discrepancies in growth rates to be increasingly exaggerated by increased size. These developmental constraints are reflected in the allometric factors described above (Deacon, in press). The fact that fate determination in mammalian brains occurs early in embryogenesis underscores the epigenetic nature of brain development.

Several features distinguish brain development from the epigenetic development of other more highly specified body tissues. In combination, these unique features allow the brain to mold itself to fit the body in which it finds itself (Deacon, 1997). Rather than being highly specified genetically, brain development is a highly interactive and contingent systemic phenomenon that depends crucially upon information from both the peripheral and central nervous system. This systemic effect follows from the long-distance connections of the axons of neurons that connect directly with distant cells and from the branching neuronal dendrites that can receive inputs from many other neurons (Deacon, 1997).

Growing axons are directed toward their targets through various mechanisms, including differences in cell adhesion, growth factors, and mechanical features of supporting tissues (Raff, 1996). Specific connections among neurons are not prespecified. Rather, they are forged through temporal correlation of firing patterns of input axons and output neurons. Multiple, nearly synchronous firing of many inputs are required to excite a receiving neuron to fire. Those structures that send the greatest number of axons to a given target in the brain will win the competition for connections. This quasi-Darwinian selective mechanism results in selective retention and elimination of neurons according to function and results in sculpting of adaptive connections among the plethora of over-produced neurons that populate the embryonic and immature brain (Deacon, 1997; Edelman, 1987).

The spatial geometry of peripheral sensory and movement systems in an organism contributes the stable connective biases whose influences propagate through the developing nervous system as their peripheral nerves fire. This means that modification of relative proportions of peripheral and central nervous structures—through developmental defects or through evolutionary change—can change connective patterns. Displacement of one system by another can occur through changes in peripheral nervous structures. So, for example, the parts of the cortex devoted to vision in typical rodents become recruited for auditory analysis in blind mole rats (Deacon, 1997).

Displacement also occurs when evolutionary changes in body size result in allometric variants. So, for example, larger-bodied creatures retain the same size eyes as their smaller relatives, and the relative amount of nervous tissue devoted to visual analysis is smaller in these species. This then frees up some brain tissue to take on other functions, for example, association functions. Hence, changes in scale can influence brain function (Deacon, 1997).

Deacon (1990a, 1997) argues that the human pattern of brain development depends upon a uniquely large allocation of brain versus body tissue in the earliest stages of embryogenesis. This is the so-called King Kong hypothesis (Deacon, 1990a) in which the embryo has a chimpanzee-size body and a *Gigantopithecus*-size brain. Deacon (1997) describes the scenario:

> Since the chimp body would only be a fraction as large as would normally carry a brain this size, the space within this brain that these inputs would recruit would be significantly reduced compared to what would happen in a normal *Gigantopithecus* body. . . . The adult brain would be quite different from either the donor or the host species' brains. Many structures and functional divisions of the transplanted brain would be smaller than expected for a *Gigantopithecus* brain, but others would be larger, if they inherited neural space from those that were more constrained by peripheral connections. (Deacon, 1997, pp. 214–215)

Figure 12-3 from Deacon (1990a) compares the ontogenetic curves of brain and body development in humans and other primates.

The implications of this model are twofold. First, the brain evolves through changes in input and output as well as changes in size rather

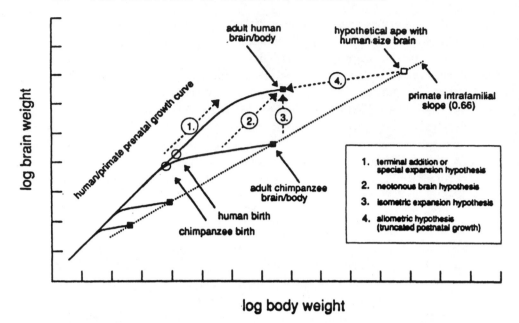

FIGURE 12-3
Ontogenetic growth curves of brain to body weight in various primates
(Deacon, 1990b). Reprinted with permission of the publisher.

than through selective addition or enlargement of parts. Second, most
cortically mediated functions are not hardwired but develop epigeneti-
cally through peripheral functioning. To the degree that modularity occurs, it is a developmental phenomenon (Elman et al., 1996; Karmiloff-
Smith, 1995). Therefore, the model of evolution of specific new cognitive
and linguistic *modules* proposed by evolutionary psychologists (Cosmides & Toobey, 1991; Pinker, 1994) seems unlikely.

BRAIN SIZE AND LIFE HISTORY

Contrasting trends in brain size and development are associated with
contrasting patterns of life history and neonatal adaptation (Harvey
et al., 1987). Large-bodied, long-lived mammals tend to mature late, have
few offspring, and invest heavily in each offspring. These so-called **K**
strategists tend to have large brains. In contrast, small-bodied, short-
lived mammals tend to mature early, have many offspring per reproductive effort, and invest little in each offspring. These so-called **r** strategists
tend to have small brains (Harvey et al., 1987; MacArthur & Wilson,
1967). Within primates, gorillas exemplify **K** strategists, whereas mouse

lemurs exemplify r strategists. (See chapter 7 for discussion of primate life histories.)

K strategists tend to have long gestation periods devoted to nurturing single offspring that are born with well-developed senses and good loco-motor capacities. These *precocial offspring* develop slowly after birth and enjoy prolonged parental investment. In contrast, r strategists tend to have short gestation periods devoted to nurturing a large litter of small offspring that are born in a fetal state with poorly developed senses and locomotor capacities. Although these *altricial offspring* enjoy intensive parental investment for a short period after birth, they develop rapidly and leave the nest or burrow as soon as they have achieved the equivalent of full gestation outside the womb. Both infants and adults of altricial species have smaller brains than infants and adults of precocial species. Altricial species experience much greater increase in brain size post-natally (from 5 to 45 times) than do precocial species (from 1.5 to 5 times). Humans are anomalous among precocial primates in increasing their brain size by a factor of 3.5 after birth (Portmann, 1990) (see chapter 7).[1]

Species differ in brain maturation at birth and postnatal brain growth and development. The percentages of growth of the brain at birth, late childhood, and late adolescence in several primate species are given in Table 12-1 from Portmann (1990).

The magnitude of multiplication of brain size after birth in various altricial and precocial mammals is given in Figure 12-4 from Portmann (1990). Note that altricial mammals increase brain size 4 times or more after birth, whereas precocial mammals do so less than 4 times.

These data clearly show a trend toward larger brain and body size from monkeys to apes to humans, as well as toward longer periods of prenatal and postnatal brain development across these taxa. They also show that human brain and body development are significantly differ-ent from those of other primates. According to various investigators, human infants might have a 12- to 21-month gestation if they followed the pattern of neonatal brain development of other primates. To put it another way, human neonates are born 3 to 12 months prematurely (Gibson, 1990; Portmann, 1990). This also suggests that adult humans

---

1. Portmann (1990) suggests that humans would have a 21-month gestation period if they were born in the state of development characteristic of other precocial mam-mals. Gibson (1990) notes, however, that 3-month-old human infants have attained a developmental state of myelination comparable to that of a neonatal rhesus macaque, a precocial primate, which suggests that they would have a 12-month gestation to be comparable.

TABLE 12-1 PROPORTION OF BRAIN SIZE ACHIEVED AT BIRTH, CHILDHOOD, AND ADOLESCENCE IN PRIMATES

Selected anthropoid primate	New born	Late childhood	Late adolescence
Macaque	40.9	79.8	91.2
Gibbon	62.7	84.3	95.5
Orangutan	40.4	90.4	96.8
Chimpanzee	45.3	83.7	98.8
Gorilla	59.4	81.4	97.9
Human	23.3	81.7	96.7
	Average		
All	—	83.55	96.1
Ape	49.74	—	—
Human	23.3	—	—

Note. From Portmann (1990). © 1990 by Columbia University Press. Reprinted with permission of the publisher.

would achieve a body size exceeding 1000 pounds if their brain-body ratio paralleled that of other primates.

COMPARATIVE MEASURES OF BRAIN SIZE

Standard measures or indices of relative brain size are prerequisite for comparative studies. Many indices have been proposed over the years. These include:

1. absolute brain size

2. ratio of brain size to body size

3. log of brain size to log of body size (Clutton-Brock & Harvey, 1980)

4. an encephalization quotient (the proportion of the brain devoted to cortex) (Jerison, 1973)

5. a neocortical progression index (the size of the cortex relative to that of a reference group) (Stephan, 1972)

6. neocoritical indices (relative neocortex volume measured as the residual of neocortex volume regressed on either body mass or the total brain mass) (Dunbar, 1992)

7. cerebral multiplier factors based on magnitude of postnatal increase in brain size (Portmann, 1990)

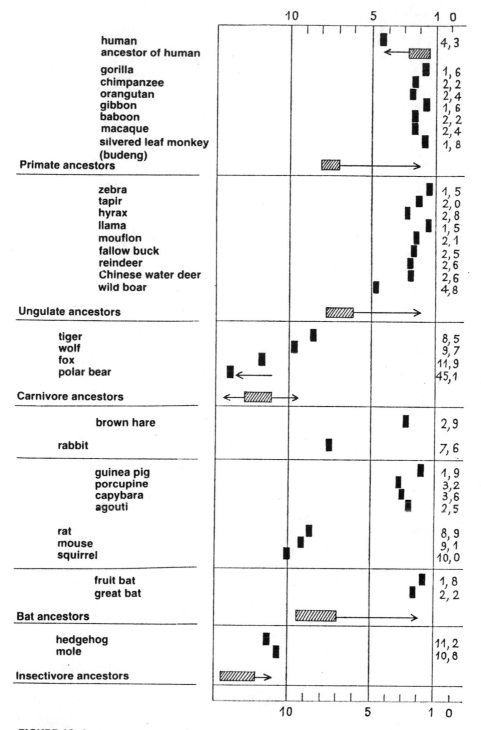

FIGURE 12-4

Multiplication of brain size after birth in various mammals (Portmann 1990)

8. comparative brain size (log of brain weight minus the sum of the elevation of the regression line for the family and the slope of regression line for the family divided by the log of body weight) (Clutton-Brock & Harvey, 1980)

9. percentage of basal metabolism (Armstrong, 1983)

Absolute brain size is significant because it reflects information-processing capacity (see discussion below) (Gibson, 1990). Brain size correlates with a variety of measures of neural complexity. Simply increasing linear brain dimensions increases the maximum size and range of axonal and cell sizes and the maximum quantity and range of myelination and decreases neural densities (cytoarchitectonic allometry). It also increases the network size through increased dendritic synaptic connections among neurons (network allometry) (Deacon, 1990a).

The relationship between brain size and body size is poorly understood. Variation of body size within genera and families is much greater than variation in brain size. Deacon (1990a) questions the common assumption that changes in brain/body ratio necessarily reflect changes in mental capacity. He notes that they may reflect, rather, selective advantages for larger body size. He suggests two reasons why brain size is more likely to determine body size than vice versa. First, the target size for brain growth is determined early in the fetal period, and brain growth ceases early in postnatal development, whereas body growth continues throughout development. Moreover, the brain, particularly the hypothalamus, interacts with extrinsic factors to regulate body growth and development, whereas the body does not regulate brain growth.

Biologists have argued that simple brain-size–to–body-size ratio is misleading because (owing to allometry) brain size does not scale proportionately with body size, and hence comparisons based on this measure tend to underestimate the brain size of larger species. An allometric formula is designed to compensate for scaling effects attendant on increasing or decreasing body size, which occurs ontogenetically in individuals within species during development and phylogenetically between species during evolution.

The *allometric exponent* is an empirical measure based on statistical analyses of many species. Earlier analyses by Jerison (1973, 1976) and others suggested that brain size scales at about 0.67 of body size (allometric exponent 0.67) where Y (brain size) = k (allometric coefficient) times X (body size) to the exponent e (scaling factor or slope of the regression line) ($Y = kX^e$ or log Y = log e [log X + log k]). Later analyses based on larger samples of animal species (Armstrong, 1983; Martin, 1981) concluded that brain size scales at about 0.75 of body size.

Some investigators believed that the earlier allometric exponent 0.67 suggested an association between brain size and body surface area (which increases as a square of the radius as compared to the volume, which increases as cube of the radius as bodies become larger) (Jerison, 1976). Others maintain that the later allometric exponent of 0.75 suggests an association between brain size and basal metabolic rate (Martin, 1981, 1983; Armstrong, 1983). These functional interpretations are based on correlations between the allometric exponents and the power functions of these two phenomena. Deacon (1990a) notes that correlation studies do not demonstrate causality. He finds both hypotheses improbable because they imply underlying causal relationships with no intervening variables.[2]

Traditional data on brain size are problematic because of small sample sizes, differences in measurement techniques, and differences in the postmortem age and condition of the specimens. The range of variation in adult brain size for five primate species is given in Table 12-2 from Schultz (1972).

Most of the common indices of brain size among living primate genera and families and fossil hominids are based on the old exponent of allometric scaling, 0.67, rather than the newer one of 0.75. This makes a substantial difference in interpretation. Figure 12-5 compares regression lines for log of brain size versus body size at various slopes determined by various allometric exponents among primates and other mammals (Deacon, 1990a).

Comparative data on brain size of living primates reveal a trend to-

---

2. The *encephalization quotient* (EQ): log (brain weight) = log (0.12) + 0.67 × log (body weight) is based on the first allometric exponent (Clutton-Brock & Harvey, 1980; Jerison, 1973). The Index of progression (IP) is the ratio between observed and expected brain weight for a species based on a regression line of brain weight on body weight for a sample of basal insectivores (Stephan, 1972). It is designed to reveal differences in the proportion of the neocortex to other parts of the brain across various vertebrates. The *comparative brain size* (CBS) index for a given genus = log (brain weight) – (elevation for family + slope for family × log [body weight]) (Clutton-Brock & Harvey, 1980). CBS is designed to overcome certain biases in slope of regression lines for species within the same family that arise because larger species in the same family and genus will tend to show relatively smaller brain sizes compared to smaller species. Clutton-Brock & Harvey designed CBS to overcome these biases by calculating separate slopes for each family.

TABLE 12-2   BRAIN SIZE OF OLD WORLD MONKEYS, APES, HOMINIDS, AND HUMANS

Species	Neonatal brain capacity (cm³)	Adult brain capacity (cm³)
Cebus	29	71 to 82
Cercopithecus (vervet)	?	59 to 79
Macaque	54 to 66	69 to 109
Papio	74	142 to 214
Gibbon	50 to 55	90 to 116 (f)
		92 to 125 (m)
Orangutan	170	276 to 431 (f)
		334 to 502 (m)
Gorilla	227	345 to 553 (f)
		410 to 715 (m)
Bonobo	?	265 to 420 (f)
		295 to 440 (m)
Chimpanzee	128	270 to 450 (f)
		322 to 503 (m)
Australopithecine	?	352 to 550
Homo erectine	?	694 to 1228
Human	384	1040 to 1615 (f)
		1160 to 1850 (m)

Note. From Schultz (1972); Harvey, Martin, and Clutton-Brock (1987); Holloway (1997). ? = No data available.

ward increasing absolute brain size in the following series of common ancestors: (1) ancestral Old World monkeys represented by macaques at about 63 cm³, (2) ancestral apes represented by gibbons at about 90 cm³, (3) ancestral great apes represented by chimpanzees at about 300 cm³, (4) ancestral *Homo* species represented by *Homo erectus* at about 1000, and (5) modern humans at about 1300 cm³. The brains of great apes are 3 to 4 times as large as those of most Old World Monkeys, those of middle Pleistocene hominids (Homo erectines) are 3 times the size of great ape brains, and those of living humans are 4 times their size (Passingham, 1982; Schultz, 1972). As we have seen in earlier chapters, the relative cognitive abilities of monkeys, great apes, and humans reflect these differences more or less exactly with macaques traversing one and a half subperiods of cognitive development, great apes traversing three and a half subperiods, and humans traversing seven subperiods (chapter 6 provides a summary). In other words, the sensorimotor intelligence of monkeys is associated with brain sizes of 100 to 200 cm³, the preoperational intelligence of great apes with 300 to 500 cm³, the concrete operations intelligence of Homo erectines with 900 to 1000 cm^3, and the formal operations intelligence of modern humans with 1250 cm^3.

COMPARING BRAIN ORGANIZATION

Considerable controversy has centered on the relative importance of differences in brain organization as compared to brain size in primate and human evolution. Brain reorganization is reflected in species differences in the proportions of various parts of the brain, particularly the cortex. Whereas brain size is relatively easy to measure in fossil as well as living forms, the proportions of various parts of the cortex are difficult to measure in fossils. This difficulty arises from the difficulty of identifying various sulci on the endocasts of fossilized brains (Holloway, 1996).

Studies of brain cases and endocasts of fossil hominids have sought to identify patterns of brain sulci that indicate the relative sizes of the cortical lobes, but have not always agreed on their interpretations. Falk (1992) argues that these features support the interpretation that the brains of Australopithecines were similar to those of living great apes, whereas those of Homo erectines were more like those of living humans. Spe-

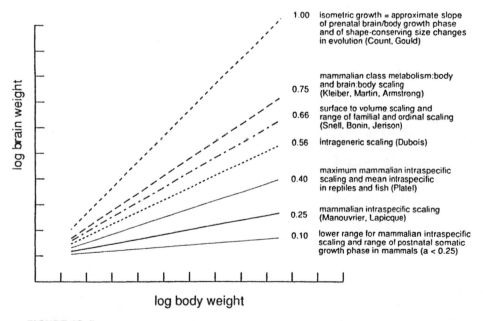

FIGURE 12-5
Comparison of slopes of regression lines of brain on body weight (Deacon, 1990a).

TABLE 12-3   HEMISPHERE AND FRONTAL LOBE VOLUMES IN PRIMATES

	Hemisphere	Frontal lobe
Macaque	62,737	17,654
Gibbon	91,385	28,438
Orangutan	268,553	94,705
Gorilla	348,336	112,912
Chimpanzee	305,521	109,800
Human	1,125,492	413,103

Notes. From Semendeferi et al. (1997). Values are given in cubic millimeters and include both hemispheres.

cifically, Australopithecines had an apelike—that is, anteriorly placed—lunate sulcus separating the occipital lobe from the rest of the brain (Falk, 1992). In contrast, Holloway (1996) argues that an increase in the parietal lobe had already occurred in Australopithecines. He argues that precise identification of the lunate sulcus in the Taung's specimen is impossible but that indirect evidence in the *Australopithecus afarensis* endocast AL 162-28 suggests a more modern position.

Several investigators have compared the sizes of various parts of human brain with those of a typical primate (Deacon, 1992; Passingham, 1982). Whereas most investigators have claimed that the frontal lobes of humans are proportionately larger than those of all other primates, recent work shows that the frontal lobe of the human cortex is proportionately somewhat larger in both humans and great apes than in other mammals. Table 12-3 compares absolute sizes of the frontal lobes of macaques, gibbons, orangutans, gorillas, chimpanzees, and humans (Semendeferi, Damasio, Frank, & Van Hoesen, 1997).

As data from volumetric studies in Table 12-3 show, in absolute terms human frontal lobes are more than 3 times the size of those of great apes, and great apes' frontal lobes are 3 times those of gibbons and 5 times those of macaques. The relative size of the frontal lobes as a ratio of the rest of the hemispheric volume increases slightly from macaques to great apes and humans: 28% in macaques, 31% in gibbons, and 32 to 36% in great apes and humans. Hence there has been a trend toward relatively larger frontal lobes from monkeys to lesser apes to great apes, but not from great apes to hominids (Semendeferi et al., 1997). It seems obvious, given cognitive differences between great apes and humans, that absolute size of the frontal lobes must be functionally significant.

Recent work on the temporal lobe of chimpanzees reveals that their left planum temporale (PT) (within Wernicke's area) is significantly larger than their right PT, a pattern previously thought to be unique to humans (Gannon, Holloway, Broadfield, & Braun, 1998). Earlier work suggested that orangutans, but not lesser apes, also show this lateral asymmetry. The authors note that this implies that this pattern must have been present in the common ancestor of chimpanzees and humans or, indeed, in the common ancestor of all the great apes. They list several hypotheses, including one that the PT provides the neural substrate for "chimpanzee language" and gave rise to human language capacity (Gannon et al., 1998). In humans, Wernicke's area is involved in understanding speech. The lateralization of the PT in great apes is consistent with evidence that their comprehension exceeds their production, as discussed in chapter 5.[3]

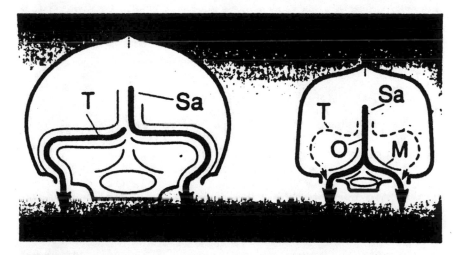

FIGURE 12-6
Blood flow routes that appear in the back of human skulls (*left*) and in Hadar hominids and robust Australopithecines (*right*). Sa, superior sagittal sinus; T, trasverse sinus; O, occipital sinus; and M, marginal sinus (Falk, 1992). Used with permission.

3. Deacon (1997) notes that brain organization is better understood in terms of brain functions rather than language functions: "The question we should ask then is . . . what is it about these areas that dictates the distribution of language functions that we observe?" (Deacon, 1997, p. 268). He notes that, as language evolved, pre-existing areas of the brain were recruited for those new information-processing functions to which they were best suited. Elman et al. (1996) make a similar point.

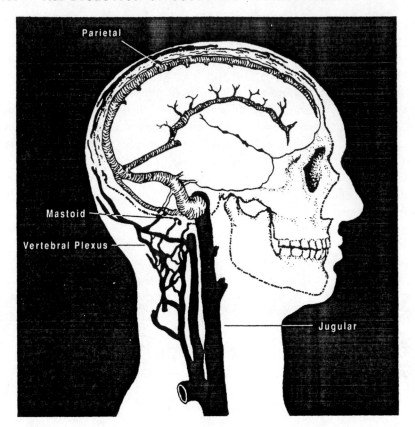

FIGURE 12-7
Side view of blood flow in humans. The radiator network of veins is extensive. Small veins on the surface of the skull enter the bone, within which they merge with other (diploic) veins. These in turn communicate with the veins on the inside of the braincase (Meningeal veins) as well as the venous sinuses. A few components of the radiator, known as emissary veins, penetrate clear through the skull via small holes known as emissary foramina (Falk, 1992). Used with permission.

COMPARISON OF BRAIN VASCULARIZATION PATTERNS

Studies of endocasts of hominid brains indicate that some Australo-pithecines (*A. afarensis* and *A. robustus*) had a pattern of venous flow in the brain that is different from that of living humans. Their occipi-tal/marginal sinuses (O/M) flowed into the sagittal sinus before exiting the cranium. This contrasts with the transverse sinus blood flow in modern humans. The transverse sinus system is associated with enlarged emissary veins in the cranium (the opthalamic, mastoid, and parietal emissary veins), which cool the brain under conditions of heat stress (Falk, 1992). Although both patterns are apparently adaptations for bipedal locomotion, the latter system is also an adaptation for temperature regulation of larger brains. (See Figures 12-6 and 12-7 for a depiction of the two patterns of blood flow.)

RELATIONS BETWEEN BRAIN DEVELOPMENT AND COGNITIVE DEVELOPMENT IN HUMANS

Perhaps the most ambitious attempt to trace the relationship between postnatal brain development and cognitive development in humans can be found in the work of Robert Thatcher. Thatcher has done large-scale studies of the development of electroencephalogram (EEG) patterns in hundreds of individuals from 2 months to 26 years of age in 3-month increments.

The EEG coherence methodology involves statistical measures of phase synchrony or shared activity between two or more neural regions. Changes in EEG coherence reflect changes in the number and synaptic strength of coriticocortical connections between neural regions, particularly in the gray matter as opposed to the white (myelinated) matter. Hence it correlates with synaptogenesis and neurotransmitter production (Thatcher, 1994).

This methodology allowed Thatcher to identify a pattern of brain growth spurts nested within 4-year cycles: cycle I from 1½ to 5 years, cycle II from 5 to 10 years, and cycle III from 10 to 14 years. Large growth spurts occurred between the ages of 5 and 7 years and between 9 and 11 years. Smaller growth spurts occurred at 1½ years and 2½ to 3 years. They were defined by phase transitions or peaks in velocity when there was a maximum increase in mean EEG coherence. These cycles differed in the left and right hemispheres (Thatcher, 1994).

Left-hemisphere cycles involved a sequential lengthening of intracortical connections between the frontal regions and the posterior parietal sensory regions governed by anterior-posterior and lateral medial gradients. Right-hemisphere cycles involved a sequential contraction of long-distance connections between the frontal regions and the posterior sensory connections. These occurred repeatedly through the life span in a sequential pattern of a left-hemisphere developmental spurt followed by a right-hemisphere developmental spurt (Thatcher, 1994).

Thatcher identified several anatomical poles of development that exhibited spatial axes and developmental gradients: (1) the anterior-posterior pole, (2) the lateral-medial pole, and (3) the left-right–hemisphere pole. As indicated above, the right and left hemispheres develop at different rates and in different ways. The left hemisphere develops from short-distance differentiation to long-distance integration, whereas the right hemisphere develops in the opposite direction, from long-distance integration to short-distance differentiation. These trends reflect specialization of the left hemisphere for analytical and sequential information processing versus the holistic and integrative information processing of the right hemisphere.

The anterior-posterior pole also shows a developmental gradient. The frontal is characterized by a sequential lengthening of connections in the left hemisphere and a sequential shortening of connections in the right hemisphere. The posterior pole is characterized by sequential lengthening of connections from the occipital and parietal regions in the left hemisphere toward the frontal region and sequential contraction of connections to the temporal and occipital lobe in the posterior direction (Thatcher, 1994).

Finally, the lateral-medial pole shows a developmental gradient. In the left hemisphere the direction is from lateral to medial, whereas in the right hemisphere the direction is from medial to lateral. The medial interhemispheric trajectories were mostly synchronous whereas the lateral interhemispheric trajectories were asynchronous (Thatcher, 1994). Figure 12-8 graphically depicts the developmental cycles, growth spurts, and poles.

Thatcher interprets these developmental cycles as a sort of spiral staircase in which cortical connections are reorganized in successive sweeps resulting in successively higher levels of neural integration (Thatcher, 1994, p. 257). He suggests that they strongly support neo-Piagetian models of stages of cognitive development recurring at higher levels of integration (Case, 1985; Fischer, 1980; Thatcher, 1994). Thatcher argues that cognitive stages are outward manifestations of underlying cycles of brain growth and that Piaget's assimilation and accommodation are reflections of the various developmental gradients in the frontal lobes. He says that, according to a predator-prey model of EEG interactions, "frontocortical predation may represent the cycle accommodation of assimilated schemata created in posterior cortical regions" (Thatcher, 1994, p. 260).

Few comparative studies of the brain have specifically addressed the relationship between brain development and cognitive development in

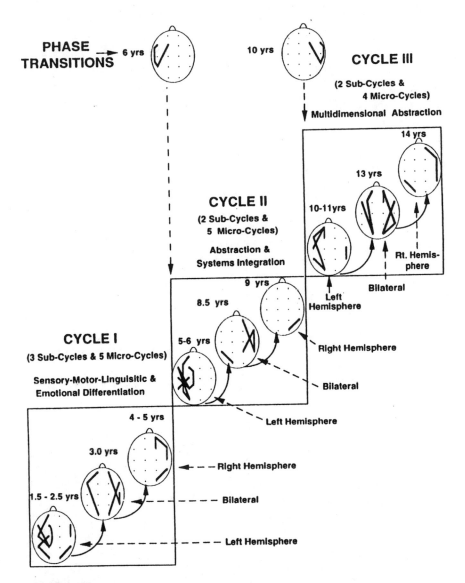

FIGURE 12-8
Diagrammatic representation of predominant developmental cycles and subcycles of neocortical reorganization (Thatcher, 1994). Reprinted with the permission of Guilford Press.

nonhuman primates. We are aware of two such studies, Gibson's (1977) study of myelination in rhesus monkeys in relation to Piagetian stages and Diamond's (1991) study of the development of object permanence in rhesus monkeys and humans. Thatcher's EEG coherence studies offer a potential tool for comparative studies of the neurodevelopmental bases for cognitive development in monkeys and apes. Such studies could reveal, for example, the number of cycles and growth spurts in other primates, which would allow investigators to see if they correlate with the number of subperiods and stages of cognitive development. If species that display more subperiods and stages of cognitive development also display more cycles of brain development, this will provide clear-cut evidence for terminal extension in the evolution of brain development.

BRAIN DEVELOPMENT AND COGNITIVE DEVELOPMENT IN PRIMATES

Brain development can be measured in a variety of ways, including gross size increase, degree of myelination, and degree of synapse formation. Of these various measures, myelination is the best studied comparatively. Although myelination is not necessary for onset of nerve function, the presence of the sheath significantly increases functional efficiency of nerve transmission and therefore correlates with behavioral development. Moreover, there is a close correspondence between myelination and increasing neuronal cell size and dendritic maturation (synaptogenesis) (Gibson, 1991b)

The sequence of *myelination,* formation of the electrical insulation covering nerve axons, is known to be very similar across vertebrates. The brain stem myelinates before the telencephalon although there is some overlap in the myelination of the two structures. Within the cortex, the various lobes myelinate in a strict sequence in the six layers. The rate of myelination and its extent at birth, however, differ significantly across taxa. In most species of altricial mammals (e.g., cats, rabbits, and rats), myelination begins after birth, whereas in species of precocial mammals (e.g., macaques and humans) it begins before birth. Altricial species experience rapid myelination, which is often complete within weeks of birth, whereas precocial species experience slower myelination, which may continue for years (Gibson, 1977, 1991b).

Gibson has compared the sequence and rate of myelination of rhesus macaque monkeys with comparable data on humans. She reports that both species begin myelination before birth. Macaques' brains are heavily myelinated at birth, particularly in the brain stem and are beginning to myelinate at the cortical level. The two species show a similar pattern of myelination at birth except that the degree of myelination is less in the brain stem and cortical myelination is not present in humans.

Cortical myelination in humans begins at 1 month postnatally and continues 6 or 8 years. In macaques, it begins prenatally and continues 1 to 2 years. Overall, postnatal myelination proceeds 3 or 4 times faster in macaques than in humans so that a 3-month-old human brain resembles that of a neonatal macaque and a 8- to 12-month-old human brain resembles that of a 3-month-old macaque in its maturation by this measure. Unfortunately, comparable data on great apes are not available, but their rate of behavioral development suggests that their myelination schedule in infancy is closer to that of humans than to that of macaques (Gibson, 1991b).

Gibson (1977, 1991b) has shown that brain myelination sequences in macaque and human infants parallel the stages of sensorimotor intellectual development, from reflexive to voluntary to differentiated and freely combinatory actions.

> Regional myelination patterns correlate with behavioral maturation, as reported by Piagetian scholars. The primary sensory and motor areas that provide fine within-modality discriminatory capacity myelinate in advance of the association areas, which provide higher-level hierarchical constructional skills. . . . Similarly, lower cortical layers, which interconnect and interact primarily with brain stem structures, mature in advance of higher cortical layers, which provide for corticocortical integration. (Gibson, 1990, p. 115)

Gibson (1990) also notes that the sequence both of myelination and of behavioral development from reflexive to voluntary coordinated actions parallels the evolution of development of sensory and motor abilities in vertebrates

In a complementary approach, Diamond has studied the neuroanatomical regions involved in object permanence and successful search for, and uncovering of, a hidden object under a second cloth in humans and rhesus monkeys. She has conclusively demonstrated localization of the requisite information processing for these tasks in the frontal lobes (Diamond, 1991). In human infants these and related abilities emerge between 9½ and 12 months, and in rhesus monkeys they emerge between 4½ and 5 months. These achievements coincide with maturational changes in the synapses in the frontal cortex in both species (Goldman-Rakic, 1987).

COMPETING MODELS FOR THE EVOLUTION OF THE HUMAN BRAIN: UNDERDEVELOPMENT VERSUS OVERDEVELOPMENT

Two competing models of human brain evolution have captured the imagination of anthropologists and psychologists, the juvenilzed ape or neoteny model and the overdeveloped ape or peramorphic model. We believe that comparative data support the latter model.

HETEROCHRONY AND BRAIN EVOLUTION

This section explains how "overdevelopment" is responsible for our most human trait, our enlarged frontal and parietal cortices. Contrary to the widespread myth, the complex behavioral and cognitive abilities of humans do not derive from the plasticity of a "juvenilized" (paedomorphic) brain. Rather, they derive from the more abundant neurons and greater synaptic complexity of an overdeveloped brain, especially the prefrontal cortex. Similarly, the juvenilized shape of the skull derives from a large brain combined with relatively small "underdeveloped" dentition.

The persistence of the myth of juvenilization is testimony to the way false but appealing ideas can become entrenched in scientific and popular culture. It is also testimony to the way that such entrenchment can distort interpretation of scientific data. The juvenilized ape notion has been woven into discussions of many disparate aspects of human behavior, including neurobiology, evolutionary psychology, and developmental psychology. This notion causes a serious distortion of evolutionary and psychological thought because the developmental and evolutionary mechanisms producing a juvenilized brain are vastly different from those producing one that is overdeveloped. Also, the personal and social implications of the two models for human evolution are very different. The cognitive deterioration that occurs when humans grow old, for example, is just the opposite of the "mental rejuvenilization" predicted by Ashley Montagu (Montagu, 1989) in his popularized view of the juvenilization model in his book *Growing Young*.

THE JUVENILIZATION MYTH OF BRAIN EVOLUTION

There are both scientific and cultural reasons for the persistence of the juvenilized ape myth. The initial scientific stimulus for the myth, according to Gould's (1977) masterful history, was the superficial morphological similarity between adult humans and living juvenile apes (see also Montagu, 1989). The relatively large, complex brain is the aspect of humans that is most commonly emphasized as the key to our evolutionary success (e.g., Allman, 1994; Desmond, 1979; Gould, 1977; Gribbin, 1988; Montagu, 1989; Wesson, 1991). Stephen Jay Gould has been an especially influential proponent of the juvenilization model and continues to promote it (Gould, 1996a).

The juvenilization model has also been incorporated into professional literature at many levels. It can be found in basic texts on developmental psychobiology (Michel & Moore, 1995), as well as advanced discussions such as those of "neural Darwinism" (Edelman, 1987). It is found in Calvin's (1993) well-known model that larger brains evolved for accurate throwing. Even the well-known developmental biologist Lewis Wolpert attributes our larger brain and enhanced learning to juvenilization (Wolpert, 1991).

There are many reasons for the persistence of the myth. In scientific circles the attraction of the juvenilization concept is abetted by basic confusion over terminology and mechanisms of developmental evolution (see chapter 8). Juvenilization is scientifically seductive because it is an economical way to explain many aspects of human anatomy and behavior. Juvenilization certainly does occur, as is well documented in domesticated animals (Mason, 1984), especially dogs (Coppinger & Schneider, 1995; Morey, 1994). In such cases, covariant suites of anatomical and behavioral traits of juveniles in the ancestral species can be readily seen in the adult of the juvenilized descendant species. Dogs are thus often characterized as retaining the puppylike appearance and behaviors of wolves (Wesson, 1991). The seductiveness of such simple scenarios for human developmental evolution is further increased by attributing juvenilization to "regulatory genes" that alter the rate or timing of development (e.g., Gould, 1977) and the apparent (but unproved) tendency for juvenilization to enhance evolutionary potential and developmental plasticity (DeBeer, 1958).

OUR OVERDEVELOPED BRAIN: PROLONGING GROWTH

Given the complex developmental origins of many heterochronies (McKinney & McNamara, 1991; Raff, 1996), it is ironic that evolution of life's most complex organ, the human brain, has a comparatively simple basis. Production of cortical neurons in mammals is limited to early prenatal developmental. Mitotic rates of neuron creation in mammals are relatively constrained, perhaps reflecting limitations on neuron mitotic processes in the prenatal environment (Sacher & Staffeldt, 1974). Thus, as discussed by Finlay and Darlington (1995), it is developmentally "easier" to produce a larger brain by extending prenatal brain growth than by altering rate of growth. In the case of modern humans, this fetal brain growth phase is extended about 25 days compared to that of living monkeys.

Extension of fetal brain growth has had profound cascading consequences that underlie all later major brain and cognitive developmental delays and complexity increase. Specifically, fetal growth extension causes the key traits (discussed below) that characterize brain overdevelopment in humans.

High Brain/Body Ratio. As noted by Shea (1989), the brain grows faster than the body during the extended fetal phase, producing a high brain/body ratio in humans. After this phase, especially postnatally, brain growth is much slower than body growth so that a lower brain/body ratio could result if body growth were accelerated as in the relatively fast-growing gorilla (McKinney & McNamara, 1991). But in humans, postnatal body growth is relatively slow, so that our bodies stay relatively small. Thus, modern humans have a brain size that, as mentioned, would be predicted for a much larger primate (Deacon, 1997).

This high brain/body ratio has, as noted, sometimes been used as evidence for juvenilization in human evolution. We suggest, however, that it results from a brain that is overdeveloped in both absolute and relative (brain/body) terms. It could be argued that humans have a juvenilized "shape," as defined by the brain/body ratio (Gould, 1977). But many workers have noted that shape is a misleading way to identify juvenilization, or any heterochrony, because the same shape can be attained by many different combinations of rate or timing changes (Raff, 1996; Shea, 1989). As Deacon (1990b, p. 270) notes, "Paedomorphism of human brain/body proportions . . . is an artificial correlate of brain size evolution. The fact that humans exhibit paedomorphic brain/stature proportions with respect to other apes . . . in no way implies any corresponding arrest of brain differentiation."

Furthermore, it makes no sense to argue that human postnatal body growth is juvenilized because humans have generally been increasing in body size over the last 5 million years as a result of progressively delayed puberty (B. H. Smith, 1992). In other words, both brain size and body size have increased, but brain size has increased relatively faster to give us a high brain/body ratio.

Greater Neural Complexity. Increased brain size alone is a coarse way to define *overdevelopment.* A finer definition would include the fact that the large number of neurons produced during prolonged fetal growth also undergoes a prolonged period of postnatal growth and maturation. Delay in offset of infancy and juvenile stages of development leads to a more complex brain because dendritic growth in the cortex is extended to 20 years or more in humans (Gibson, 1991b). Similarly, glial cell growth and synaptogenesis are all also prolonged and thus extend beyond ancestral developmental patterns (Gibson, 1990, 1991b). This extension of dendritic and synaptic growth explains why neurons in the human brain have many more dendritic and synaptic interconnections with each other than they have in other primates (Purves, 1988).

Brain myelination is also prolonged in humans. Because it promotes more effective nerve transmission, myelination plays a large role in improvements in memory, intelligence, and language skills during development (Case, 1992b; Gibson, 1991b). Myelination is highly canalized; that is, it follows the same specific sequence among primates. So regional delays are readily seen in humans. In both monkeys and humans, myelination begins in the brain stem and proceeds through subcortical areas, with neocortex and especially prefrontal cortex being among the last to myelinate (Gibson, 1991b). But rhesus monkeys show myelination only up to about 3½ years, whereas in humans, myelination continues until well over 12 years of age (Gibson, 1991b). Even peak synaptic complexity is delayed in humans. Dendritic complexity in the human brain peaks at about 2 years of age and begins to decline thereafter as "pruning" begins (Edelman, 1987). In contrast, in apes and monkeys, dendritic pruning begins well before 2 years old (Purves, 1988).

More Neocortex and Prefrontal Cortex. Finlay and Darlington (1995) show that increase in brain size, from shrews to primates, accounts for an enormous amount (over 96%) of the species variation in size of individual brain regions. Of special relevance to humans is that the order of events in neurogenesis is phylogenetically highly conserved. As a result, evolutionary brain-size increase has occurred by disproportionate growth in areas of the brain, such as the neocortex, that are generated late in development (Finlay and Darlington, 1995). Because the neocortex (or isocortex) is a "general-purpose integrator," increasing neocortical size produces an increasing capacity to process information of all kinds (Finlay & Darlington, 1995). Tool use, language, and social behavior have all increased in complexity as the neocortex increased in size because, as Gibson (1993) notes, they all share common neocortical substrates that promote mental constructional skills.

The area of the neocortex long considered crucial for cognitive function is the prefrontal cortex. Significantly, the prefrontal cortex is the center for short-term memory ("central executive system"), in which information is temporarily stored and manipulated (Case, 1992b). This function has been confirmed by magnetic resonance imaging of prefrontal brain activity during various cognitive tasks (D'Esposito et al., 1995). The phyletic increase in this area and the parietal cortex has been much greater than in any other area of the brain. Other areas of the cortex, including the occipital and olfactory lobes, have concomitantly decreased (see above).

On the basis of such evidence, Gibson (1991b) concludes that "on neurological grounds . . . there is nothing paedomorphic about the adult human brain." In direct contrast to the infant primate, adult human brains are "large, highly fissurated and myelinated with complex synaptic morphology."

OUR OVERDEVELOPED COGNITIVE SKILLS: PROLONGED ACCELERATION

Human mental development follows the same general pattern as that of the brain, that is, prolongation with no reduction in rate of development, ultimately producing an overdeveloped adult. Humans have a prolonged period of cognitive development relative to great apes while following the same general cognitive sequence (chapter 6). Modern human cognitive development thus goes beyond that of other primates. It is also accelerated relative to that of great apes in that the same stages of late sensorimotor and early preoperations are attained at a younger age in humans. Accelerated cognition is ultimately attributable to having a larger endowment of neurons, dendrites, and synapses, which allows storing and manipulating information at a faster rate, especially in the prefrontal area.

MECHANISMS OF BRAIN OVERDEVELOPMENT

The mechanisms that result in developmental evolution can be described at many levels, from genetic to behavioral (Raff, 1996). At a coarse level, the evolution of human cognitive overdevelopment resulted from the combination of extended and accelerated cognitive development associated with prolonged childhood and life history in general. But this accelerated and prolonged cognitive development can be traced back one step further, to the prolongation of fetal growth, which extends the duration of neuron mitosis (Finlay & Darlington, 1995). This extended fetal growth not only produces the large number of cortical neurons characteristic of the human brain, it also has cascading effects on neuronal complexity by prolonging development of individual neurons, allowing more complex dendritic and synaptic outgrowths and connections.

Furthermore, many aspects of brain complexity seem to originate as cascading effects that ultimately derive from prolongation of fetal brain growth. As Deacon (1990b, p. 270) notes,

> In all the major indicators of mammalian brain development and maturation, including the level of differentiation of brain structures, the morphological maturation of neurons, the level of myelination of axons, the rates and specificity of neurotransmitter synthesis, . . . as well as many other measures, the adult human brain has at least achieved the level of . . . mature mammalian brains and typically has carried these trends much further.

That many aspects of brain size and complexity are overdeveloped at so many scales, from the cellular to the gross morphological, is indicative of a highly integrated system. This high integration agrees with the highly covariant, constrained patterns found by Finlay and Darlington (1995) in mammalian brain evolution. Or, as Deacon (1990b, p. 277) points out in his overview of brain ontogeny and phylogeny, "Brain development parameters are remarkably similar in different species and across a phenomenal range of sizes."

Can the sequence of causation be taken back one step further to specify causes of prolonged fetal brain growth? Deacon (in press) has suggested that the prolonged fetal brain growth occurs because of changes in segmentation in very early embryonic development. Changes in homeotic gene expression may alter the initial proportion of late-maturing embryonic stem (parent) neurons. If so, then the "ultimate" heterochronic event underlying human brain evolution would be traceable to mutations in homeotic genes that produce *predisplacement*, the heterochronic term referring to "starting off with more" (McKinney & McNamara, 1991).

It is unclear that the prolonged fetal growth of humans is also responsible for our prolonged childhood and life span. For example, a variety of mechanisms are involved in aging, many of which are poorly understood (Wolpert, 1991). Is our long life span traceable all the way back to a resetting of the "mitotic clock" during fetal growth? Or has selection acted upon a different set of mechanisms (e.g., endocrine timing) that are activated postnatally?

It is important to note that the early fetal developmental mechanisms that produce overdeveloped brain and cognition in humans have little to do with the known mechanisms that produced the juvenilized morphology and behaviors of domesticated dogs. Wayne (1986) shows that the juvenilized morphology of dogs is produced by growth rate changes occurring mainly during late fetal development. Furthermore, the juvenilized social behaviors of domesticated dogs, when compared to wolves (Goodwin, Bradshaw, & Wickens, 1997), are apparently unaccompanied by any change in cognitive abilities.

WHAT, EXACTLY, IS RECAPITULATED?

The idea that human mental evolution is recapitulated during ontogeny was popular at the turn of the nineteenth century. The idea that the development of children may repeat the mental development of early humans was favored by Darwin himself and many early psychologists such as Baldwin and G. Stanley Hall (Morss, 1990). It is a theme in the writings of Freud and Piaget and continues to be proposed today (Ekstig, 1994). Recently, at least a few anthropologists have begun to interpret human cultural evolution in terms of mental ontogeny (Mithen, 1996; Parker & Gibson, 1979; Wynn, 1989).

It is tempting to interpret brain overdevelopment as a result of terminal addition that produces evolutionary recapitulation in modern human ontogeny. There are at least two caveats to this interpretation, however. First, brain overdevelopment is not a simple process of terminal addition. Second, although general aspects of brain and cognitive evolution are recapitulated, all the fine details of brain and especially cognitive evolution are not repeated. It is important therefore to specify exactly what is being recapitulated.

TERMINAL EXTENSION, NOT TERMINAL ADDITION

Human cognitive overdevelopment has been produced largely by prolonged fetal brain growth. This prolongation produces not only the addition of more neurons at the end of brain growth but also overdevelopment in many aspects of brain (and cognitive) development, such as neuron, dendritic, and synaptic complexity and connectedness. For this reason, the term *terminal extension* is more accurate than *terminal addition* because it better describes the process whereby many aspects of the developmental trajectory, from cellular to brain morphology to behavior, are extended (McKinney, in press-a, in press-b).

GENERAL BRAIN AND COGNITIVE RECAPITULATION

MacLean's (1990) "triune brain" hypothesis is the most popularized expression of the overdeveloped brain concept (e.g., Damasio, 1994; Sagan, 1977). The triune brain refers to three grades of brain evolution, represented by the protoreptilian (R-complex), the paleomammalian (limbic system), and the neomammalian (neocortex) structures (MacLean, 1990).

The details of the triune brain hypothesis have been strongly criticized (Deacon, 1990a, 1990b). But the evidence that brain development is highly covariant and brain evolution has been highly constrained (Deacon, 1990a, 1990b, 1997, in press; Finlay & Darlington, 1995) implies that brain ontogeny should recapitulate general evolutionary patterns. Raff (1996) has described such patterns as the result of "phylotypic" embedding, in which certain ontogenetic traits become entrenched among higher taxa. Traits that develop late in ontogeny are "modularized" and can be modified independently by selection. Traits that develop earlier, during organogenesis, are less liable to modification because modularization has not yet occurred. In this light, the "triune brain" appears to describe phylotypic embedding representing reptilian, early mammal, and late mammal brain evolution, respectively.

At a finer scale, there is evidence that the origination and maturation (e.g., synaptogenesis, myelination) of various areas of the brain roughly correspond to the order of their evolutionary appearance (see especially Gibson, 1990, 1991b). Konner (1991) discusses how such brain recapitulation may translate into behavioral recapitulation. Ritualized motor displays similar to those found in reptiles are among the first displays to appear in human infants due to maturation of the basal ganglia. In mammals, these are followed by bonding, attachment, and other more complex emotional behaviors evoked by maturation of the limbic system (Konner, 1991). Finally, maturation of neocortical areas underpins cognitive development, a process that lasts for a substantial part of the human life span (Gibson, 1991b).

This scenario of recapitulation is very general. It focuses mainly on the fact that neocortical areas were the last brain areas to originate (Deacon, 1990a, 1990b) and enlarge (Finlay & Darlington, 1995) and are the last to mature during development (Gibson, 1990, 1991b). It is unclear to what degree finer details of development of areas, within the human neocortex, recapitulate their evolution. Human cognition rests largely on the prefrontal area, which has increased relative to an average living anthropoid (Deacon, 1990a, 1990b). But the exact evolutionary sequence that produced that increase is unknown because fossil crania do not preserve the details of neocortical structure. Evidence suggests, however, that the steadily increasing brain size of human evolution largely translated into steadily increasing prefrontal area. Fossil crania, for instance, show disproportionate increase of the prefrontal cavity during human evolution (Holloway, 1996).

General recapitulation in brain structure is reflected in brain function. Cognitive abilities are related to neocortical size and complexity, which permit increasingly complex mental constructions (Gibson, 1990, 1991b). Such mental constructions include increasingly complex combinations of objects, words, and, ultimately, ideas that are created and manipulated (Gibson, 1993). It seems that, as our complex brain matures (e.g., synaptogenesis, myelination), especially the prefrontal cortex (Case, 1992b), our general ability to "think" by manipulating ideas (and objects and words) recapitulates the evolution of those abilities of our ancestors (Ekstig, 1994). This is the outcome of the increasing mental abilities of a series of hominoid ancestors, which resulted in the cognitive recapitulation discussed in chapter 9.

CONCLUSION: EVOLUTION OF COGNITIVE COMPLEXITY CAN BE "EASY"

The evidence discussed here invalidates the juvenilization scenario for the origins of human cortical hypertrophy. The human brain is generally overdeveloped when compared to ancestral primate brains. This overdevelopment is manifested in many ways, ranging from our larger relative brain size down to the fine details of our greater complexity of neurons, dendrites, and synapses. Of special note is the relatively greater size and neural complexity of the prefrontal and parietal areas, which are crucial for the most complex cognitive processes. As a result of this brain overdevelopment humans attain higher levels of cognitive development, accelerating cognitive development during each stage (Langer, 1996; Parker, 1996b). These developmental extensions of brain size and complexity are exactly the opposite of those predicted by the popular juvenilization myth of human evolution (e.g., Gould, 1996a).

The relatively simple developmental mechanism underlying the evolution of complex brains implies that neurological overdevelopment (McNamara, 1997) and higher intelligence can readily evolve once certain prerequisites are in place. The genetic and developmental mechanisms that produced brain overdevelopment are surprisingly simple, especially given their enormous impact on brain size and complexity, cognition, and behavior. All of these may be consequences of a single underlying cause, predisplacement, in which a mutation or mutations in homeotic gene expression allot proportionately more late-maturing embryonic stem neurons in the developing human fetus (Deacon, 1997; in press). Increased size and complexity of the neocortex, including disproportionate increase in the prefrontal and parietal areas, are produced by this initial change as cascading consequences of the highly conservative development of the mammalian brain, which produces disproportionately more growth in late-maturing areas (Finlay & Darlington, 1995). Later maturation not only increases the general size of those areas but greatly increases complexity at the synaptic level by extending neuronal and dendritic growth (Gibson, 1990, 1991b).

As life's most complex organ (Katz, 1987), the human brain has been at the center of heated debates about evolution and progress (Gould, 1977; Morss, 1990). In addition to promoting the juvenilization scenario of human evolution, Stephen Jay Gould has spent many years attempting to refute the concept of evolutionary progress by arguing that evolution is dominated by "luck" (e.g., Gould, 1996b). This argument is countered in the final chapter.

CHAPTER 13
COGNITIVE COMPLEXITY AND
PROGRESS IN EVOLUTION

It is important to view the evolution of human cognition in the larger context of the history of life. To use Darwin's famous metaphor, the "tree" of human evolution is just a tiny branch on the much larger evolutionary tree of life. Examination of this entire tree shows that, for all our uniqueness, humans represent the culmination of many general patterns of evolution found in the biosphere. In particular, humans are the result of a general trend toward increasing complexity, both morphological and behavioral, that has characterized the history of life.

The concept of evolutionary progress is unpopular in many scientific circles. Perhaps the British paleontologist Simon Conway Morris (1995) said it best when he noted that "It is now distinctly unfashionable to talk about evolutionary progress." But why is this so? One set of reasons is cultural. Many commentators have noted how scientific notions are influenced by their cultural context. The popularity of the idea of evolutionary progress in the nineteenth and early twentieth centuries may have been a response to the great economic and technological expansion occurring in Europe during that time. Conversely, the recent decline in popularity of the idea of evolutionary progress has been a reaction against the social Darwinism, racism, and colonialism associated with that expansion (Gould, 1977; Lovejoy, 1960). Unlike the turn of the nineteenth century, the turn of the twentieth century is marked by somber prognostications. Indeed, Matthew Nitecki's (1988) excellent book of readings on evolutionary progress begins with the sentence, "Pessimism prevails in our world." Adding to such background influences on our perception of the idea of progress, are vocal critics who actively argue against the concept. Stephen Jay Gould (1988a), for example, has written that "progress is a noxious, culturally embedded, untestable, nonoperational intractable idea that must be replaced."

In addition, there are scientific reasons for the declining fashion of the idea of evolutionary progress. Many of the earlier assertions about evolutionary progress are now known to be grossly inaccurate. Simplistic notions of orthogenesis and teleology are now readily dismissed in a few lines in even the most basic textbooks. Predictably, the pendulum of scientific opinion has swung in the other direction, and many observers now see evolution as largely directionless and random (Gould, 1989, 1995) and dominated by catastrophic events that wipe clean the evolutionary slate (Raup, 1991).

Despite these cultural and scientific influences, the concept of evolutionary progress is far from dead. Michael Ruse (1993) has suggested that the idea of evolutionary progress keeps resurfacing because, at least in part, it is seductive. In this chapter, we argue that it keeps resurfacing because evolutionary progress is real at many scales of observation. Furthermore, we argue in favor of a much-maligned heresy: that there is an ultimate, absolute evolutionary progress. Even the most ardent critics often acknowledge that there are many local progressive trends that affect only a few lineages, such as increasing body size or shell thickness (Gould, 1988a; Hull, 1988). Such progress is relatively well documented (Ruse, 1993).

But here we argue for what Ruse (1993) has called *absolute progress*, meaning increase on a scale of fixed value. The scale of fixed value we use is *complexity*. Complexity is an enormously complicated topic that has recently produced a huge literature. As we show below, complexity has the important advantage of being measurable in a number of ways. This is crucial if we are to make an objective assessment of progressive change in evolution.

Complexity has another advantage: It is a general trait that, unlike, for example, brain size, can be measured in many kinds of organisms from bacteria to humans. Also, as the opposite of organizational simplicity, which was the starting point of life, it is the logical metric for global progress of the biosphere as a whole. In contrast to the local kinds of progress noted earlier, we can attempt to identify any general trend in complexity in the history of life. In Daniel Dennett's (1995) terms, complexity would represent global progress, or in David Hull's (1988) terms, it represents one "big" direction as opposed to the many "little" directional trends that are restricted in time and space to a few evolutionary lineages and limited periods of time.

EVOLUTION OF COMPLEXITY

This chapter reviews four major points concerning evolutionary increase in complexity:

1. The upper limit of both morphological and behavioral complexity in the biosphere has increased through geological time.

2. A key mechanism driving this complexity increase is natural selection of modified ontogenies, that is, of mutations affecting late ontogeny that can extend the developmental trajectory.

3. The progressive increase in the upper limit of complexity has not been a simple monotonic increase, but is a statistical diffusion process into more complex kinds of morphology and behavior. It is characterized by many reversals and delays.

4. Morphological complexity has slowed its rate of increase, but unlike morphology, behavioral and cognitive complexity have accelerated their rate of increase. We suggest that this reflects a qualitative difference between morphological and mental adaptations. Behavioral and cognitive adaptations represent ways of overcoming the severe physical limitations inherent in morphological adaptation. Behavior and cognition are ultimately much less constrained.

Individually, none of these ideas is new. John Bonner (1988), for instance, has argued extensively in favor of most of them. But theoretical and empirical advances in the last few years, especially in paleontology, further validate these points. These recent advances will argue that the evolution of complexity is not "boring," as Maynard-Smith (1988) once labeled it. His is a sentiment expressed by a number of workers who point out that, because life began as simple organisms, then of course it must evolve complexly because there is nowhere else to go. This view has assumed various names, such as the "nowhere but up" process (McKinney, 1990; McShea, 1994) and Gould's (1988b) "expansion of variance" process discussed later. These views are correct, but they are hardly boring. Instead, they pose many intriguing questions. For example, what was driving the expansion of variance? Was it intrinsic genetic or developmental innovation or was it extrinsic environmental selection?

WHAT IS COMPLEXITY?

Complexity is a difficult but essential concept. Many kinds of detailed definitions have each proved inadequate in some way (Horgan, 1995). The effort by Brooks and Wiley (1988) to measure the evolution of complexity in terms of information theory has also been subjected to the familiar criticisms that it is too abstract and therefore difficult to test.

To minimize such problems of abstraction, we discuss complexity in fairly specific ways. Usually complexity refers to more types of something (O'Neill, DeAngeles, Wiede, & Allen, 1986). So the general implication for morphological complexity is that there is an increase in the number of morphological parts and hence in their potential interactions. Similarly, increased behavioral complexity implies more kinds of behaviors, although the fossil record obviously records morphological evolution better.

One of the most widely discussed metrics of measuring morphological complexity has been the number of cell types. Bonner (1988), for example, plotted the number of cell types displayed by the most complex organisms alive at a given time in geologic history and found that the upper limit of biosphere complexity had increased. This approach was recently refined by Valentine and others (Valentine, Collins, & Meyer, 1994), who documented an average rate of origin of one new cell type per 3 million years. Cell type number is obviously a very crude approximation of complexity. Among many other things, it gives equal weight to all differences of cell type, as Michael Ruse (1993) has noted. But it is better than body size (e.g., Bonner, 1988)—reducing all organizational complexity to pounds or kilograms is an even less discriminating way of giving equal weight to very large differences.

Another approach has been to estimate complexity by tabulating morphological parts. For example, John Cisne (1974) found that the average number of appendages in aquatic arthropods followed a logistic pattern, with a rapid rise in the Paleozoic era and relatively little change since the early Mesozoic. Such a logistic pattern of complexity evolution is what is predicted from many theoretical approaches that model development as a system of interacting parts that rather quickly become "congealed" or canalized so that further significant changes are more difficult (see chapter 8).

Complexity, however we define it, tends to positive feedback toward greater complexity as it is constructed during ontogeny and as it evolves phylogenetically. The basic idea is that the more parts something has, then the faster it will evolve because there are more parts to alter. Vermeij (1973) has called this the "law of independent parts," but it is also well known in many theories of hierarchy (O'Neill et al., 1986). Many workers, such as Stanley (1990), have noted that increasing morphological and behavioral complexity correlates with increasing rates of evolution. This includes both increasing rates of speciation and extinction with increasing complexity.

DEVELOPMENTAL EVOLUTION BY TERMINAL EXTENSION

Just as past abuses of the notion of evolutionary progress confound modern discussions, the same is true of past abuses of developmental evolution. Biologists now generally acknowledge that universal Haeckelian recapitulation does not occur (McNamara, 1997; Raff, 1996). But conservation of early ontogenetic processes does promote a statistical tendency toward late-acting developmental modifications, including terminal extension. Such terminal extensions can lead to generalized recapitulatory patterns in development, as well as the evolution of complexity. Developmental mechanisms that cause terminal extension are often heterochronic, involving rate or timing changes in late development (as discussed in chapter 8). Examples include increased rate of development or increased duration of development (McNamara, 1997).

Given that universal Haeckelian terminal addition does not occur, we cannot expect that ontogeny will recapitulate phylogeny in every detail. However, if early development is generally conserved and terminal extension is a statistically favored pattern, then we can expect to see general patterns of recapitulation at very coarse scales of the phylogenetic tree of life. We can also expect to see glimpses of recapitulation at finer scales, but must specify what is being recapitulated: which cells, which tissues, which behaviors. Recapitulation is not an "all or nothing" proposition. Mosaic evolution has long been acknowledged (Levinton, 1988); this alone tells us that some organs and developmental pathways can be truncated or extended whereas others can be unaffected.

EVIDENCE FOR COMPLEXITY EVOLUTION VIA TERMINAL EXTENSION

It is clear that development influences both the rate and direction of evolution (Hall, 1992; Levinton, 1988; McKinney & McNamara, 1991;

McNamara, 1997; Raff, 1996). The arguments for terminal extension have traditionally been based on the fact that early development is highly constrained by contingencies of interaction. Our goal here is to briefly review the evidence that early development is indeed highly constrained and that terminal extension of development has produced morphological, cognitive, and behavioral complexity in evolution.

FOSSIL EVIDENCE

Levinton (1988), McKinney and McNamara (1991), Hall (1992), Erwin (1993), and Raff (1996) are among those who review the considerable developmental and paleontological evidence that the defining traits of a group (the *body plans*) have gradually congealed after an initial period of relative plasticity. Higher taxonomic levels, ostensibly representing the most novel ontogenetic and evolutionary variants, are overwhelmingly clustered in the early Paleozoic era.

This pattern is also seen within the evolution of specific groups. Morphological analysis of blastozoan echinoderms (Foote, 1992) and crinoids (Foote, 1995) shows that these basic phylogenetic groups originated in a burst of accelerated morphological diversification, making relatively large jumps in morphospace in a brief period of time in the early Paleozoic. Following this burst, morphological diversification was much slower and essentially consisted of lower-ranked subclades within the blastozoans and crinoids gradually occupying smaller areas of morphospace that had already been "carved out" in the initial burst. Similar patterns of accelerated early diversification, with more minor modifications later, have been documented in marine arthropods (Wills et al., 1994). Also, Hughes (1991) presented data that Cambrian trilobites showed less developmental canalization than Devonian trilobites. Wagner (1995) infers that increasing phylogenetic constraints played a greater role in the early evolution of gastropods than increasing ecological constraints.

DATA FROM LIVING ORGANISMS

Defining traits of higher taxonomic groups (e.g., phyla and classes) alive today tend to appear early in ontogeny and show reduced levels of variation compared to traits that appear later in ontogeny (overviews in Hall, 1992; Raff, 1996). This is evidence for an evolutionary increase in ontogenetic constraint.

Kirschner (1992) reviews the evolution of the cell in development and notes that the basic developmental patterns have been extremely conserved in vertebrates and insects. Indeed, nearly all metazoa share many basic developmental processes that date back to common ancestry in the late Precambrian (Valentine et al., 1996). Ontogenetic conservativism among higher taxa is also becoming evident at the genetic and subcellular level. Recent evidence shows that evolution in metazoa has largely involved regulatory changes in expression of highly conserved arrays of homeotic genes and the many developmental genes that they regulate (Carroll, 1995; Valentine et al., 1996). Homeotic genes are early-acting genes that sculpt the basic body plan of these animals via pattern formation processes just noted.

In agreement with "von Baer's law" (Hall, 1992), ontogenetic conservatism is even greater among more closely related taxa. In other words, lower (more closely related) taxa share similar developmental patterns even later into development than do higher taxa (which share the basic body plan patterns just noted). For example, fish, birds, and mammals all have the same initial forelimb bud formation, including the same molecular pattern formation process controlled by the same homeobox genes (Davidson, Peterson, & Cameron, 1995). Only in later development are the limbs modified into fins, wings, legs, and so on among more closely related species within each specific groups. Sordino and others (Sordino et al., 1995) have shown how terminal changes in limb appendages (via minor regulatory gene change) can explain the evolutionary origin of fingers and toes from fish fins, by unequal proliferation of cells. Similar patterns are seen in the formation of legs and wings among lower taxa of insects (Davidson et al., 1995).

COMPLEXITY DIFFUSION AND THE MYTH OF CHANCE

It seems from the evidence that terminal extension of ontogenies has progressively increased the upper limit of morphological (and behavioral) complexity in the biosphere. We see this whether we measure morphological complexity by number of cells, number of body parts, or many other metrics. Similarly, in this book, we have seen how behavioral and cognitive complexity has also increased via (sequential) terminal extensions of brain and life history developmental stages. As noted below, increasing cognitive and behavioral complexity is a way of overcoming the limits on morphological complexity increase.

But we want to re-emphasize that this process of increasing complexity is only a statistical trend in the biosphere. It is far from the deterministic terminal addition that has been associated with recapitulationism. In other words, the intrinsic and extrinsic forces that drive evolution cause many other developmental changes besides terminal extension. Early developmental changes can occur in evolution, as reported by Wray (1992) among a recently evolved echinoid species (see also Raff, 1996). Shubin (1994) discusses examples from the tetrapod limb. Furthermore, even late developmental alterations are often not terminal additions or extensions. Examples include preterminal insertions, late developmental substitutions, and terminal truncations (Hall, 1992; Mabee & Humphries, 1993; Raff, 1996).

On the other hand, there is evidence that late-acting terminal change is indeed common. Mabee's (1993) thorough phylogenetic study of living species finds that terminal addition may account for up to 51.9% of character state evolution in centrarchid fishes. She also shows that, when other terminal changes are included (e.g., terminal deletions, substitutions), up to 75% of change is accounted for.

EVOLUTIONARY DIFFUSION

An excellent way to quantify and visualize evolution is as a process of diffusion (McKinney, 1990). Using fossil data and computer models, McShea (1994) shows that evolutionary diffusion can follow two basic patterns, passive diffusion and directionally biased diffusion. In passive diffusion, species diversification is seen as random movement in morphospace. Species move about in morphospace in a way analogous to random gas particles in an expanding cloud. An example would be the "expansion of variance" trends of Stanley (1973) and Gould (1988b) in body size and complexity within clades. Alternatively, directional driving forces can bias the movement of individual gas particles, which in this case are species in morphospace. An example would be some of the comparative progressive trends discussed by Ruse (1993), such as the arms race. In these, there is an external systematic force, such as increasingly efficient predators, that acts on a large number of species in the clade to bias their movement in morphospace.

THE MYTH OF CHANCE IN EVOLUTION

Gould has been persistent critic of the idea of evolutionary progress, arguing that evolution is instead characterized by episodic upheavals —punctuations—and is generally dominated by chance (Gould, 1988b, 1995). But Gould's usage of *chance* has always been nonrigorous and informal and rarely employs statistical testing. As Raup (1977) has noted, random patterns do not reflect a lack of causation. Rather, they indicate that all the processes causing the patterns are not known and may involve many complex interactions that preclude our having full knowledge of them.

In other words, perceived randomness or chance is often an artifact of ignorance of the observer (McKinney, 1990). Our perception of chance in such cases says nothing about the true forces driving evolutionary patterns. To say that progressive complexity follows a diffusive model of random particle movement does not mean that particle movement lacks specific intrinsic and extrinsic forces pushing them. Rather, it means that the total forces determining morphospace evolution, including terminal extension, cannot be entirely specified.

Furthermore, our degree of ignorance depends on the scale of our observation. The scale at which we analyze evolutionary patterns plays a huge role in our perceptions of them. As Raup (1988) noted, if one views the history of life from a great distance, there is an obvious directionality. And, as we have noted, there are also coarse patterns of developmental evolution, such as in the origin of phyla. But at finer scales, the patterns of evolutionary progress become much less clear. And so it is with developmental constraints: At fine evolutionary scales, such as in the origin of modern species, including humans, there is clearly no simple deterministic process of evolution by universal terminal addition to ontogeny. Instead, we have argued for a statistical view of developmental change that emphasizes probabilistic biases in the way ontogeny evolves. These are evident only when viewed over long time scales that include many speciation events within a group.

BRAIN AND COGNITIVE EVOLUTION BY SEQUENTIAL TERMINAL EXTENSION

Evolution of the primate and human lineage shows many typical patterns of the history of life. For example, overdevelopment via terminal extension of ontogenies has produced many progressive statistical trends, including increases in size and complexity in many lineages (McNamara, 1997). In chapter 12, we discussed the specific case of the modern human brain, which was produced by sequential terminal extension of brain developmental stages (see also McKinney, in press-a, in press-b).

In this chapter, we suggest that sequential terminal extension of brain and cognitive ontogeny is a major mechanism of overcoming limitations on morphological complexity. In the perpetual struggle to find new ways of surviving in the evolving biotic environment, increased brain and cognitive complexity has many advantages. Most important perhaps are the physical limitations on evolving morphological complexity and the generally slower pace of morphological evolution.

PROSPECTS FOR BEHAVIORAL AND BRAIN EVOLUTION

Speculations about the future must consider the dynamics of the past. Future patterns of complex phenomena, whether of stock markets or evolutionary trends, are rarely simple extrapolations of the past. But the past does provide measurable clues about the probability of events (Casti, 1995). In this case, even though the dynamics of the biosphere (e.g., speciation and extinction) are enormously complex, in the long term they have produced a nonrandom statistical pattern at the very coarse scale of geological time. There is a tendency for life on Earth to become more complex. Furthermore, we can identify a main driving force of this complexity evolution as that of development: Ontogenies evolve so that late-acting developmental traits are often favored. Terminal extension of development has been common trend because of this.

But virtually all trends reach an upper limit. The evolution of complexity in hundreds of psychological, social, and technological phenomena show an S-shaped curve (Casti, 1995). The apparent reason is that any novelty or innovation initially shows a period of rapid exponential growth but design limitations eventually cause a reduction of the complexity trend. Such limitations usually reflect some ultimate structural constraint on the phenomenon (O'Neill et al., 1986). One way to circumvent such limitations is to innovate and create a fundamentally new kind of design.

In the case of morphological evolution, this S-shaped pattern is found when the upper limit of ontogenetic complexity in the biosphere is measured by cell type (Valentine et al., 1994), body size (Bonner, 1988), and genomic complexity (Brooks and Wiley, 1988). Thus, Figure 13-1 shows why brain and cognitive evolution is so important to evolutionary biology. A basic premise of Darwinian selection is that it favors "new ways of doing things." This reduces competition and allows exploitation of new resources; becoming more complex is one way of doing something new. This is an extrinsic factor in promoting the diffusive trend toward complexity in the biosphere. In the case of genetic and morphological complexity, complexity evolution via terminal extension seems to have encountered the kind of structural limitations referred to above, as shown by the S-shaped patterns observed.

The only way to overcome the structural constraints intrinsic to such

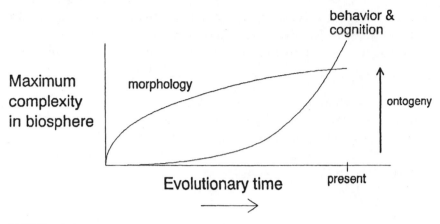

FIGURE 13-1
Maximum complexity of the biosphere versus evolutionary time. Terminal
extensin of ontogeny has increased maximum complexity of the morphology in
the biosphere through evolutionary time. But morphological complexification
has leveled off because of intrinsic constructional, functional, and other
constraints. Behavioral and cognitive complexification, as yet, show no signs of
slowing down.

a functional design is through a fundamentally new design. Behavior
and cognition represent a fundamentally new design that overcomes
the structural constraints inherent in the morphological (cellular) assembly of an individual. Enhanced cognitive ability, for instance, allows
apes and humans to overcome morphological limitations by the use of
tools. Complex behavior represents new ways of utilizing the morphological variation of the biosphere because behavior is what organisms do
with their morphological traits. In contrast to morphological evolution,
behavioral and cognitive evolution shows a pattern of increasing complexity (e.g., discussions by Bonner, 1988, and Dennett, 1995). Empirical
evidence for this is that, in general, relative brain size in mammals has
increased throughout most of the Cenozoic (Bonner, 1988). Furthermore,
examination of Jerison's brain-size data appears to indicate that, at least
for carnivores and ungulates, this was a trend that was driven by an
"arms race" of biotic interactions (Gould, 1988b).

A simple extrapolation of the cognitive trend of Figure 13-1 would
seem to imply that cognitive complexity will continue its exponential
increase. Wills (1993), for example, speculates that the current human
population probably contains a substantial number of genes that would

promote prolongation of the human life span and, presumably, pro-longed gestation, childhood, and so on. But whereas genes for longer gestation (and thus greater neuron proliferation) do likely exist, infants with larger brains would produce even greater stress for the mother's birth canal, which is already near its apparent maximum (Trevathan, 1987). This would seem to represent the biological design limitation for human cognitive evolution, and we might be tempted to predict an "S-shaped" leveling off of the cognitive curve in Figure 13-1. But given the fact that humans now have the ability to exceed biological designs in a variety of ways (e.g., recombinant DNA, cesarean births, and potentially even extrauterine births), it would seem that no immediate constraint on cognition is visible. Culturally transmitted technological innovations have already emancipated humans from their physical limitations. This emancipation is even more profound in the case of computerized technology. Donald (1991), among others, describes how future cognitive complexity evolution will involve the use of computers and other non-biological information storage devices.

SUMMARY AND CONCLUSIONS

Paleontological and developmental evidence supports the following major points concerning complexity increase in biological evolution.

1. The upper limit of both morphological and behavioral complexity in the biosphere has increased through geological time.

2. A key mechanism driving this complexity increase is natural selection of modified ontogenies, especially mutations affecting late ontogeny, which can extend the developmental trajectory.

3. The progressive increase in the upper limit of complexity has not been a simple monotonic increase, but is a statistical diffusion process into more complex kinds of morphology and behavior. It is thus characterized by many reversals and delays, but evolution has an episodic trend toward complexity.

4. Morphological complexity has slowed its rate of increase, but unlike morphology, behavioral complexity has accelerated its rate of increase. This reflects a qualitative difference between morphological and behavioral adaptations. Behavioral adaptations represent ways of overcoming the severe physical limitations inherent in morphological adaptation. Behavior and cognition are ultimately much less constrained in their evolutionary potential for complexification.

5. Human cognitive evolution is a predictable outcome of the evolutionary tendency to produce behavioral complexity by modifying late ontogeny. The mechanism by which this occurred in humans, by sequential terminal extension of ontogenetic stages, is common in the history of life.

REFERENCES

Abitbol, M. M. (1995). Lateral view of *Australopithecus afarensis:* Primitive aspects of bipedal positional behavior in the earliest hominids. *Journal of Human Evolution, 28,* 211–229.

Alberch, P. (1990). Natural selection and developmental constraints versus internal determinants of order in nature. In C. J. DeRousseau (Ed.), *Primate life history and evolution.* New York: Wiley-Liss.

Allman, W. F. (1994). *The stone age present.* New York: Simon & Schuster.

Altmann, J. (1974). Observational study of behaviour: Sampling methods. *Behaviour, 49,* 227–265.

Alp, R. (1997). "Stepping-sticks" and "seat-sticks": New types of tools used by wild chimpanzees *(Pan troglodytes)* in Sierra Leone. *American Journal of Primatology, 41,* 45–52.

Anderson, J. R. (1994). The monkey in the mirror: A strange conspecific. In S. T. Parker, R. W. Mitchell, & M. L. Boccia (Eds.), *Self-awareness in animals and humans* (pp. 315–329). New York: Cambridge University Press.

Antinucci, F. (Ed.). (1989). *Cognitive structure and development in nonhuman primates.* Hillsdale, NJ: Erlbaum.

Antinucci, F., Spinozzi, G., Visalberghi, E., & Volterra, V. (1982). Cognitive development in a Japanese macaque *(Macaca fuscata).* Ann. Ist Super. Sanita, 18 (2), 177–184.

Armstrong, D. F., Stokoe, W. C., & Wilcox, S. E. (1995). *Gesture and the nature of language.* Cambridge, UK: Cambridge University Press.

Armstrong, E. (1983). Relative brain size and metabolism in mammals. *Science, 220,* 1302–1304.

Astington, J., Harris, P., & Olson, D. R. (Eds.). (1988). *Developing theories of mind.* New York: Cambridge University Press.

Baillargeon, R. (1987). Young infants' reasoning about the physical and spatial properties of a hidden object. *Cognitive Development, 2,* 179–220.

Baillargeon, R., Spelke, E. S., & Wasserman, S. (1985). Object permanence in 5 month old infants. *Cognition, 20,* 191–208.

Baldwin, J. M. (1897). *Social and ethical interpretations of mental development.* New York: Macmillan.

Baldwin, J. M. (1902). *Development and evolution.* New York: Macmillan.

Baldwin, J. M. (1906). *Mental development of the child and the race* (3rd ed.). New York: Macmillan.

Bard, K. (1995). Sensorimotor cognition in young feral orangutans (*Pongo pygmaeus*). *Primates, 36* (3), 297–321.

Barkow, J., Cosmides, L., & Toobey, J. (Eds.). (1992). *The adapted mind.* New York: Oxford University Press.

Baron-Cohen, S. (1995). *Mind blindness: An essay on autism and theory of mind.* Cambridge, MA: M. I. T. Press.

Bates, E. (1976). *Language in context: The acquisition of pragmatics.* New York: Academic Press.

Bates, E. (1993). Comprehension and production in early language development, *Language comprehension in ape and child* (Vol. 58, pp. 22–241). Chicago: University of Chicago Press.

Bates, E., Benigni, L., Bretherton, I., Camaioni, L., & Volterra, V. (1979). *The emergence of symbols.* New York: Academic Press.

Bates, E., & Goodman, J. C. (1996). The inseparability of grammar and the lexicon: Evidence from acquisition, aphasia and real-time processing. *Technical Report of the Center for Research in Language, University of California at San Diego, 9603.*

Bates, E., Thal, D., & Marchman, V. (1991). Symbols and syntax: A Darwinian approach to language development. In N. Krasnegor, D. Rumbaugh, R. Schiefelbusch, & M. Studdert-Kennedy (Eds.), *Biological and behavioral determinants of language development* (pp. 30–65). Hlllsdale, NJ: Erlbaum.

Bayley, N. (1969). *Bayley Scales of Infant development.* San Antonio, TX: Psych. Corp.

Beach, F. (1965). The snark was a boojum. In T. E. McGill (Ed.), *Readings in animal behavior* (pp. 3–14). New York: Holt, Rinehart & Winston.

Begun, D., & Walker, A. (1993). The endocast. In A. Walker & R. Leakey (Eds.), *The Nariokotome* Homo erectus *skeleton* (pp. 326–358). Cambridge MA: Harvard University Press.

Bertanthal, B. L., & Fischer, K. W. (1978). Development of self-recognition in the infant. *Developmental Psychology, 14,* 44–50.

Boccia, M. L. (1994). Mirror behavior in macaques. In S. T. Parker, R. W. Mitchell, & M. L. Boccia (Eds.), *Self-awareness in animals and humans* (pp.350–360). New York: Cambridge University Press.

Boesch, C. (1991a). Symbolic communication in wild chimpanzees. *Human Evolution, 6* (1), 81–90.

Boesch, C. (1991b). Teaching among wild chimpanzees. *Animal Behaviour, 41,* 530–532.

Boesch, C. (1993). Aspects of transmission of tool use in wild chimpanzees. In

K. R. Gibson & T. Ingold (Eds.), *Tools, language and cognition in human evolution* (pp. 171–183). Cambridge, UK: Cambridge University Press.

Boesch, C., & Boesch, H. (1984). Mental map in wild chimpanzees: An analysis of hammer transports for nut cracking. *Primates, 25* (2), 160–170.

Boesch, C., & Boesch, H. (1989). Hunting behavior of wild chimpanzees in the Tai National Park. *American Journal of Physical Anthropology, 78* 547–573.

Bogartz, R., Shinskey, J., & Speaker, C. (in press). Interpreting infant looking. *Developmental Psychology.*

Bogin, B. (1988). *Patterns of human growth.* Cambridge, UK: Cambridge University Press.

Bogin, B. (1997). Evolutionary hypotheses for human childhood. *Yearbook of Physical Anthropology, 40,* 1–27.

Bonner, J. T. (1988). *The evolution of complexity by means of natural selection.* Princeton, NJ: Princeton University Press.

Bonvillian, J. D., & Patterson, F. G. (1993). Early sign language acquisition in children and gorillas: Vocabulary content and sign iconicity. *First Language, 13* 315–338.

Borchert, C., & Zihlman, A. (1990). The ontogeny and phylogeny of symbolizing. In M. L. Foster (Ed.), *The life of symbols* (pp. 15–44). Boulder, CO: Westview.

Bower, T. G. R. (1979). *Human development.* New York: Freeman.

Boysen, S. T., & Berntson, G. (1990). The development of numerical skills in the chimpanzee *(Pan troglodytes).* In S. T. Parker & K. Gibson (Eds.), *"Language" and intelligence in monkeys and apes* (pp. 435–450). New York: Cambridge University Press.

Boysen, S. T., Berntson, G., & Pentice, J. (1987). Simian scribbles: A reappraisal of drawing in the chimpanzee *(Pan troglodytes). Journal of Comparative Psychology, 101* (1), 82–89.

Boysen, S. T., Berntson, G., Shreyer, T., & Quiqley, K. (1993). Processing of ordinality and transitivity by chimpanzees. *Journal of Comparative Psychology, 107* (2), 208–215.

Boysen, S. T., Bryan, K. M., & Shreyer, T. A. (1994). Shadows and mirrors: Alternative avenues to the development of self-recognition in chimpanzees. In S. T. Parker, R. W. Mitchell, & M. L. Boccia (Eds.), *Self-awareness in animals and humans* (pp. 227–240). New York: Cambridge University Press.

Boysen, S. T., & Capaldi, E. J. (Eds.). (1993). *The development of numerical competence: Animal and human models.* Hillsdale, NJ: Erlbaum.

Braggio, J. P., Hall, A. D., Buchanan, J. P., & Nadler, R. D. (1979). Cognitive capacities of juvenile chimpanzees on a Piagetian-type multiple-classification task? *Psychological Reports, 44* 1087–1097.

Brainerd, C. (1974). Training and transfer of transitivity, conservation and class inclusion in length. *Child Development, 45,* 324–334.

Brakke, K., & Savage-Rumbaugh, S. E. (1995). The development of language skills in bonobos and chimpanzees. *Language and Communication, 15* (2), 121–148.

Bretherton, I. (1984). Representing the social world in symbolic play: Reality and fantasy. In I. Bretherton (Ed.), *Symbolic play: The development of social understanding* (pp. 32–41). New York: Academic Press.

Bretherton, I., & Bates, E. (1984). The development of representation from 10 to

28 months. In R. Emde & R. Harmon (Eds.), *Continuities and discontinuities in development*. New York: Plenum.

Brewer, S., & McGrew, W. (1990). Chimpanzee use of a tool-set to get honey. *Folia Primatologica, 54*, 100–104.

Brooks, D., & McLennan, D. (1991). *Phylogeny, ecology, and behavior*. Chicago: University of Chicago Press.

Brooks, D. R., & Wiley, E. O. (1988). *Evolution as entropy*. Chicago: University of Chicago Press.

Brown, F. H., Harris, J. M., Leakey, R. E. F., & Walker, A. (1985). Early *Homo erectus* skeleton from West Lake Turkana, Kenya. *Nature, 316*, 788–792.

Brown, R. (1973). *A first language: The early stages*. Cambridge, MA: Harvard University Press.

Bryant, P., & Trabasso, T. (1971). Transitive inference and memory in young children. *Nature, 232*, 456–458.

Butler, A. B., & Hodos, W. (1996). *Comparative vertebrate neuroanatomy*. New York: Liss-Wiley.

Butterworth, G. (1991). The ontogeny and phylogeny of joint visual attention. In A. Whiten (Ed.), *Natural theories of mind* (pp. 223–232). Oxford, UK: Oxford University Press.

Byrne, R. W. (1995). *The thinking ape: Evolutionary origins of intelligence*. Oxford, UK: Oxford University Press.

Byrne, R. W. (1997). What's the use of anecdotes? Distinguishing psychological mechanisms in primate tactical deception. In R. W. Mitchell, N. S. Thompson, & H. L. Miles (Eds.), *Anthropomorphism, anecdotes, and animals* (pp. 134–150). Albany, NY: SUNY Press.

Byrne, R. W., & Byrne, J. M. E. (1991). Hand preferences in the skilled gathering tasks of mountain gorillas (*Gorilla g. beringei*). *Cortex, 27*, 521–546.

Byrne, R. W., & Russon, A. (in press). Learning by imitation: A hierarchical approach. *Behavioral and Brain Sciences*.

Byrne, R., & Whiten, A. (1992). Cognitive evolution in primates: Evidence from tactical deception. *Man, 27* (3), 609–627.

Call, J., & Rochat, P. (1996). Liquid conservation in orangutans (*Pongo pygmaeus*) and humans (*Homo sapiens*): Individual differences and perceptual strategies. *Journal of Comparative Psychology, 110* (3), 219–232.

Call, J., & Tomasello, M. (1995). Use of social information in the problem solving of orangutans (*Pongo pygmeaus*) and human children (*Homo sapiens*). *Journal of Comparative Psychology, 109* (3), 308–320.

Calvin, W. H. (1993). The unitary hypothesis: A common neural circuitry for novel manipulations, language, plan-ahead and throwing? In K. R. Gibson & T. Ingold (Eds.), *Language, tools, and cognition in human evolution* (pp. 230–250). Cambridge, UK: Cambridge University Press.

Cameron, P. A., & Gallup, G. G., Jr. (1988). Shadow self-recognition in human infants. *Infant Behavior and Development, 11*, 465–471.

Carey, S., & Spelke, E. (1994). Domain-specific knowledge and conceptual change. In L. A. Hirshfield & S. A. Gelman (Eds.), *Mapping the mind: Domain specificity in cognition and culture* (pp. 169–299). Cambridge, UK: Cambridge University Press.

Caro, T. M., & Hauser, M. D. (1992). Is there teaching in nonhuman animals? *The Quarterly Review of Biology, 67* (2), 151–174.

Carroll, S. B. (1995). Homeotic genes and the evolution of arthropods and chordates. *Evolution, 28,* 479–485.

Case, R. (1985) *Intellectual development from birth to adulthood,* Orlando, FL: Academic Press.

Case, R. (1987). Neo-Piagetian theory: Retrospect and prospect. *International Journal of Psychology, 22,* 773–791.

Case, R. (1991). Stages in the development of the young child's first sense of self. *Developmental Review, 11,* 210–230.

Case, R. (Ed.). (1992a). *The mind's staircase: Exploring the conceptual underpinnings of children's thought and knowledge.* Hillsdale, NJ: Erlbaum.

Case, R. (1992b). The role of the frontal lobes in the regulation of cognitive development. *Brain and Cognition, 20,* 51–73.

Case, R. (1996). Introduction: Reconceptualizing the nature of children's conceptual structures and their development in middle childhood. In R. Case & Y. Okamoto (Eds.), *The role of central conceptual structures in the development of children's thought* (Series No. 246, Vol. 61, pp. 1–26). Chicago: University of Chicago Press.

Case, R., & Khanna, F. (1981). The missing links: Stages in children's progression from sensorimotor to logical thought. In K. W. Fischer (Ed.), *Cognitive development* (Vol. 12, pp. 21–32). New York: Jossey Bass.

Case, R., & Okamoto, Y. (Eds.). (1996). *The role of central conceptual structures in the development of children's thought* (Series No. 246, Vol. 61). Chicago: University of Chicago Press.

Casti, J. L. (1995). *Complexification.* New York: Harper.

Chance, M. R. A., & Mead, A. P. (1953). Social behavior and primate evolution. *Evolution, 7,* 395–439.

Chapman, M., & Lindenberger, U. (1988). Functions, operations, and decalage in the development of transitivity. *Developmental Psychology, 24,* 542–551.

Cheney, D., & Seyfarth, R. (1990). *How monkeys see the world.* Chicago: University of Chicago Press.

Cheney, D., & Seyfarth, R. (1991). Reading minds or reading behaviour? Tests for a theory of mind in monkeys. In A. Whiten (Ed.), *Natural theories of mind.* Oxford, UK: Basil Blackwell.

Chevalier-Skolnikoff, S. (1976). The ontogeny of primate intelligence and its implications for communicative potential: A preliminary report. *Annals of the New York Academy of Sciences, 280* (Origins and evolution of language and speech), 173–211.

Chevalier-Skolnikoff, S. (1977). A Piagetian model for comparing the socialization of monkey, ape, and human infants. In S. Chevalier-Skolnikoff & F. Poirier (Eds.), *Primate biosocial development* (pp. 159–188). New York: Garland.

Chevalier-Skolnikoff, S. (1983). Sensorimotor development in orangutans and other primates. *Journal of Human Evolution, 12,* 545–561.

Chevalier-Skolnikoff, S., Galdikas, B., & Skolnikoff, A. (1982). The adaptive significance of higher intelligence in orangutans: A preliminary report. *Journal of Human Evolution, 11,* 639–652.

Chevalier-Skolnikoff, S., & Poirier, F. (Eds.). (1977). *Primate biosocial development*. New York: Garland.

Chomsky, N. (1957). A review of B. F. Skinner's *Verbal Behavior*. *Language, 35*, 26–58.

Cisne, J. L. (1974). Evolution of the world fauna of aquatic free-living arthropods. *Evolution*, 337–366.

Clark, W. L. G. (1964). *The fossil evidence for human evolution*. Chicago: University of Chicago Press.

Clutton-Brock, T., & Harvey, P. H. (1980). Primates, brains and ecology. *Journal of Zoology (London), 190*, 309–323.

Coddington, J. A. (1988). Cladistic tests of adaptational hypotheses. *Cladistics, 4*, 3–22.

Connor, R. C., Smolker, R., & Richards, A. F. (1992). Two levels of alliance formation among male bottlenose dolphins. *Proceedings of the National Academy of Sciences, 89*, 987–990.

Cook-Gumprez, J. (1992). Gendered contexts. In P. Auer & A. d. Luzio (Eds.), *The contextualization of language* (pp. 177–198). Philadelphia: John Benjamin.

Cooley, C. H. (1983). *Human nature and the social order* (rev. ed.). New Brunswick, NJ: Transaction Books.

Coppinger, R., & Schneider, R. (1995). Evolution of working dogs. In J. Serpell (Ed.), *The domestic dog: Its evolution, behavior, and interactions with people* (pp. 21–47). Cambridge, UK: Cambridge University Press.

Corman, H., & Escalona, S. (1969). Stages of sensorimotor development: A replication study. *Merrill-Palmer Quarterly, 15*, 351.

Cosmides, L., & Toobey, J. (1991). Cognitive adaptations for social exchange. In J. Barkow, L. Cosmides, & J. Toobey (Eds.), *The adapted mind* (pp. 163–228). New York: Oxford University Press.

Cronin, J. E., Sarich, V. M., & Ryder, O. (1984). Molecular perspectives on the evolution of the lesser apes. In H. Preuschoft, D. J. Chivers, W. Y. Brockelman, & N. Creel (Eds.), *The lesser apes* (pp. 468–485). Edinburgh: Edinburgh University Press.

Custance, D., & Bard, K. (1994). The comparative and developmental study of self-recognition and imitation: The importance of social factors. In S. T. Parker, R. W. Mitchell, & M. L. Boccia (Eds.), *Self-awareness in animals and humans* (pp. 207–226). New York: Cambridge University Press.

Custance, D., Whiten, A., & Bard, K. (1995). Can young chimpanzees *(Pan troglodytes)* imitate arbitrary actions? Hayes and Hayes (1952) revisited. *Behaviour, 132*, 839–858.

D'Esposito, M., Detre, J., Alsop, D., Shin, R., Atlas, S., & Grossman, M. (1995). The neural basis of the central executive system of working memory. *Science, 378*, 279–281.

Damasio, A. (1994). *Decartes' error*. New York: Avon.

Darwin, C. (1930). *The descent of man*. New York: Appleton and Co.

Davidson, E. H., Peterson, K., & Cameron, R. (1995). Origin of bilateral body plans: Evolution of developmental regulatory systems. *Science, 270*, 1319–1324.

Davidson, P. E. (1914). *The recapitulation theory and human infancy*. New York: Teachers College, Columbia University.

Davis, H., & Perusse, R. (1988). Numerical competence in animals: Definitional issues, current evidence, and a new research agenda. *Behavioral and Brain Sciences, 11,* 561–615.

Dawkins, R. (1986). *The blind watchmaker*. New York: W. W. Norton.

Dawson, B. V., & Foss, B. M. (1965). Observational learning in budgerigars. *Animal Behaviour, 13,* 470–474.

De Blois, S., & Novak, M. (1994). Object permanence in rhesus monkeys (*Macaca mulatta*). *Journal of Comparative Psychology, 108* (4), 318–327.

de Queiroz, K. (1985). The ontogenetic method for determining character polarity and its relevance to phylogenetic systematics. *Systematic Zoology, 34,* 280–299.

de Waal, F. (1983). *Chimpanzee politics*. New York: Harper and Row.

de Waal, F. (1989). *Peacemaking among primates*. Cambridge, MA: Harvard University Press.

de Waal, F. (1991). Complementary methods and convergent evidence in the study of primate social cognition. *Behaviour, 118,* 297–320.

de Waal, F. (1996a). Conflict as negotiation. In W. McGrew, L. Marchant, & T. Nishida (Eds.), *Great ape societies* (pp. 159–172). Cambridge, UK: Cambridge University Press.

de Waal, F. (1996b). *Good natured: The origins of right and wrong in humans and other animals*. Cambridge, MA: Harvard University Press.

Deacon, T. (1990a). Fallacies of progression in theories of brain-size evolution. *International Journal of Primatology, 11,* 193–236.

Deacon, T. W. (1990b). Problems of ontogeny and phylogeny in brain size evolution. *International Journal of Primatology, 11,* 237–282.

Deacon, T. (1992). The human brain. In S. Jones, R. Martin, & D. Pilbeam (Eds.), *The Cambridge encyclopedia of human evolution* (pp. 115–123). Cambridge, UK: Cambridge University Press.

Deacon, T. (1997). *The symbolic species*. New York: W. W. Norton.

Deacon, T. (in press). Is there a cellular-molecular basis for hetcrochrony? A view from neuroembryology. In S. T. Parker, M. L. McKinney, & J. Langer (Eds.), *The evolution of development: Biology, brain, and behavior*. Santa Fe, NM: School of American Research.

DeBeer, G. (1958). *Embryos and ancestors*. Oxford, UK: Oxford University Press.

DeLoache, J. S. (1990). Young children's understanding of models. In R. F. Hudson & J. Hudson (Eds.), *Knowing and remembering in young children*. New York: Cambridge University Press.

Dennett, D. C. (1988). The intentional stance in theory and practice. In R. Byrne & A. Whiten (Eds.), *Machiavellian Intelligence: Social expertise and the evolution of intelligence in monkeys, apes, and humans* (pp. 180–202). Oxford, UK: Oxford University Press.

Dennett, D. C. (1995). *Darwin's dangerous idea*. New York: Simon & Schuster.

Desmond, A. (1979). *The ape's reflexion*. London: Quartet Books.

Diamond, A. (1991). Frontal lobe involvement in cognitive changes during the first year of life. In K. R. Gibson & A. Peterson (Eds.), *Brain maturation and cognitive development* (pp. 127–180). Chicago: Aldine.

DiLisi, R. (1987). A cognitive-developmental model of planning. In S. Friedman, E. Scholnick, & R. Cocking (Eds.), *Blueprints for thinking* (pp. 79–109). Cambridge, UK: Cambridge University Press.

Donald, M. (1991). *Origins of the modern mind*. Cambridge, MA: Harvard University Press.

Doran, D. M. (1992). The ontogeny of chimpanzee and pygmy chimpanzee locomotor behavior: A case study of paedomorphism and its behavioral correlates. *Journal of Human Evolution, 23*, 139–157.

Doran, D. M. (1997). Ontogeny of locomotion in mountain gorillas and chimpanzees. *Journal of Human Evolution, 32*, 323–344.

Dore, F., & Dumas, C. (1987). Psychology of animal cognition: Piagetian studies. *Psychological Bulletin, 102* (2), 242–248.

Dunbar, R. I. M. (1992). Neocortex size as a constraint on group size in primates. *Journal of Human Evolution, 20*, 409–493.

Dunbar, R. I. M. (1993). Coevolution of neocortical size, group size and language in humans. *Behavioral and Brain Sciences, 16*, 681–735.

Edelman, G. (1987). *Neural darwinism*. New York: Basic Books.

Ekman, P., & Friesen, W. (1969). The repertoire of nonverbal behavior—categories, origins, usage, and coding. *Semiotica, 1*, 49–98.

Ekstig, B. (1994). Condensation of developmental stages and evolution. *Bioscience, 44*, 158–164.

Elman, J., Bates, E., Johnson, M., Karmiloff-Smith, A., Pariisi, D., & Plunkett, K. (1996). *Rethinking innateness*. Cambridge, MA: M. I. T. Press.

Erwin, D. H. (1993). The origin of metazoan development: A paleobiological perspective. *Biol. J. Linnean Soc., 50*, 255–274.

Falk, D. (1992). *Braindance: New discoveries about human brain evolution*. New York: Henry Holt & Co.

Ferster, C. B. (1958). Intermittent reinforcement of a complex response in the chimpanzee. *Journal of the Experimental Analysis of Behavior, 1*, 163–165.

Finlay, B. L., & Darlington, R. B. (1995). Linked regularities in the development and evolution of mammalian brains. *Science, 286*, 1578–1584.

Fischer, K. W. (1980). A theory of cognitive development: The control and construction of hierarchies of skills. *Psychological Review, 87* (6), 477–531.

Fischer, K. W., & Bidell, T. R. (1991). Constraining nativist inferences about cognitive capacities. In S. Carey & R. Gelman (Eds.), *The epigenesis of mind: Essays on biology and mind* (pp. 199–235). Hillsdale, NJ: Erlbaum.

Fischer, K. W., Bullock, D. H., Rotenberg, E. J., & Raya, P. (1993). The dynamics of competence: How context contributes directly to skill. In R. Wozniak & K. W. Fischer (Eds.), *Development in context* (pp. 93–117). Hillsdale, NJ: Erlbaum.

Fischer, K. W., Pipp, S. L., & Bullock, D. (1984). Detecting developmental discontinuities. In R. Emde & Harmon (Eds.), *Continuities and discontinuities in development* (pp. 95–121). New York, NY: Plenum.

Flavell, J. H., Botkin, P. T., Fry, C. L., Wright, J. W., & Jarvis, P. E. (1968). *The development of role-taking and communication skills in children*. New York: Wiley.

Fleagle, J. G. (1988). *Primate adaptation and evolution.* New York: Academic Press.

Fodor, J. A. (1983). *Modularity of mind.* Cambridge, MA: M. I. T. Press.

Fong, D. W., Kane, T. C., & Culver, D. C. (1995). Vestigualization and loss of non-functional characters. *Annual Review of Ecology & Systematics, 26,* 249–288.

Foote, M. (1992). Rarefaction analysis of morphological and taxonomic diversity. *Paleobiology, 18,* 1–16.

Foote, M. (1995). Morphological diversification of Paleozoic crinoids. *Paleobiology, 21* 273–299.

Fouts, R. (1973). Acquisition and testing of gestural signs in four young chimpanzees. *Science, 180,* 978–80.

Fouts, R., Fouts, D., & Schoenfeld. (1984). Sign language conversational interactions between chimpanzees. *Sign Language Studies, 34,* 1–12.

Fouts, R., Fouts, D., & Van Cantfort, T. (1989). The infant Loulis learns signs from cross-fostered chimpanzees. In R. A. Gardner, B. T. Gardner, & T. E. Van Cantfort (Eds.), *Teaching sign language to chimpanzees* (pp. 293–307). Albany, NY: SUNY Press.

Fox, E., & van Schaik, C. (in press). Intelligent tool use in wild Sumatran orang-utans. In S. T. Parker, H. L. Miles, & R. W. Mitchell (Eds.), *The mentality of gorillas and orangutans.* Cambridge, UK: Cambridge University Press.

Fruth, B., & Hohmann, G. (1996). Nest building behavior in the great apes: The great leap forward? In W. McGrew, L. Marchant, & T. Nishida (Eds.), *Great ape societies* (pp. 225–240). New York: Cambridge University Press.

Furness. W. (1916). Observations on the mentality of chimpanzees. *Proceedings of the American Philosophical Society. 55,* 281–290.

Galdikas, B. (1995). *Reflections of Eden: My years with the orangutans of Borneo.* New York: Little, Brown.

Galef, B. G., Jr. (1988). Imitation in animals: History, definition, and interpretation of data. In T. Zentall & B. C. Galef Jr. (Eds.), *Social learning* (pp. 3–28). Hillsdale, NJ: Erlbaum.

Gallup, G. G., Jr. (1970). Chimpanzees: Self-recognition. *Science, 167,* 86–87.

Gallup, G. G., Jr. (1977). Self-recognition in primates. *American Psychologist, 32,* 329–338.

Gamble, C. (1976). Interaction and alliance in Palaeolithic society. *Man, 17,* 92–107.

Gannon, P. J., Holloway, R. L., Broadfield, D. C., & Braun, A. R. (1998). Asymmetry of chimpanzee planum temorale: Humanlike pattern of Wernicke's brain language area homolog. *Science, 279,* 220–222.

Gardner, B. T., & Gardner, R. A. (1980). Two comparative psychologists look at language acquisition. In K. Nelson (Ed.), *Children's language* (Vol. 2, pp. 331–369).

Gardner, B. T., & Gardner, R. A. (1989a). Prelinguistic development of children and chimpanzees. *Human Evolution, 4* (6), 433–460.

Gardner, B. T., & Gardner, R. A. (1989b). A test of communication. In R. A. Gardner, B. T. Gardner, & T. E. Van Cantfort (Eds.), *Teaching sign language to chimpanzees* (pp. 181–197). Albany, NY: SUNY Press.

Gardner, B. T., Gardner, R. A., & Nichols, S. G. (1989). The shapes and uses of signs in a cross-fostering laboratory. In R. A. Gardner, B. T. Gardner, & T. E. Van Cantfort (Eds.), *Teaching sign language to chimpanzees* (pp. 55–180). Albany, NY: SUNY Press.

Gardner, H. (1987). *The mind's new science.* New York: Basic Books.

Gardner, R. A., & Gardner, B. T. (1969). Teaching sign language to a chimpanzee. *Science, 165,* 664–672.

Gardner, R. A., & Gardner, B. T. (1989). A cross-fostering laboratory. In R. A. Gardner, B. T. Gardner, & T. E. Van Cantfort (Eds.), *Teaching sign language to chimpanzees* (pp. 1–28). Albany, NY: SUNY Press.

Gardner, R. A., Gardner, B. T., & Van Cantfort, T. E. (Eds.). (1989). *Teaching sign language to chimpanzees.* Albany, NY: SUNY Press.

Gelman, R., & Gallistel, C. R. (1978). *The child's understanding of number.* Cambridge, MA: Harvard University Press.

Gergely, G. (1994). From self-recognition to theory of mind. In S. T. Parker, R. W. Mitchell, & M. L. Boccia (Eds.), *Self-awareness in animals and humans* (pp. 51–60). New York: Cambridge University Press.

Gesell, A. (1945). *How a baby grows.* New York: Harper and Brothers.

Gesell, A., Halverson, H., Thompson, H., Ilg, F., Castner, B., Ames, L., & Amatruda, C. (1940). *The first five years of life: A guide to the study of the preschool child.* New York: Harper.

Ghiselin, M. (1969). *The triumph of the Darwinian method.* Berkeley, CA: University of California Press.

Gibson, K. R. (1977). Brain structure and intelligence in macaques and human infants from a Piagetian perspective. In S. Chevalier-Skolnikoff & F. Poirier (Eds.), *Primate biosocial development* (pp. 113–157). New York: Garland.

Gibson, K. R. (Ed.). (1986). *Cognition, brain size and the extraction of embedded food resources.* Cambridge, UK: Cambridge University Press.

Gibson, K. R. (1990). New perspectives on instincts and intelligence: Brain size and the emergence of hierarchical mental construction skills. In S. T. Parker & K. R. Gibson (Eds.), *"Language" and intelligence in monkeys and apes* (pp. 97–128). New York: Cambridge University Press.

Gibson, K. R. (1991a). Basic neuroanatomy for the nonspecialist. In K. R. Gibson & A. C. Peterson (Eds.), *Brain maturation and cognitive development: Comparative and cross-cultural perspectives* (pp. 13–27). New York: Aldine Gruyter.

Gibson, K. R. (1991b). Myelination and behavioral development: A comparative perspective on questions of neoteny, altriciality, and intelligence. In K. R. Gibson & A. C. Peterson (Eds.), *Brain maturation and cognitive development: Comparative and cross-cultural perspectives* (pp. 29–64). New York: Aldine Gruyter.

Gibson, K. R. (1993). Tool use, language and social behavior in relationship to information processing capacities. In K. R. Gibson & T. Ingold (Eds.), *Tools, language and cognition in human evolution* (pp. 251–270). Cambridge, UK: Cambridge University Press.

Gibson, K. R. (1995). Hypermorphosis in hominid brain evolution. Paper presented at American Association for the Advancement of Science, Baltimore, MD.

Gibson, K. R., & Ingold, T. (Eds.). (1993). *Tools, language and cognition in human evolution.* Cambridge, UK: Cambridge University Press.

Gillan, D. J. (1981). Reasoning in the chimpanzee: II. Transitive inference. *Journal of Experimental Psychology: Animal Behavior Processes, 7* (2), 150–164.

Goldin-Meadow, S. (1993). When does gesture become language? A study of gesture as a primary communication system by deaf children of hearing parents. In K. R. Gibson & T. Ingold (Eds.), *Tools, language, and cognition in human evolution* (pp. 63–85). Cambridge, UK: Cambridge University Press.

Goldman-Rakic, P. S. (1987). Development of cortical circuitry and cognitive function. *Child Development, 58,* 601–622.

Gomez, J. C. (1990). The emergence of intentional communication as a problem-solving strategy in the gorilla. In S. T. Parker, R. W. Mitchell, & M. L. Boccia (Eds.), *"Language" and intelligence in monkeys and apes* (pp. 333–355). New York: Cambridge University Press.

Gomez, J. C. (1991). Visual behaviour as a window for reading the minds of others in primates. In A. Whiten (Ed.), *Natural theories of mind* (pp. 195–208). London: Blackwell.

Goodall, J. (1986). *Chimpanzees of the Gombe.* Cambridge, MA: Harvard University Press.

Goodall, J. (1990). *Through a window: My thirty years with the chimpanzees of Gombe.* New York: Houghton Mifflin.

Goodenough, W. (1981). *Language, culture, and society* (2nd ed.). Menlo Park, CA: Benjamin Cummings.

Goodwin, D., Bradshaw, J. W. S., & Wickens, S. M. (1997). Paedomorphosis affects agonistic visual signals of dogs. *Animal Behaviour, 53,* 297–304.

Gopnik, A., & Meltzoff, A. (1987). Early semantic developments and their relationship to object permanence, means-ends understanding and categorization. In K. E. Nelson & A. v. Kleeck (Eds.), *Children's language* (pp. 191–212). Hillsdale, NJ: Erlbaum.

Gopnik, A., & Meltzoff, A. (1994). Minds, bodies and persons: Young children's understanding of the self and others as reflected in imitation and theory of mind research. In S. T. Parker, R. W. Mitchell, & M. L. Boccia (Eds.), *Self-awareness in animals and humans* (pp. 166–185). New York: Cambridge University Press.

Gould, S. J. (1977). *Ontogeny and phylogeny.* Cambridge, MA: Harvard University Press.

Gould, S. J. (1988a). On replacing the idea of progress with an operational notion of directionality. In M. H. Nitecki (Ed.), *Evolutionary progress* (pp. 319–339). Chicago: University of Chicago Press.

Gould, S. J. (1988b). Trends as changes in variance. *Journal of Paleontology, 62,* 319–329.

Gould, S. J. (1989). *Wonderful life.* New York: W. W. Norton.

Gould, S. J. (1995). The evolution of life on earth. *Scientific American, 272,* 85–91.

Gould, S. J. (1996a). Creating the creators. *Discover, 17,* 42–57.

Gould, S. J. (1996b). *Full house.* New York: Harmony Books.

Gould, S. J., & Lewontin, R. (1979). The spandrels of San Marcos and the Panglossian paradigm: A critique of the adaptationist programme. *Proceedings of the Royal Society of London, 205B*, 581–598.

Greenfield, P. M. (1992). Language, tools and brain: The ontogeny and phylogeny of hierarchically organized sequential behavior. *Behavioral and Brain Sciences, 14*, 531–595.

Greenfield, P. M., & Savage-Rumbaugh, E. S. (1990). Grammatical combination in *Pan paniscus:* Processes of learning and invention in the evolution and development of language. In S. T. Parker & K. R. Gibson (Eds.), *"Language" and intelligence in monkeys and apes* (pp. 540–548). New York: Cambridge University Press.

Greenfield, P. M., & Savage-Rumbaugh, E. S. (1993). Comparing communicative competence in child and chimp: The pragmatics of repetition. *Journal of Child Language, 20*, 1–26.

Greenfield, P. M., & Smith, J. H. (1976). *The structure of communication in early language development.* New York: Academic Press.

Gribbin, J. M. (1988). *The one percent advantage.* New York: Blackwell.

Guillaume, P. (1971). *Imitation in children* (E. Halperin, Trans., 2nd ed.). Chicago: University of Chicago Press.

Hall, B. K. (1992). *Evolutionary developmental biology.* London: Chapman & Hall.

Hall, G. S. (1897). A study of fears. *American Journal of Psychology, 8*, 147–249.

Hall, G. S. (1904). *Adolescence, its psychology and its relations to physiology, anthropology, sociology, sex, crime, religion and education* (Vols. 1 & 2). New York: Appleton.

Hallock, M. B., & Worobey, J. (1984). Cognitive development in infant chimpanzees. *Journal of Human Evolution, 13*, 441–447.

Hanken, J., & Wake, D. B. (1993). Miniaturization of body size: Organismal consequences and evolutionary significance. *Annual Reviews of Ecology and Systematics, 24*, 501–520.

Harcourt, A. (1988). Alliances in contests and social intelligence. In R. Byrne & A. Whiten (Eds.), *Machiavellian intelligence* (pp. 132–151). Oxford, UK: Oxford University Press.

Hart, D., & Fegley, S. (1994). Social imitation and the emergence of a mental model of self. In S. T. Parker, R. W. Mitchell, & M. L. Boccia (Eds.), *Self-awareness in animals and humans* (pp. 149–165). New York: Cambridge University Press.

Harvey, P. H. (1990). Life-history variation: Size and mortality patterns. In C. J. DeRousseau (Ed.), *Primate life history and evolution* (pp. 81–88). New York: Wiley-Liss.

Harvey, P., Martin, R. D., & Clutton-Brock, T. (1987). Life histories in comparative perspective. In B. Smuts, D. Cheney, R. Seyfarth, R. Wrangham, & T. Struhsaker (Eds.), *Primate societies* (pp. 181–196). Chicago: University of Chicago Press.

Harvey, P. H., & Pagel, M. D. (1991). *The comparative method in evolutionary biology.* Oxford, UK: Oxford University Press.

Hauser, M. (1988). Invention and social transmission: New data from wild vervet monkeys. In R. Byrne & A. Whiten (Eds.), *Machiavellian intelligence* (pp. 327–343). Oxford, UK: Oxford University Press.

Hayes, C. (1952). *Ape in our house.* New York: Harper.

Hayes, K. J., & Hayes, C. (1951). Imitation in a home-raised chimpanzee. *Journal of Comparative Physiological Psychology, 45,* 450–459.

Hayes, K. J., & Nissen, C. H. (1971). Higher mental functions of a home-raised chimpanzee. In A. M. Schrier & F. Stollnitz (Eds.), *Behavior of nonhuman primates* (pp. 59–115). New York: Academic Press.

Hennig, W. (1979). *Phylogenetic systematics.* Urbana, IL: University of Illinois Press.

Heyes, C. (1995). Self-recognition in mirrors: Further reflections create a hall of mirrors. *Animal Behaviour, 50,* 1533–1542.

Holloway, R. (1996). Evolution of the human brain. In A. Lock & C. Peters (Eds.), *Handbook of human symbolic evolution* (pp. 74–125). Oxford, UK: Clarendon Press.

Horgan, J. (1995). From complexity to perplexity. *Scientific American, 272,* 104–109.

Hughes, N. C. (1991). Morphological plasticity and genetic flexibility in a Cambrian trilobite. *Geology, 19,* 913–916.

Hull, D. (1988). Progress in ideas of progress. In M. H. Nitecki (Ed.), *Evolutionary progress* (pp. 27–48). Chicago: University of Chicago Press.

Humphrey, N. K. (1976). The social function of intellect. In P. P. G. Bateson & R. A. Hinde (Eds.), *Growing points in ethology* (pp. 303–317). Cambridge, UK: Cambridge University Press.

Huxley, T. H. (1959). *Man's place in nature.* Ann Arbor, MI: University of Michigan Press.

Hyatt, C., & Hopkins, W. (1994). Self-awareness in bonobos and chimpanzees: A comparative perspective. In S. T. Parker, R. W. Mitchell, & M. L. Boccia (Eds.), *Self-awareness in animals and humans* (pp. 248–253). New York: Cambridge University Press.

Ingmanson, E. (1996). Tool-using behavior in wild *Pan paniscus:* Social and ecological considerations. In A. Russon, K. Bard, & S. T. Parker (Eds.) *Reaching into thought: The minds of the great apes* (pp. 190–210). Cambridge, UK: Cambridge University Press.

Inhelder, B., & Piaget, J. (1964). *The early growth of logical thinking.* New York: Routledge, Kegan Paul.

Itakura, S. (1994a). Mother-infant interaction of chimpanzees during the presentation of unfamiliar objects: Evidence for social referencing in chimpanzees. *Bulletin of Oita Prefectural College of Arts and Culture, 32,* 72–80.

Itakura, S. (1994b). Recognition of line-drawing representations by a chimpanzee. *The Journal of General Psychology, 12* (3), 189–197.

Itani, J., & Nishimura, A. (1973). The study of infrahuman culture in Japan. In E. W. Menzel (Ed.), *Protocultural primate behavior.* Basel, Switzerland: Karger.

Jensvold, M. L. A., & Fouts, R. (1993). Imaginary play in chimpanzees (*Pan troglodytes*). *Human Evolution, 8,* 217–227.

Jerison, H. (1973). *The evolution of the brain and intelligence.* New York: Academic Press.

Jerison, H. (1976). Principles of the evolution of brain and behavior. In R. B. Masterton, W. Hodos, & H. Jerison (Eds.), *Evolution, brain, and behavior.* Hillsdale, NJ: Erlbaum.

Jolly, A. (1964). Choice of cue in prosimian learning. *Animal Behaviour, 12* (4), 571–577.

Jolly, A. (1966). Lemur social behaviour and primate intelligence. *Science, 153,* 501–506.

Jolly, A. (1972). *The evolution of primate behavior.* New York: McGraw Hill.

Karmiloff-Smith, A. (1995). *Beyond modularity.* Cambridge, MA: M. I. T. Press.

Katz, M. J. (1987). Is evolution random? In R. Raff & E. Raff (Eds.), *Development as an evolutionary process* (pp. 181–224). New York: De Gruyter.

Kaufman, S. (1993). *Origins of order.* Oxford, UK: Oxford University Press.

Kellogg, W., & Kellogg, L. (1933). *Development of ape and child.* New York: McGraw Hill.

Kessen, W. (1965). *The child.* New York: John Wiley & Sons.

King, B. J. (1994). *The information continuum.* Santa Fe, NM: School of American Research.

Kirschner, M. (1992). Evolution of the cell. In P. R. Grant & H. S. Horn (Eds.), *Molds, molecules, and metazoa* (pp. 99–126). Princeton, NJ: Princeton University Press.

Köhler, W. (1927). *The mentality of apes.* New York: Vintage.

Konner, M. (1976). Maternal care, infant behavior and development among the !Kung. In R. B. Lee & I. DeVore (Eds.), *Kalahari hunter-gatherers: Studies of the !Kung San and their neighbors* (pp. 218–245). Cambridge, MA: Harvard University Press.

Konner, M. (1991). Universals of behavioral development in relation to brain myelination. In K. R. Gibson & A. Petersen (Eds.), *Brain maturation and cognitive development* (pp. 181–224). New York: De Gruyter.

Kottler, M. J. (1985). Charles Darwin and Alfred Russell Wallace: Two decades of debate over natural selection. In D. Kohn (Ed.), *The Darwinian heritage* (pp. 367–432). Princeton, NJ: Princeton University Press.

Kuhlmeier, V., & Boysen, S. T. (1997). Understanding of the representational nature of a scale model by chimpanzees *(Pan troglodytes).* Paper delivered at the American Society of Primatologists, Annual Meeting, San Diego, CA.

Kummer, H. (1967). Tripartite relations in hamadryas baboons. In S. A. Altmann (Ed.), *Social communication among primates* (pp. 63–71). Chicago: University of Chicago Press.

Kummer, H., & Dasser, V. (1990). Exploring primate social cognition: Some critical remarks. *Behaviour, 112* (1–2), 84–98.

Kuroda, S. (1989). Developmental retardation and behavioral characteristics of pygmy chimpanzees. In P. G. Heltne & L. A. Marquardt (Eds.), *Understanding chimpanzees* (pp. 184–193). Cambridge, MA: Harvard University Press.

Langer, J. (1980). *The origins of logic: Six to twelve months.* New York: Academic Press.

Langer, J. (1986). *The origins of logic: One to two years.* New York: Academic Press.

Langer, J. (1996). Heterochrony and the evolution of primate cognitive development. In A. Russon, K. Bard, & S. T. Parker (Eds.), *Reaching into thought: The minds of great apes* (pp. 257–277). Cambridge, UK: Cambridge University Press.

Langer, J. (1998). Phylogenetic and ontogenetic origins of cognition: Classification. In J. Langer & M. Killen, *Piaget, Evolution, and Development* (pp. 33–54). Mahwah, NJ: Erlbaum.

Langer, J. (in press). The heterochronic evolution of primate cognitive development. In S. T. Parker, J. Langer, & M. McKinney (Eds.), *The evolution of behavioral ontogeny.* Santa Fe, NM: School of American Research.

Ledbetter, D. H., & Basen, J. A. (1982). Failure to demonstrate self-recognition in gorillas. *American Journal of Primatology, 2,* 307–310.

Leslie, A. M. (1988). Some implications of pretense for mechanisms underlying the child's theory of mind. In J. Astington, P. Harris, & D. Olsen (Eds.), *Developing theories of mind* (pp. 19–46). Cambridge, UK: Cambridge University Press.

Levinton, J. S. (1988). *Genetics, paleontology, and macroevolution.* Cambridge, UK: Cambridge University Press.

Lewis, M. (1994). Myself and me. In S. T. Parker, R. W. Mitchell, & M. L. Boccia (Eds.), *Self-awareness in animals and humans* (pp. 20–34). New York: Cambridge University Press.

Lewis, M., & Brooks-Gunn, J. (1979). *Social cognition and the acquisition of self.* New York: Plenum.

Lewis, M., Sullivan, M. W., Stanger, C., & Weiss, M. (1989). Self development and self-conscious emotions. *Child Development, 60,* 146–156.

Lewontin, R. (1990). Evolution of cognition. In D. Osherson & E. E. Smith (Eds.), *Thinking: An invitation to cognitive science* (Vol. 3, pp. 229–246). Cambridge, MA: M. I. T. Press.

Lillard, A. S. (1993a). Pretend play skills and the child's theory of mind. *Child Development, 64,* 348–371.

Lillard, A. S. (1993b). Young children's conceptualization of pretense: Action or mental representational state? *Child Development, 64,* 372–386.

Limongelli, L., Boysen, S. T., & Visalberghi, E. (1995). Comprehension of cause-effect relations in a tool-using task by chimpanzees (Pan troglodytes). *Journal of Comparative Psychology, 109* (1), 18–26.

Lindburg, D. G. (Ed.). (1980). *The macaques: Studies in ecology, behavior and evolution.* New York: Van Nostrand Reinhold.

Linn, A., Bard, K., & Anderson, J. R. (1992). Development of self-recognition in chimpanzees. *Journal of Comparative Psychology, 106,* 120–127.

Lorenz, K. (1950). The comparative method in studying innate behavior patterns. *Symposium of the Society for Experimental Biology, 42,* 166–181.

Lorenz, K. (1965). *The evolution and modification of behavior.* Chicago: University of Chicago Press.

Lourenco, O., & Machado, A. (1996). In defense of Piaget's theory: A reply to 10 common criticisms. *Psychological Review, 102* (1), 143–164.

Lovejoy, A. (1960). *The great chain of being.* New York: Harper Torchbooks.

Luria, A. R. (1973). *The working brain: An introduction to neuropsychology.* New York: Penguin Books.

Mabee, P. M. (1993). Phylogenetic interpretation of ontogenetic change: Sorting out the actual and the artifactual in an empirical case study of centrarchid fishes. *Zoological Journal of the Linnean Society, 107*, 175–291.

Mabee, P., & Humphries, J. (1993). Coding polymorphic data: Examples from allozymes and ontogeny. *Systematic Zoology, 42*, 166–181.

MacArthur, R. H., & Wilson, E. O. (1967). *Theory of island biogeography*. Princeton, NJ: Princeton University Press.

MacLean, P. D. (1990). *The triune brain in evolution*. New York: Plenum.

Maddison, D., & Maddison, W. (1992). *MacClade*. Sunderland: Sinaur Associates.

Mandler, J. (1988). How to build a baby: On the development of an accessible representation system. *Cognitive Development, 3*, 113–136.

Maple, T., & Hoff, M. (1982). *Gorilla behavior*. New York: Van Nostrand Reinhold.

Martens, K., & Psarakos, S. (1994). Evidence of self-awareness in the bottle nose dolphin. In S. T. Parker, R. W. Mitchell, & M. L. Boccia (Eds.), *Self-awareness in animals and humans* (pp. 361–379). New York: Cambridge University Press.

Martin, P., & Bateson, P. (1993). *Measuring behaviour*. (2nd ed.). Cambridge, UK: Cambridge University Press.

Martin, R. D. (1981). Relative brains size and basal metabolic rate in terrestrial vertebrates. *Nature, 293*, 57–60.

Martin, R. D. (1983). *Human brain evolution in an ecological context*. New York: American Museum of Natural History.

Martin, R. D., & MacLarnon, A. M. (1990). Reproductive patterns in primates and other mammals: The dichotomy between altricial and precocial offspring. In C. J. DeRousseau (Ed.), *Primate life history and evolution*. New York: Wiley-Liss.

Mason, H. (Ed.). (1984). *Evolution of domesticated animals*. London: Longman.

Mathieu, M., & Bergeron, G. (1981). Piagetian assessment on cognitive development in chimpanzee (*Pan troglodytes*). In A. B. Chiarelli & R. S. Corruccini (Eds.), *Primate behavior and sociobiology* (pp. 142–147). New York: Springer-Verlag.

Mathieu, M., Bouchard, M., Granger, L., & Hersovitch, J. (1976). Piagetian object permanence in *Cebus capucinus, Lagothrica flavicauda* and *Pan troglodytes*. *Animal Behaviour, 24*, 575–588.

Matsuzawa, T. (1985a). Color naming and classification in a chimpanzee. *Journal of Human Evolution, 14*, 283–291.

Matsuzawa, T. (1985b). Use of numbers by a chimpanzee. *Nature, 315* (6014), 57–59.

Matsuzawa, T. (1990). Spontaneous sorting in human and chimpanzee. In S. T. Parker & K. R. Gibson (Eds.), *"Language" and intelligence in monkeys and apes* (pp. 451–468). New York: Cambridge University Press.

Matsuzawa, T. (1994). Field experiments on use of stone tools in the wild. In R. Wrangham, W. C. McGrew, F. B. M. de Waal, P. Heltne, & L. A. Marquardt (Eds.), *Chimpanzee cultures* (pp. 351–370). Cambridge, MA: Harvard University Press.

Matsuzawa, T. (1995). *Chimpanzee being* (excerpt, T. Matsuzawa, Trans.). Tokyo: Iwanami-shoten.

Matsuzawa, T. (1996). Chimpanzee intelligence in nature and in captivity: Isomorphism of symbol use and tool use. In W. McGrew, L. Marchant, & T. Nishida (Eds.), *Great ape societies* (pp. 196–209). Cambridge, UK: Cambridge University Press.

Matsuzawa, T. (1997). The death of an infant chimpanzee at Bossou, Guinea. *Pan African News, 4* (1), 4–6.

Maynard-Smith, J. R. (1988). Evolutionary progress and levels of selection. In M. H. Nitecki (Ed.), *Evolutionary progress* (pp. 219–230). Chicago: University of Chicago Press.

Maynard-Smith, J., Burian, R., Kaufman, S., Alberch, J., Campbell, J., Goodwin, B., Lande, R., Raup, D., & Wolpert, L. (1985). Developmental constraints and evolution. *Quarterly Review of Biology, 60,* 265–287.

Mayr, E. (1994). Recapitulation reinterpreted: The somatic program. *Quarterly Review of Biology, 69* (2), 223–232.

McGonigle, B. O., & Chalmers, M. (1977). Are monkeys logical? *Nature, 267,* 694–696.

McGrew, W. (1992). *Chimpanzee material culture.* New York: Cambridge University Press.

McHenry, H. (1986). The first bipeds: A comparison of the *A. afarensis* and *A. africanus* postcranium and implications for the evolution of bipedalism. *Journal of Human Evolution, 15,* 177–191.

McKinney, M. L. (1990). Classifying and analyzing evolutionary trends. In K. J. McNamara (Ed.), *Evolutionary trends* (pp. 28–58). Tucson, AZ: University of Arizona Press.

McKinney, M. L. (in press-a). Biological evolution of behavior and cognition. In M. K. J. Langer (Ed.), *Piaget, development and evolution,* Hillsdale, NJ: Erlbaum.

McKinney, M. L. (in press-b). The juvenilized ape myth: Our overdeveloped brain. *Bioscience.*

McKinney, M. L., & Gittleman, J. G. (1995). Ontogeny and phylogeny: Tinkering with covariation in life history, behavior, and morphology. In K. McNamara (Ed.), *Evolutionary change through heterochrony* (pp. 15–31). New York: Wiley.

McKinney, M., & McNamara, K. (1991). *Heterochrony: The Evolution of Ontogeny.* New York: Plenum.

McLennan, D. A., Brooks, D. R., & McPhail, J. D. (1988). The benefits of communication between comparative ethology and phylogenetic systematics: A case study using gasterosteid fishes. *Canadian Journal of Zoology, 66,* 2177–2190.

McNamara, K. (1997). *The shapes of time.* Baltimore: The Johns Hopkins University Press.

McPeek, M. A. (1995). Testing hypotheses about evolutionary change on single branches of a phylogeny using evolutionary contrasts. *The American Naturalist, 145* (5), 686–703.

McShea, D. W. (1994). Investigating mechanisms of large-scale evolutionary trends. *Evolution, 48,* 1747–1763.

Mead, G. H. (1970). *Mind, self and society.* Chicago: University of Chicago Press.

Mellors, P., & Stringer, C. (Eds.). (1989). *The human revolution.* Princeton, NJ: Princeton University Press.

Meltzoff, A. (1985). The roots of social and cognitive development: Models of man's original nature. In T. M. Field & N. A. Fox (Eds.), *Social perception in infants* (pp. 1–30). Norwood, NJ: Ablex.

Meltzoff, A. (1990). Foundations for developing a concept of self: The role of imitation in relating self to other and the value of social mirroring, social modeling, and self practice in infancy. In D. Cicchetti & M. Beeghly (Eds.), *The self in transition: Infancy to childhood.* Chicago: University of Chicago Press.

Meltzoff, A., & Gopnik, A. (1993). The role of imitation in understanding persons and developing a theory of mind. In S. Baron-Cohen, H. Tager-Flusberg, & D. J. Cohen (Eds.), *Understanding other minds* (pp. 335–366). Oxford, UK: Oxford University Press.

Meltzoff, A., & Moore, M. K. (1997). Imitation of facial and manual gestures by human neonates. *Science, 198,* 75–78.

Meltzoff, A., & Moore, M. K. (1983). Newborn infants imitate adult facial gestures. *Child Development, 54,* 702–709.

Meltzoff, A., & Moore, M. K. (1989). Imitation in newborn infants: Exploring the range of gestures imitated and the underlying mechanisms. *Developmental Psychology, 25,* 954–962.

Menzel, E. W., Jr., Premack, D., & Woodruff, G. (1978). Map reading by chimpanzees. *Folia primatologica, 29,* 241–249.

Michel, G. F., & Moore, C. L. (1995). *Developmental psychobiology.* Cambridge, MA: M.I.T. Press.

Mignault, C. (1985). Transition between sensorimotor and symbolic activities in nursery-reared chimpanzees (*Pan troglodytes*). *Journal of Human Evolution, 14,* 747–758.

Miles, H. L. (1983). Apes and language: The search for communicative competence. In J. d. Luce & H. T. Wilder (Eds.), *Language in primates: Perspectives and implications* (pp. 43–61). New York: Springer-Verlag.

Miles, H. L. (1990). The cognitive foundations for reference in a signing orangutan. In S. T. Parker & K. R. Gibson (Eds.), *"Language" and intelligence in monkeys and apes* (pp. 511–539). New York: Cambridge University Press.

Miles, H. L. (1994). Me Chantek: The development of self-awareness in a signing gorilla. In S. T. Parker, R. W. Mitchell, & M. L. Boccia (Eds.), *Self-awareness in animals and humans* (pp. 254–272). New York: Cambridge University Press.

Miles, H. L. (1997). Anthropomorphism, apes, and language. In R. W. Mitchell, N. S. Thompson, & H. L. Miles (Eds.), *Anthropomorphism, anecdotes, and animals* (pp. 383–404). Albany, NY: SUNY Press.

Miles, H. L., Mitchell, R., & Harper, S. (1996). Simon says: The development of imitation in an enculturated orangutan. In A. Russon, K. Bard, & S. T. Parker (Eds.), *Reaching into thought: The minds of great apes* (pp. 278–299). Cambridge UK: Cambridge University Press.

Milton, K. (1981). Distribution patterns of tropical plant foods as an evolutionary stimulus to primate mental development. *American Anthropologist, 83,* 534–548.

Milton, K. (1988). Foraging behaviour and the evolution of primate intelligence. In R. Byrne & A. Whiten (Eds.), *Machiavellian intelligence* (pp. 285–305). Oxford, UK: Oxford University Press.

Mitchell, R. W. (1986). A framework for discussing deception. In R. W. Mitchell & N. Thompson (Eds.), *Deception: Perspectives on human and nonhuman deceit* (pp. 3–40). Albany, NY: SUNY Press.

Mitchell, R. W. (1987). A comparative-developmental approach to understanding imitation. In P. P. G. Bateson & P. H. Klopfer (Eds.), *Perspectives in ethology*, (Vol. 7, pp. 183–215). New York: Plenum.

Mitchell, R. W. (1991). Deception and hiding in captive lowland gorillas (*Gorilla gorilla gorilla*). *Primates, 32* (4), 523–527.

Mitchell, R. W. (1993). Mental models of mirror-self-recognition: Two theories. *New Ideas in Psychology, 11*, 295–325.

Mitchell, R. W. (1994). The evolution of primate cognition: Simulation, self-knowledge, and knowledge of other minds. In D. Quiatt & J. Itani (Eds.), *Hominid culture in primate perspective* (pp. 177–232). Boulder, CO: University of Colorado Press.

Mitchell, R. W. (1997). Anthropomorphic anecdotalism as method. In R. W. Mitchell, N. S. Thompson, & H. L. Miles (Eds.), *Anthropomorphism, anecdotes, and animals* (pp. 151–169). Albany, NY: SUNY Press.

Mitchell, R. W. (in press). Deception and concealment as strategic script violations in great apes and humans. In S. T. Parker, R. W. Mitchell, & H. L. Miles (Eds.), *Mentalities of gorillas and orangutans*. Cambridge, UK: Cambridge University Press.

Mitchell, R. W., & Anderson, J. (1997). Communicative and deceptive pointing by capuchin monkeys (*Cebus apella*). *Animal Behaviour, 111*, 351–361.

Mitchell, R. W., Thompson, N., & Miles, H. L. (Eds.). (1997). *Anthropomorphism, anecdotes and animals*. Albany, NY: SUNY Press.

Mitchell, R., & Thompson, N. (Eds.). (1986). *Deception: Perspectives on human and nonhuman deceit*. Albany, NY: SUNY Press.

Mithen, S. (1993). Tool-use and language in apes and humans. *Cambridge Archeological Journal, 3* (2), 285–300.

Mithen, S. (1996). *The prehistory of the mind*. New York: Thames and Hudson.

Moerk, E. L. (1978). The fuzzy set called "imitations." In G. E. Speidel & K. E. Nelson (Eds.), *The many faces of imitation* (pp. 277–303). New York: Springer-Verlag.

Montagu, A. (1989). *Growing young* (2nd ed.). Granby, MA: Bergin and Garvey.

Moore, J. (1996). Savanna chimpanzees, referential models and the last common ancestor. In W. McGrew, L. Marchant, & T. Nishida (Eds.), *Great ape societies* (pp. 275–292). Cambridge, UK: Cambridge University Press.

Morey, D. F. (1994). The early evolution of the domestic dog. *American Scientist, 82*, 336–347.

Morell, V. (1994). Will primate genetics split one gorilla into two? *Science, 265*, 1661.

Morgan, C. L. (1895). *An introduction to comparative psychology*. London: Walter Scott.

Morin, P. A., Moore, J. J., Chakraborty, R., Jin, L., Goodall, J., & Woodruff, D. S. (1994). Kin selection, social structure, gene flow, and the evolution of chimpanzees. *Science, 265*, 1193–1201.

Morris, D. (1962). *Biology of art*. London: Methuen and Co.

Morris, S. C. (1995). Ecology in deep time. *Trends in Ecology and Evolution, 10,* 290–294.

Morss, J. R. (1990). *The biologising of childhood: Developmental psychology and the Darwinian myth.* Hillsdale, NJ: Erlbaum.

Mounoud, P., & Vinter, A. (1981). Representation and sensorimotor development. In G. Butterworth (Ed.), *Infancy and epistemology,* Brighton, UK: Harvester Press.

Muller, U., Sokol, B., & Overton, W. F. (in press). Reframing a constructivist model of the development of mental representation. *Developmental Review.*

Nagell, K., Olguin, R., & Tomasello, M. (1993). Processes of social learning in the tool use of chimpanzees (*Pan troglodytes*) and human children (*Homo sapiens*). *Journal of Comparative Psychology, 107* (2), 174–186.

Natale, F. (1989a). Causality II: The stick problem. In F. Antinucci (Ed.), *Cognitive structure and development in nonhuman primates* (pp. 121–135). Hillsdale, NJ: Erlbaum.

Natale, F. (1989b). Patterns of object manipulation. In F. Antinucci (Ed.), *Cognitive structure and development in nonhuman primates* (pp. 121–133). Hillsdale, NJ: Erlbaum.

Natale, F. (1989c). Stage 5 object-concept. In F. Antinucci (Ed.), *Cognitive structure and development in nonhuman primates* (pp. 89–96). Hillsdale, NJ: Erlbaum.

Natale, F. (1989d). Stage 6 object-concept and representation. In F. Antinucci (Ed.), *Cognitive structure and development in nonhuman primates* (pp. 97–112). Hillsdale, NJ: Erlbaum.

Nelson, K. (1983). The derivation of concepts and categories from event representations. In E. Scholnick (Ed.), *New trends in conceptual representation: Challenges to Piagetian theory* (pp. 129–149). Hillsdale, NJ: Erlbaum.

Nelson, K., & Gruendel, J. (1986). Children's scripts. In K. Nelson (Ed.), *Event knowledge* (pp. 21–45). Hillsdale, NJ: Erlbaum.

Nelson, K., & Seidman, s. (1984). Playing with scripts. In I. Bretherton (Ed.), *Symbolic play* (pp. 45–72). New York: Academic Press.

Nishida, T. (1986). Local traditions and cultural transmission. In B. Smuts, D. Cheney, R. Seyfarth, R. Wrangham, & T. Struhsaker (Eds.), *Primate societies* (pp. 462–474). Chicago: University of Chicago Press.

Nitecki, M. H. (Ed.). (1988). *Evolutionary progress.* Chicago: University of Chicago Press.

Noack, R. A. (1995). A radical reversal in cortical information flow as the mechanism for human cognitive abilities: The frontal feedback model. *The Journal of Mind and Behavior, 16* (1), 281–304.

Noble, W., & Davidson, I. (1996). *Human evolution, language, and mind.* Cambridge UK: Cambridge University Press.

O'Neill, R., DeAngeles, D., Wiede, J., & Allen, T. (1986). *A hierarchical concept of ecosystems.* Princeton, NJ: Princeton University Press.

O'Sullivan, C., & Yeager, C. P. (1989). Communicative context and linguistic competence: The effects of social setting on a chimpanzee's conversational skill. In R. A. Gardner, B. Gardner, & T. Van Cantfort (Eds.), *Teaching sign language to chimpanzees* (pp. 269–279). Albany, NY: SUNY Press.

Oakley, K. P. (1959). *Man the tool-maker.* Chicago: Phoenix Books, University of Chicago Press.

Parker, S. T. (1977). Paiget's sensorimotor series in an infant macaque: A model for comparing unstereotyped behavior and intelligence in human and non-human primates. In S. Chevalier-Skolnikoff & F. Poirier (Eds.), *Primate biosocial development* (pp. 43–112). New York: Garland.

Parker, S. T. (1984). Playing for keeps: An evolutionary perspective on human games. In P. K. Smith (Ed.), *Play in animals and humans* (pp. 271–293). London: Blackwell.

Parker, S. T. (1987). A sexual selection model for hominid evolution. *Human Evolution, 2* (3), 235–53.

Parker, S. T. (1990a). The origins of comparative developmental evolutionary studies of primate mental abilities. In S. T. Parker & K. R. Gibson (Eds.), *"Language" and intelligence in monkeys and apes* (pp. 3–64). New York: Cambridge University Press.

Parker, S. T. (1990b). Why big brains are so rare: Energy costs of intelligence and brain size in anthropoid primates. In S. T. Parker & K. R. Gibson (Eds.), *"Language" and intelligence in monkeys and apes* (pp. 129–154). New York: Cambridge University Press.

Parker, S. T. (1991). A developmental approach to the origins of self-awareness in great apes and human infants. *Human Evolution, 6,* 435–449.

Parker, S. T. (1993). Imitation and circular reactions as evolved mechanisms for cognitive construction. *Human Development, 36,* 309–323.

Parker, S. T. (1996a). Apprenticeship in tool-mediated extractive foraging: The origins of imitation, teaching, and self-awareness in great apes. In A. Russon, K. Bard, & S. T. Parker (Eds.), *Reaching into thought: The minds of great apes* (pp. 348–370). Cambridge, UK: Cambridge University Press.

Parker, S. T. (1996b). Using cladistic analysis of comparative data to reconstruct the evolution of cognitive development in hominids. In E. Martins (Ed.), *Phylogenies and the comparative method in animal behavior* (pp. 361–398). New York: Oxford University Press.

Parker, S. T. (1997). Anthropomorphism is the null hypothesis and recapitulation is the Bogey man in comparative developmental evolutionary studies. In R. Mitchell, N. Thompson, & H. L. Miles (Eds.), *Anthropomorphism, anecdotes and animals* (pp. 348–362). Albany, NY: SUNY Press.

Parker, S. T. (in press). Development of intelligence and social roles in the play of an infant gorilla. In S. T. Parker, R. W. Mitchell, & H. L. Miles (Eds.), *The mentalities of gorillas and orangutans,* Cambridge, UK: Cambridge University Press.

Parker, S. T., & Baars, B. (1990). How scientific usages reflect implicit theories: Adaptation, development, instinct, learning, cognition, and intelligence. In S. T. Parker & K. R. Gibson (Eds.), *"Language" and intelligence in monkeys and apes* (pp. 65–96). New York: Cambridge University Press.

Parker, S. T., & Gibson, K. R. (1977). Object manipulation, tool use, and sensorimotor intelligence as feeding adaptations in cebus monkeys and great apes. *Journal of Human Evolution, 6,* 623–641.

Parker, S. T., & Gibson, K. R. (1979). A developmental model for the evolution of language and intelligence in early hominids. *Behavioral and Brain Sciences,* 2, 367–408.

Parker, S. T., & Gibson, K. G. (Eds.). (1990). *"Language" and intelligence in monkeys and apes.* New York: Cambridge University Press.

Parker, S. T., & Milbrath, C. (1993). Higher intelligence, propositional language, and culture as adaptations for planning. In K. R. Gibson & T. Ingold (Eds.), *Tools, language, and cognition in human evolution* (pp. 314–33). Cambridge, UK: Cambridge University Press.

Parker, S. T., & Milbrath, C. (1994). Contributions of imitation and role-playing games to the construction of self in primates. In S. T. Parker, R. W. Mitchell, & M. L. Boccia (Eds.), *Self-awareness in animals and humans* (pp. 108–128). New York: Cambridge University Press.

Parker, S. T., Markowitz, H., Gould, J., & Kerr, M. (in press). A survey of tool use in zoo gorillas. In S. T. Parker, R. W. Mitchell, & H. L. Miles (Eds.), *The mentality of gorillas and orangutans.* Cambridge, UK: Cambridge University Press.

Parker, S. T., Mitchell, R. W., & Boccia, M. L. (Eds.). (1994). *Self-awareness in animals and humans.* New York: Cambridge University Press.

Parker, S. T., Mitchell, R. W., & Miles, H. L., (Eds.). (in press). *The mentalities of gorillas and orangutans.* Cambridge, UK: Cambridge University Press.

Parker, S. T., & Poti', P. (1990). The role of innate motor patterns in ontogenetic and experiential development of intelligent use of sticks in cebus monkeys. In S. T. Parker & K. R. Gibson (Eds.), *"Language" and intelligence in monkeys and apes* (pp. 219–245). New York: Cambridge University Press.

Passingham, R. (1982). *The human primate.* San Francisco: W. H. Freeman.

Passingham, R. E. (1975). Changes in the size and organization of the brain in man and his ancestors. *Brain, Behavior and Evolution, 11,* 73–90.

Patterson, F. (1980). Innovative use of language by gorilla: A case study. In K. Nelson (Ed.), *Children's language,* (Vol. 2, pp. 497–561). New York: Gardner.

Patterson, F., & Cohn, R. (1994). Self-recognition and self-awareness in lowland gorillas. In S. T. Parker, R. W. Mitchell, & M. L. Boccia (Eds.), *Self-awareness in animals and humans* (pp. 273–290). New York: Cambridge University Press.

Patterson, F., & Linden, E. (1981). *The education of Koko.* New York: Holt, Rinehart & Winston.

Perner, J. (1991). *Understanding the representational mind.* Boston: M.I.T. Press.

Piaget, J. (1928). *Judgment and reasoning in the child* (M. Warden, Trans.). London: Routledge, Kegan Paul.

Piaget, J. (1929). *The child's conception of the world* (J. and A. Tomlinson, Trans.). London: Routledge, Kegan Paul.

Piaget, J. (1952). *The origins of intelligence in children.* New York: International Universities Press.

Piaget, J. (1954) *The construction of reality in the child.* New York: Basic Books.

Piaget, J. (1962) *Play, dreams and imitation in childhood.* New York: W. W. Norton.

Piaget, J. (1965a). *The child's conception of number.* New York: W. W. Norton.

Piaget, J. (1965b). *The moral judgment of the child.* New York: Free Press.

Piaget, J. (1966). *Psychology of intelligence.* New York: Littlefield Adams & Co.

Piaget, J. (1968). *Six psychological studies.* New York: Random House.

Piaget, J. (1970a). *Genetic epistemology.* New York: Columbia University Press.

Piaget, J. (1970b). *Structuralism.* New York: Harper Colophon Books.

Piaget, J., & Garcia, R. (1974). *Understanding causality* (D. and M. Miles, Trans.). New York: W. W. Norton.

Piaget, J., & Garcia, R. (1991). *Toward a logic of meanings.* Hillsdale, NJ: Erlbaum.

Piaget, J., Grize, J.-B., Szeminska, A., & Bang, V. (1977). *Epistemology and psychology of functions* (J. and A. Tomlinson, Trans.). London: Routledge, Kegan Paul.

Piaget, J., & Inhelder, B. (1967). *The child's conception of space.* New York: W. W. Norton.

Piaget, J., & Inhelder, B. (1969). *The psychology of the child.* New York: Basic Books.

Piaget, J., & Inhelder, B. (1971). *Mental imagery in the child: A study of the development of mental representation* (P. A. Chilton, Trans.). New York: Basic Books.

Piaget, J., & Inhelder, B. (1974). *The child's construction of quantities.* New York: Routledge, Kegan Paul.

Piaget, J., Inhelder, B., & Szeminska, A. (1960). *The child's conception of geometry.* London: Routledge, Kegan Paul.

Pinker, S. (1994). *The language instinct.* New York: Harper Collins.

Portmann, A. (1990). *A zoologist looks at humankind* (J. Schaefer, Trans.). New York: Columbia University Press.

Poti', P. (1996). Spatial aspects of spontaneous object grouping by young chimpanzees (*Pan troglodytes*). *International Journal of Primatology, 17,* 101–116.

Poti', P. (1997). Logical structures of young chimpanzees: Spontaneous object grouping. *International Journal of Primatology, 18,* 33–59.

Poti', P., & Antinucci, F. (1989). Logical operations. In F. Antinucci (Ed.), *Cognitive structures and development in nonhuman primates* (pp. 189–228). Hillsdale, NJ: Erlbaum.

Poti', P., & Spinozzi, G. (1994). Early sensorimotor development in chimpanzees (*Pan troglodytes*). *Journal of Comparative Psychology, 108* (1), 93–103.

Povinelli, D. (1994). How to create a self-recognizing gorilla (but don't try it on macaques). In S. T. Parker, R. W. Mitchell, & M. L. Boccia (Eds.), *Self-awareness in animals and humans* (pp. 291–300). New York: Cambridge University Press.

Povinelli, D. J., & Cant, J. G. H. (1995). Arboreal clambering and the evolution of self conception. *Quarterly Review of Biology, 70* (4), 393–421.

Povinelli, D. J., & Eddy, T. J. (1996). *What young chimpanzees know about seeing* (Vol. Series 247, Vol. 61). Chicago: University of Chicago Press.

Povinelli, D. J., Parks, K. A., & Novak, M. A. (1992). Role reversal by rhesus monkeys, but no evidence of empathy. *Animal Behaviour, 44,* 269–281.

Povinelli, D. J., Nelson, K. E., & Boysen, S. T. (1992). Comprehension of role reversal in chimpanzees: Evidence of empathy? *Animal Behaviour, 43* 633–640.

Povinelli, D., Rulf, A., Landau, K., & Bierschwale, D. (1993). Self-recognition in chimpanzees (*Pan troglodytes*): Distribution, ontogeny, and patterns of emergence. *Journal of Comparative Psychology, 107* (4), 347–372.

Premack, D. (1976) *Intelligence in apes and man,* Hillsdale, NJ: Erlbaum.

Premack, D. (1983). The codes of man and beast. *Behavioral and Brain Sciences, 6,* 123–167.

Premack, D. (1988a). 'Does the chimpanzee have a theory of mind?' revisited. In R. Byrne & A. Whiten (Eds.), *Machiavellian intelligence: Social expertise and the evolution of intellect in monkeys, apes, and humans* (pp. 160–178). Oxford, UK: Oxford University Press.

Premack, D. (1988b). Minds with and without language. In L. Weiskrantz (Ed.), *Thought without language* (pp. 46–65). New York: Oxford University Press.

Premack, D., & Premack, A. (1983). *The mind of an ape.* New York: W.W. Norton.

Premack, D., & Woodruff, G. (1978). Does the chimpanzee have a theory of mind? *Behavioral and Brain Sciences, 4,* 515–526.

Preuschoft, H., Chivers, D. J., Brockelman, W. Y., & Creel, N. (Eds.). (1984). *The lesser apes: Evolutionary and behavioural biology.* Edinburgh: Edinburgh University Press.

Purves, D. (1988). *Body and brain: A trophic theory of neural connections.* Cambridge, MA: Harvard University Press.

Raemaekers, J. (1984). Large vs. small gibbons: Relative role of bioenergetics and competition in their ecological segregation in sympatry. In H. Preuscroft, D. J. Chivers, & N. Creel (Eds.), *The lesser apes* (pp. 209–218). Edinburgh: Edinburgh University Press.

Raff, R. A. (1996). *The shape of life.* Chicago: University of Chicago Press.

Raup, D. M. (1977). Stochastic models in evolutionary paleontology. In A. Hallam (Ed.), *Patterns of evolution* (pp. 59–78). Amsterdam: Elsevier.

Raup, D. M. (1988). Testing the fossil record for evolutionary progress. In M. Nitecki (Ed.), *Evolutionary progress* (pp. 293–317). Chicago: University of Chicago Press.

Raup, D. M. (1991). *Extinction: Bad genes or bad luck?* New York: W. W. Norton.

Redshaw, M. (1978). Cognitive development in human and gorilla infants. *Journal of Human Evolution, 7,* 113–141.

Reed, J. (1987). Robert M. Yerkes and the comparative method. In E. Tobach (Ed.), *Historical perspectives and the international status of comparative psychology* (pp. 91–102). Hillsdale, NJ: Erlbaum.

Reynolds, P. C. (1993). The complementation theory of language and tool use. In K. R. Gibson & T. Ingold (Eds.), *Tools, language and cognition in human evolution* (pp. 407–428). Cambridge, UK: Cambridge University Press.

Rice, S. (1997). The analysis of ontogenetic trajectories: When a change in size or shape is not heterochrony. *Proceedings National Academy of Sciences, 94,* 907–912.

Richard, A. (1985). *Primates in nature.* New York: W. H. Freeman.

Richards, R. J. (1987). *Darwin and the emergence of evolutionary theories of mind and behavior.* Chicago: University of Chicago Press.

Ridley, M. (1986). *Evolution and classification.* New York: Longman.

Rilling, M. (1993). Invisible counting animals: A history of contributions from comparative psychology, ethology, and learning theory. In S. T. Boysen & E. J. Capaldi (Eds.), *The development of numerical competence, animal and human models* (pp. 3–37). Hillsdale, NJ: Erlbaum.

Rimpau, J. B., Gardner, R. A., & Gardner, B. T. (1989). Expression of person, place and instrument in ASL utterances of children and chimpanzees. In R. A.

Gardner, B. T. Gardner, & T. E. Van Cantfort (Eds.), *Teaching sign language to chimpanzees* (pp. 240–268). Albany, NY: SUNY Press.

Rivera, S. M., Wakeley, A., & Langer, J. (in press). The drawbridge phenomenon: Representational reasoning or perceptual preference? *Developmental Psychology.*

Roback, A. A. (1964). *A history of American psychology* (rev. ed.). New York: Collier Books.

Romanes, G. (1882). *Animal intelligence.* London, UK: Kegan Paul, French.

Romanes, G. (1888). *Mental evolution in animals.* London: Kegan Paul, French.

Romanes, G. (1898). *Mental evolution in man: Origins of human faculty.* New York: Appleton.

Rumbaugh, D. (Ed.). (1977). *Language learning by a chimpanzee.* New York: Academic Press.

Rumbaugh, D., & Washburn, D. A. (1993). Counting by chimpanzees and ordinality judgments by macaques in video-formatted tasks. In S. T. Boysen & E. J. Capaldi (Eds.), *The development of numerical competence, animals and human models.* Hillsdale, NJ: Erlbaum.

Ruse, M. (1993). Evolution and progress. *Trends in ecology and evolution, 8,* 55–59.

Russon, A. (1996). Imitation in everyday use: Matching and rehearsal in the spontaneous imitation of rehabilitant orangutans (*Pongo pygmaeus*). In A. Russon, K. Bard, & S. T. Parker (Eds.), *Reaching into thought: The minds of great apes* (pp. 152–176). Cambridge, UK: Cambridge University Press.

Russon, A. (in press). Imitation of tool use in orangutans. In S. T. Parker, H. L. Miles, & R. W. Mitchell (Eds.), *The mentalities of gorillas and orangutans.* Cambridge, UK: Cambridge University Press.

Russon, A., Bard, K., & Parker, S. T. (Eds.). (1996). *Reaching into thought: The minds of great apes.* Cambridge, UK: Cambridge University Press.

Russon, A., & Galdikas, B. (1992). *The cognitive complexity of object manipulations in free-ranging rehabilitant orangutans.* Paper presented at the 14th Congress of the International Primatological Society, Strasbourg, France.

Russon, A., & Galdikas, B. (1993). Imitation in free-ranging rehabilititant orangutans (*Pongo pygmaeus*). *Journal of Comparative Psychology, 107* (2), 147–161.

Russon, A., & Galdikas, B. (1995). Constraints on great ape imitation: Model and action selectivity in rehabilitant orangutans (*Pongo pygmaeus*) imitation. *Journal of Comparative Psychology, 109,* 5–17.

Ruvolo, M. (1994). Molecular evolutionary processes and conflicting gene trees: The hominoid case. *American Journal of Physical Anthropology, 94,* 89–113.

Sacher, G. A. (1982). Relation of lifespan to brain weight and body weight in mammals. In E. Armstrong & D. Falk (Eds.), *Primate brain evolution: Methods and concepts.* New York: Plenum.

Sacher, G. A., & Staffeldt, E. (1974). Relation of gestation time to brain weight for placental mammals. *American Naturalist, 108,* 593–615.

Sagan, C. (1977). *The dragons of Eden.* New York: Random House.

Sarmiento, E. E. (1995). Cautious climbing and folivory: A model of hominoid differentiation. *Human Evolution, 10* (4), 289–321.

Savage-Rumbaugh, E. S., Murphy, J., Sevcik, R., Brakke, K. E., Williams, S. L., & Rumbaugh, D. M. (1993). *Language comprehension in ape and child* (Vol. 58). Chicago: University of Chicago Press.

Savage-Rumbaugh, E. S., Romski, M. A., Hopkins, W. D., & Sevcik, R. (1989). Symbol acquisition and use by *Pan troglodytes, Pan paniscus,* and *Homo sapiens.* In P. G. Heltne & L. A. Marquardt (Eds.), *Understanding chimpanzees* (pp. 266–295). Cambridge, MA: Harvard University Press.

Savage-Rumbaugh, E. S., Wilkerson, B., & Bakeman, R. (1977). Spontaneous gestural communication among conspecifics in pygmy chimpanzees. In G. Bourne (Ed.), *Progress in ape research* (pp. 287–309). New York: Academic Press.

Savage-Rumbaugh, S. (1995,). *Symbolic communication in bonobos.* Paper presented at the Jean Piaget Society, Berkeley, CA.

Savage-Rumbaugh, S., & Lewin, R. (1994). *Kanzi: The ape at the brink of the human mind.* New York: Wiley.

Savage-Rumbaugh, S., & MacDonald, K. (1988). Deception and social manipulation in symbol-using apes. In R. Byrne & A. Whiten (Eds.), *Machiavellian Intelligence: Social expertise and the evolution of intellect in monkeys, apes, and humans* (pp. 224–237). Oxford, UK: Oxford University Press.

Schaller, G. (1963). *The mountain gorilla: Biology and behavior.* Chicago: University of Chicago Press.

Schiller, P. (1952). Innate constituents of complex responses in primates. *Psychological Review, 59* (3), 177–191.

Schiller, P. (1957). Innate motor actions as a basis for learning. In P. Schiller (Ed.), *Instinctive behavior* (pp. 264–287). New York: International Universities Press.

Schultz, A. (1972). *The life of primates.* New York: Universe Books.

Sebeok, T. A., & Rosenthal, R. (Eds.). (1981). *The Clever Hans phenomenon: Communication with horses, whales, apes, and people* (Vol. 364). New York: New York Academy of Sciences.

Semaw, S., Renne, P., Harris, J. W. K., Feibel, C. S., Bernor, R. L., Fesseha, N., & Mowbray, K. (1997). 2.5-million-year-old stone tools from Gona, Ehtiopia. *Nature, 385,* 333–336.

Semendeferi, K. (in press). The frontal lobes of the great apes with a focus on the gorilla and orangutan. In S. T. Parker, R. W. Mitchell, & H. L. Miles (Eds.), *The mentality of gorillas and orangutans.* Cambridge, UK: Cambridge University Press.

Semendeferi, K., Damasio, H., Frank, R., & Van Hoesen, G. W. (1997). The evolution of the frontal lobes: A volumetric analysis based on three-dimensional reconstructions of magnetic resonance scans of human and ape brains. *Journal of Human Evolution, 32,* 375–388.

Shapiro, G. (1982). Sign acquisition in a home reared/free ranging orangutan: Comparisons with other signing apes. *American Journal of Primatology, 3,* 121–129.

Shea, B. (1989). Heterochrony in human evolution: The case for human neoteny. *Yearbook of Physical Anthropology, 32,* 69–101.

Shea, B. (1992). Developmental perspective on size change and allometry in evolution. *Evolutionary Anthropology, 1,* 125–133.

Shea, B. T. (1990). Dynamic morphology: Growth, life history, and ecology in primate evolution. In C. J. DeRousseau (Ed.), *Primate life history and evolution* (pp. 325–352). New York: Wiley.

Shubin, N. H. (1994). The phylogeny of development and the origin of homology. In L. Grande & O. Rieppel (Eds.), *Interpreting the hierarchy of nature* (pp. 201–226). New York: Academic Press.

Silk, J. B. (1978). Patterns of food sharing among mother and infant chimpanzees at Gombe National Park, Tanzania. *Folia primatologica, 29,* 129–141.

Sinclair, H. (1970). The transition from sensory-motor behaviour to symbolic activity. *Interchange, 1,* 119–126.

Skelton, R. R., McHenry, H. M., & Drawhorn, G. (1986). Phylogenetic analysis of early hominids. *Current Anthropology, 27* (1), 21–43.

Slatkin, M. (1987). Quantitative genetics of heterochrony. *Evolution, 41,* 799–811.

Smith, B. H. (1992). Life history and the evolution of human maturation. *Evolutionary Anthropology, 1,* 134–142.

Smith, B. H. (1993). The physiological age of KNM-WT 15000. In A. Walker & R. Leakey (Eds.), *The Nariokotome* Homo erectus *skeleton* (pp. 195–220). Cambridge, MA: Harvard University Press.

Smith, B. H., Crummett, T. L., & Brandt, K. L. (1994). Ages of eruption of primate teeth: A compendium for aging individuals and comparing life histories. *Yearbook of Physical Anthropology, 37,* 177–231.

Smith, J. M., Burian, R., Kaufman, S., Alberch, P., Campbell, J., Godwin, B., Lande, R., Raup, D., & Wolpert, L. (1985). Developmental constraints and evolution. *Quarterly Review of Biology, 60* (3), 265–285.

Smith, R. J., Gannon, P. J., & Smith, B. H. (1995). Ontogeny of Australopithecines and early Homo evidence from cranial capacity and dental eruption. *Journal of Human Evolution, 29,* 155–168.

Snow, C. E. (1990). Building memories: The ontogeny of autobiography. In D. Cicchetti & M. Beeghly (Eds.), *The self in transition: Infancy to childhood* (pp. 213–42). Chicago: University of Chicago Press.

Sordino, P., Hoeven, F., & Duboule, D. (1995). Hox gene expression in teleost fins and the origin of vertebrate digits. *Nature, 375* (678–681).

Spelke, E., Breinlinger, K., Macomber, J., & Jacobson, K. (1992). Origins of knowledge. *Psychological Review, 99* (4), 605–612.

Spinozzi, G. (1993). Development of spontaneous classificatory behavior in chimpanzees (*Pan troglodytes*). *Journal of Comparative Psychology, 107,* 193–200.

Spinozzi, G., & Natale, F. (1989). Classification. In F. Antinucci (Ed.), *Cognitive structure and development in nonhuman primates* (pp. 163–187). Hillsdale, NJ: Erlbaum.

Stanley, S. (1996). *Children of the ice age.* New York: Harmony Books.

Stanley, S. M. (1973). An explanation for Cope's rule. *Evolution, 27,* 1–26.

Stanley, S. M. (1990). The general correlation between rate of speciation and rate of extinction: Fortuitious causal linkages. In R. Ross & W. Allmon (Eds.), *Causes of evolution: A paleontological perspective* (pp. 103–127). Chicago: University of Chicago Press.

Stanley, S. M. (1992). An ecological theory for the origin of Homo. *Paleobiology, 18,* 237–257.

Starkey, P. (1992). The early development of numerical reasoning. *Cognition, 43,* 93–126.

Starkey, P., Spelke, E., & Gelman, R. (1990). Numerical abstractions by human infants. *Cognition, 36,* 97–127.

Stearns, S. C. (1992). *The evolution of life histories.* Oxford, UK: Oxford University Press.

Stephan, H. (1972). Evolution of primate brains: A comparative anatomical investigation. In R. Tuttle (Ed.), *The functional and evolutionary biology of primates* (pp. 155–174). New York: Aldine.

Stern, D. (1977). *The first relationship.* Cambridge, MA: Harvard University Press.

Straight, D. S., Grine, F. E., & Moniz, M. A. (1997). A reappraisal of early hominid phylogeny. *Journal of Human Evolution, 32,* 17–82.

Sugiyama, Y. (1997). Social tradition and the use of tool-composites by wild chimpanzees. *Evolutionary Anthropology, 6,* 23–27.

Susman, R. (1988). Hand of *Paranthropus robustus* from Member 1, Swartkrans: Fossil evidence for tool behavior. *Science, 240,* 781–784.

Swartz, K., & Evans, S. (1991). Not all chimpanzees (*Pan troglodytes*) show self-recognition. *Primates, 32,* 483–496.

Swartz, K., & Evans, S. (1994). Social and cognitive factors in chimpanzee and gorilla mirror behavior and self-recognition. In S. T. Parker, R. W. Mitchell, & M. L. Boccia (Eds.), *Self-awareness in animals and humans* (pp. 189–205). New York: Cambridge University Press.

Tanner, J., & Byrne, R. W. (1993). Concealing facial evidence of mood: Perspective-taking in a captive gorilla? *Primates, 34* (4), 451–457.

Tanner, J. E., & Byrne, R. W. (1996). Representation of action through iconic gesture in a captive lowland gorilla. *Current Anthropology, 37* (1), 162–173.

Tanner, J., & Byrne, R. W. (in press). Spontaneous gestural communication in captive lowland gorillas. In S. T. Parker, R. W. Mitchell, & H. L. Miles (Eds.), *Mentalities of gorillas and orangutans.* Cambridge, UK: Cambridge University Press.

Tattersall, I. (1986). Species recognition in human paleontology. *Journal of Human Evolution, 15,* 165–175.

Tayler, C. K., & Saayman, G. S. (1973). Imitative behaviour by Indian Ocean bottlenose dolphins (*Tursiops aduncus*) in captivity. *Behaviour, 44,* 286–298.

Teleki, G. (1974). Chimpanzee subsistence technology: Materials and skills. *Journal of Human Evolution, 3,* 575–594.

Terrace, H. (1979). *Nim: A chimpanzee who learned sign language.* New York: Knopf.

Terrace, H., Petitito, L., Saunders, R. J., & Bever, T. (1979). Can an ape create a sentence? *Science, 206,* 891–902.

Thatcher, R. (1994). Cyclic cortical reorganization: Origins of human cognitive development. In G. Dawson & K. Fischer (Eds.), *Human behavior and the developing brain* (pp. 232–266). New York: Guilford.

Thomas, R., & Peay, L. (1976). Length judgments by squirrel monkeys: Evidence for conservation? *Developmental Psychology, 12* (4), 349–352.

Thompson, R. K. R., Boysen, S. T., & Oden, D. L. (1997). Language-naive chimpanzees (*Pan troglodytes*) judge relations between relations in a conceptual matching-to-sample task. *Journal of Experimental Psychology, 23* (1), 31–43.

Thompson, R. L., & Boatright-Horowitz, S. (1994). The question of mirror-mediated self-recognition in apes and monkeys: Some new results and reservations. In S. T. Parker, R. W. Mitchell, & M. L. Boccia (Eds.), *Self-awareness in animals and humans* (pp. 330–349). New York: Cambridge University Press.

Thorpe, W. H. (1956). *Learning and instinct in animals.* London: Methuen.

Tomasello, M., & Call, J. (1994). Social cognition of monkeys and apes. *Yearbook of Physical Anthropology, 37,* 273–305.

Tomasello, M., & Call, J. (1997). *Primate cognition.* New York: Oxford University Press.

Tomasello, M., Gust, D., & Frost, G. T. (1989). A longitudinal investigation of gestural communication in young chimpanzees. *Primates, 30* (1), 35–50.

Tomasello, M., Kruger, A. C., & Ratner, H. H. (1993). Cultural learning. *Behavioral and Brain Sciences, 16,* 495–552.

Tooby, J., & Cosmides, L. (1992). The psychological foundations of culture. In J. Barkow, L. Cosmides, & J. Tooby (Eds.), *The adapted mind* (pp. 19–136). New York: Oxford University Press.

Tooby, J., & DeVore, I. (1987). The reconstruction of hominid behavioral evolution through strategic modeling. In W. G. Kinzey (Ed.), *The evolution of human behavior: Primate models* (pp. 183–287). Albany, NY: SUNY Press.

Torigoe, T. (1985). Comparison of object manipulation among 74 species of nonhuman primates. *Primates, 26* (2), 182–194.

Trevarthan, C. (1973). Behavioral embryology. In E. C. Carterette & M. P. Friedman (Eds.), *Handbook of perception* (Vol. 3, pp. 89–117).

Trevathan, W. (1987). *Human birth: An evolutionary perspective.* Hawthorne, NY: Aldine de Gruyter.

Trivers, R. (1971). The evolution of reciprocal altruism. *Quarterly Review of Biology, 46,* 35–57.

Trivers, R. (1972). Parental investment and sexual selection. In B. Campbell (Ed.), *Sexual selection and the descent of man* (pp. 136–179). New York: Aldine.

Trivers, R. (1974). Parent -offspring conflict. *American Zoologist, 14,* 249–264.

Trivers, R. (1985). *Social evolution.* Menlo Park, CA: Benjamin Cummings.

Tuttle, R. (1986). *Apes of the world: Their socialization, behavior, mentality, and ecology.* Park Ridge, NJ: Hayes.

Uzgiris, I., & Hunt, M. (1975). *Assessment in infancy: Ordinal scales of psychological development.* Urbana, IL: University of Illinois Press.

Valentine, J. W., Collins, A., & Meyer, C. P. (1994). Morphological complexity increase in metazoans. *Paleobiology, 20,* 131–142.

Valentine, J. W., Erwin, D. H., & Jablonski, D. (1996). Developmental evolution of metazoan bodyplans: The fossil evidence. *Developmental Biology, 173,* 373–381.

Van Cantfort, T. E., Gardner, B., & Gardner, R. A. (1989). Developmental trends in replies to Wh-questions by children and chimpanzees. In R. A. Gardner, B. T. Gardner, & T. E. Van Cantfort (Eds.), *Teaching sign language to chimpanzees* (pp. 198–239). Albany, NY: SUNY Press.

van Schaik, C. P., Fox, E. A., & Sitompul, A. F. (1996). Manufacture and use of tools in wild Sumatran orangutans: Implications for human evolution. *Naturwissenschaften, 83*, 186–188.

Vauclair, J. (1982). Sensorimotor intelligence in humans and nonhuman primates. *Journal of Human Evolution, 11*, 257–264.

Vauclair, J. (1996). *Animal cognition.* Cambridge, MA: Harvard University Press.

Vauclair, J., & Bard, K. (1983). Development of manipulations with objects in ape and human infants. *Journal of Human Evolution, 12*, 631–645.

Vaughter, R. M., Smotherman, W., & Ordy, J. M. (1972). Development of object permanence in an infant squirrel monkey. *Developmental Psychology, 7*, 34–38.

Vermeij, G. J. (1973). Morphological patterns in high intertidal gastropods. *Marine Biology, 20*, 319–346.

Visalberghi, E., & Fragaszy, D. (1990). Do monkeys ape? In S. T. Parker & K. R. Gibson (Eds.), *"Language" and intelligence in monkeys and apes* (pp. 247–273). New York: Cambridge University Press.

Visalberghi, E., & Fragaszy, D. (1996). Pedagogy and imitation in monkeys: Yes, no, or maybe? In D. Olsen (Ed.), *Handbook of psychology in education: New models of learning, teaching, and schooling* (pp. 277–301). New York: Blackwell.

Visalberghi, E., Fragaszy, D., & Savage-Rumbaugh, S. E. (1995). Performance in a tool-using task by common chimpanzees (*Pan troglodytes*), bonobos (*Pan paniscus*), and orangutan (*Pongo pygmeaus*), and capuchin monkeys (*Cebus apella*). *Journal of Comparative Psychology, 109* (1), 52–60.

Visalberghi, E., & Limongelli, L. (1994). Lack of comprehension of cause-effect relations in tool-using capuchin monkeys (*Cebus apella*). *Journal of Comparative Psychology, 108* (1), 15–22.

Visalberghi, E., & Limongelli, L. (1996). Acting and understanding: Tool use revisited through the minds of capuchins monkeys. In A. Russon, K. Bard, & S. T. Parker (Eds.), *Reaching into thought: The minds of the great apes* (pp. 57–79). Cambridge, UK: Cambridge University Press.

Volterra, V. (1987). From single communicative signal to linguistic combinations in hearing and deaf children. In J. Montangero, A. Tyrphon, & S. Dionnet (Eds.), *Symbolism and knowledge*, (Vol. 8, pp. 87–105). Geneva: Jean Piaget Archives Foundation.

Wagner, P. J. (1995). Testing evolutionary constraint hypotheses for early gastropods. *Paleobiology, 21*, 248–272.

Wallace, A. R. (1864). The origin of human races and the antiquity of man deduced from the theory of "natural selection." *Anthropological Review, 2* (civiii-cixx).

Watson, J. B. (1913). Psychology as a behaviorist sees it. *Psychological Review, 20*, 158–177.

Watson, J. S. (1972). Smiling, cooing, and "the game." *Merrill Palmer Quarterly, 18*, 323–339.

Watson, J. S. (1994). Detection of self: The perfect algorithm. In S. T. Parker, R. W. Mitchell, & M. L. Boccia (Eds.), *Self-awareness in animals and humans* (pp. 131–148). New York: Cambridge University Press.

Watson, M. W. (1981). The development of social roles: A sequence of social-cognitive development. In K. W. Fischer (Ed.), *New directions for child development*, (Vol. 12, pp. 33–41). New York: Jossey Bass, Inc.

Watson, M. W. (1984). Development of social role understanding. *Developmental Review, 4*, 192–213.

Watson, M. W. (1990). Aspects of self development as reflected in children's role playing. In D. Cicchetti & M. Beeghly (Eds.), *The self in transition* (pp. 265–280). Chicago: University of Chicago Press.

Watson, M. W., & Fischer, K. W. (1977). A developmental sequence of agent use in infancy. *Child Development, 48*, 828–36.

Watson, M. W., & Fischer, K. W. (1980). Development of social roles in elicited and spontaneous behavior during preschool years. *Developmental Psychology, 16* (5), 483–94.

Watts, E. (1985). Adolescent growth and development of monkeys, apes, and humans. In E. Watts (Ed.), *Nonhuman primate models for human growth and development* (pp. 41–65). New York: Alan R. Liss.

Watts, E. (1990). A comparative study of neonatal skeletal development in cebus and other primates. In D. M. Fragaszy, J. G. Robinson, & E. Visalberghi (Eds.), *Adaptation and adaptability of capuchin monkeys*, (Vol. 54, pp. 217–224). Basel, Switzerland: Karger.

Wayne, R. K. (1986). Cranial morphology of domestic and wild canids: The influence of development on morphological change. *Evolution, 40*, 243–261.

Weiss, M. (1987). Nucleic acid evidence bearing on hominoid relationships. *Yearbook of Physical Anthropology, 30*, 41–73.

Wellman, H. M. (1990). *The child's theory of mind.* Cambridge, MA: M. I. T. Press.

Werner, E. E., & Gilliam, J. F. (1984). The ontogenetic niche and species interactions in size-structured populations. *Annual Review of Ecology and Systematics, 15*, 393–425.

Wesson, R. (1991). *Beyond natural selection.* Cambridge, MA: M. I. T. Press.

West-Eberhard, M. J. (1983). Sexual selection, social competition, and speciation. *Quarterly Review of Biology, 58* (2), 155–183.

Westergaard, G. (1988). Lion-tailed macaques (*Macaca silenus*) manufacture and use tools. *Journal of Comparative Psychology, 102*, 152–159.

Westergaard, G. C. (1992). Object manipulation and use of tools by infant baboons. *Journal of Comparative Psychology, 106*, 398–403.

Westergaard, G. C. (1993). Development of combinatorial manipulation in infant baboons. *Journal of Comparative Psychology, 107*, 34–38.

Westergaard, G., & Suomi, S. (1993). Use of a tool-set by capuchin monkeys. *Primates, 34* (4), 459–462.

Westergaard, G., & Suomi, S. (1994). Hierarchical complexity of combinatorial manipulation in capuchin monkeys (*Cebus apella*). *American Journal of Primatology, 32*, 171–176.

Whiten, A. (Ed.). (1991). *Natural theories of mind: Evolution, development and simulation of everyday mindreading.* Cambridge, MA: Basil Blackwell.

Whiten, A., & Byrne, R. (Eds.). (1988). *Machiavellian intelligence.* London: Oxford University Press.

Whiten, A., & Custance, D. (1996). Studies of imitation in chimpanzees and children. In B. G. Galef & C. M. Heyes (Eds.), *Social learning in animals: The roots of culture* (pp. 291–318). New York: Academic Press.

Whiten, A., Custance, D., Gomez, J.-C., Teixidor, P., & Bard, K. A. (1996). Imitative learning of artificial fruit processing in children (*Homo sapiens*) and chimpanzees (*Pan troglodytes*). *Journal of Comparative Psychology, 110* (1), 3–14.

Whiten, A., & Ham, R. (1992). On the nature and evolution of imitation in the animal kingdom: Reappraisal of a century of research. In P. J. B. Slater, J. S. Rosenblatt, C. Beer, & M. Milinski (Eds.), *Advances in the study of behavior* (Vol. 21, pp. 239–283). New York: Academic Press.

Wiley, E. O. (1981). *Phylogenetics: The theory and practice of phylogenetic systematics.* New York: John Wiley.

Williams, G. (1966). *Adaptation and natural selection.* Princeton, NJ: Princeton University Press.

Williams, G. C. (1992). *Natural selection: Domains, levels, and challenges.* New York: Oxford University Press.

Wills, C. (1993). *The runaway brain.* New York: Basic Books.

Wills, M. A., Briggs, D., & Fortney, R. A. (1994). Disparity as an evolutionary index: A comparison of Cambrian and recent arthropods. *Paleobiology, 20,* 93–130.

Wise, K. L., Wise, L. A., & Zimmerman, R. R. (1974). Piagetian object permanence in the infant rhesus monkey. *Developmental Psychology, 10,* 429–437.

Wolpert, L. (1991). *The triumph of the embryo.* Oxford, UK: Oxford University Press.

Wood, B. (1992). Origins and evolution of the genus *Homo. Nature, 355,* 783–790.

Wood, D., Bruner, J., & Ross, G. (1976). The role of tutoring in problem solving. *Journal of Child Psychology and Psychiatry, 17,* 89–100.

Wood, S., Moriarty, K., Gardner, B., & Gardner, A. (1980). Object permanence in child and chimpanzee. *Animal Learning and Behavior, 8* (1), 3–9.

Woodruff, G., & Premack, D. (1979). Intentional communication in the chimpanzee: The development of deception. *Cognition, 7,* 333–362.

Woodruff, G., Premack, D., & Kennel, K. (1978). Conservation of liquid and solid quantity by the chimpanzee. *Science, 202,* 991–994.

Wrangham, R. (1995). Ape cultures and missing links. *Symbols: The Peabody Museum and the Harvard University (Spring 1995),* 2–20.

Wrangham, R., & Peterson, D. (1996). *Demonic males: Ages and the origins of human violence.* Boston: Houghton Mifflin.

Wray, G. A. (1992). Rates of evolution in developmental processes. *American Zoologist, 32,* 123–134.

Wright, R. V. S. (1972). Imitative learning of a flaked-tool technology—the case of an orangutan. *Mankind, 8,* 296–306.

Wynn, T. (1989). *The evolution of spatial competence.* Urbana, IL: University of Illinois Press.

Wynn, T., & McGrew, W. (1989). An ape's view of the Oldowan. *Man, 24,* 388–397.

Yerkes, R. M. (1916). *The mental life of monkeys and apes.* New York: Holt.

Yerkes, R. M., & Yerkes, A. (1929). *The great apes.* New Haven, CT: Yale University Press.

Zazzo, R. (1982). The person: Objective approaches. In W. W. Hartup (Ed.), *Review of child development research* (Vol. 6, pp. 247–290). Chicago: University of Chicago Press.

INDEX

abstract codes vs. imaginal codes, 83, 85

acceleration of development, 244, 246, 255, 256, 269, 273, 340

accommodation, 4, 27, 107

Acheulian tools, 201, 285

active intermodal matching (AIM), 109

adaptation, criteria for
  convergence criterion, 299, 300, 307, 309
  correlation criterion, 309, 310, 311
  phyletic criterion, 299, 300, 302, 304, 305, 307, 309

adaptation, definitions of, 294–95

adaptation, models for hominid cognitive
  apprenticeship model, 277, 278, 281, 305
  declarative planning model, 277, 286, 287
  joint attention model, 277, 285
  ontogenetic niche model, 305–7

adaptation, models for primate cognitive
  atavistic model, 282, 305

  ecological models, 277, 278, 303, 304
  locomotor models, 305–7
  socialization models, 277, 304, 305
  social models, 277, 278, 301–3

adaptive arrays of primates, 222

adaptive radiations of primates, 222

adolescence, evolution of, 220

adultified apes, 237

agency (in causality), 50

agent, in grammar, 164

Ai, symbol-trained chimpanzee, 69, 88, 156

alliances, 287, 290

allometric exponent, 324

allometry, 249, 257, 318, 324, 325

Altmann, J., 18

altricial vs. precocial offspring, 223, 321, 334

American Sign Language (ASL), 112, 174–79

Anderson, J., 149, 150

anecdotes, 11

antecedent heterochrony, 256

anthropologists, 9, 275, 309

anthropomorphism, 10–11

LIBRARY OF CONGRESS CATALOGING-IN-PUBLICATION DATA

Parker, Sue Taylor.
Origins of intelligence : the evolution of cognitive development in monkeys, apes, and humans / Sue Taylor Parker and Michael L. McKinney.
p. cm.
Includes bibliographical references and index.
ISBN 0-8018-6012-1 (hardcover : alk. paper)
1. Cognition. 2. Cognition in animals. 3. Animal intelligence. 4. Genetic psychology. 5. Psychology, Comparative. I. McKinney, Michael L. II. Title.
BF311.P31363 1999
156′.3—dc21   98-37428
                                                              CIP